CAMBRIDGE TRACTS IN MATHEMATICS

General Editors

106 Arithmetic of quadratic forms

YOSHIYUKI KITAOKA
Department of Mathematics, Nagoya University

Arithmetic of quadratic forms

CAMBRIDGE
UNIVERSITY PRESS

Published by the Press Syndicate of the University of Cambridge
The Pitt Building, Trumpington Street, Cambridge CB2 1RP
40 West 20th Street, New York, NY 10311-4211, USA
10 Stamford Road, Oakleigh, Victoria 3166, Australia

First published 1993

Printed in Great Britain at the University Press, Cambridge

Library of Congress cataloguing in publication data available
British Library cataloguing in publication data available

ISBN 0 521 40475 4 hardback

Contents

Preface vii

Notation ix

Chapter 1 General theory of quadratic forms 1
- 1.1 Symmetric bilinear forms 1
- 1.2 Quadratic forms 3
- 1.3 Quadratic forms over finite fields 12
- 1.4 The Clifford algebra 20
- 1.5 Quaternion algebras 24
- 1.6 The spinor norm 29
- 1.7 Scalar extensions 32

Chapter 2 Positive definite quadratic forms over $\mathbb{R}$ 33
- 2.1 Reduction theory 33
- 2.2 An estimate of Hermite's constant 42

Chapter 3 Quadratic forms over local fields 47
- 3.1 p-adic numbers 47
- 3.2 The quadratic residue symbol 52
- 3.3 The Hilbert symbol 54
- 3.4 The Hasse invariant 56
- 3.5 Classification of quadratic spaces over p-adic number fields 60

Chapter 4 Quadratic forms over $\mathbb{Q}$ 64

Chapter 5 Quadratic forms over the p-adic integer ring 70
5.1 Dual lattices 70
5.2 Maximal and modular lattices 71
5.3 Jordan decompositions 79
5.4 Extension theorems 86
5.5 The spinor norm 92
5.6 Local densities 93

Chapter 6 Quadratic forms over $\mathbb{Z}$ 129
6.1 Fundamentals 129
6.2 Approximation theorems 134
6.3 Genus, spinor genus and class 147
6.4 Representation of codimension 1 151
6.5 Representation of codimension 2 157
6.6 Representation of codimension ≥ 3 164
6.7 Orthogonal decomposition 169
6.8 The Minkowski-Siegel formula 173

Chapter 7 Some functorial properties of positive definite quadratic forms 189
7.1 Positive lattices of E-type 190
7.2 A fundamental lemma 199
7.3 Weighted graphs 217
7.4 The tensor product of positive lattices 222
7.5 Scalar extension of positive lattices 239

Notes 240
References 263
Index 268

Preface

The purpose of this book is to introduce the reader to the arithmetic of quadratic forms. Quadratic forms in this book are mainly considered over the rational number field or the ring of rational integers and their completions. It is of course possible to discuss quadratic forms over more general number theoretic fields and rings. But it is not hard to understand the general theory once readers have learned the simpler theory in this book. This book is self-contained except for the following: The theory of quadratic fields is used to prove the first approximation theorem in Chapter 6. In Chapter 7, the relative theory is treated and so the use of algebraic number fields is inevitable. Dirichlet's theorem on prime numbers in arithmetic progressions is also used without proof.

Several exercises are given. The answers are known anyway. Some of them are easy and some of them are not. Some problems are also given in the Notes. Problems mean questions for which I do not know the answer and want to know it. They are very subjective and arbitrary. I do not say that they have citizenship in the world of quadratic forms.

Inheriting pioneering works of Eisenstein, Smith, Minkowski, Hardy and others, Hasse established the so-called Minkowski-Hasse principle on quadratic forms and Siegel gave the so-called Siegel formula and other great works. Some of them were formulated from the view point of algebraic groups, and stimulated and developed them. One of the achievements of Eichler was to introduce the important concept of spinor genus. Kneser showed that group-theoretic consideration was fruitful. Their method gave a unified perspective to sporadic results. Now there seem to be no famous outstanding problems. But it does not mean that there remains nothing to search out. The important view point is again to look at quadratic forms as diophantine equations. Although it may be hard, it must be interesting and deep.

It must be emphasized that this book does not cover even all of the arithmetic part of quadratic forms, of course, and I did not touch Witt rings, hermitian forms, analytic theory, etc. At first I intended to give a brief introduction to analytic theory, in particular the theory of modular forms, but I gave up as the book was already too long. I say only that analytic theory is a fertile part of quadratic forms.

I wish to thank Mr. D. Tranah, who gave me the chance to write this book, and Professor M. Kneser who gave useful suggestions and finally I am cordially grateful to Professor J. S. Hsia who examined in full detail the first version of the manuscript mathematically and linguistically.

Notation

$\mathbb{Z}$ and $\mathbb{Q}$ denote the ring of integers and the field of rational numbers, respectively, $\mathbb{Z}_p$, $\mathbb{Q}_p$ are their p-adic completion, and $\mathbb{R}$ is the field of real numbers.

$A := B$ means that A is defined by B and sometimes B by A.

For any two sets X, Y, $X \subset Y$ denotes the inclusion, so $X = Y$ is possible.

$X \backslash Y$ denotes $\{x \in X \mid x \notin Y\}$.

For a set Z, $\sharp Z$ denotes the cardinality of Z.

Almost all elements of a set X means all but a finite number of elements of X.

For a commutative ring R, $M_{m,n}(R)$ is the set of $m \times n$ matrices with entries in R, and we put $M_m(R) := M_{m,m}(R)$. 1_m denotes the unit matrix of degree m.

$GL_m(R)$ is the group of all invertible elements of $M_m(R)$.

For $X \in M_{m,n}$, ${}^tX$ means the transposed matrix of X, and for matrices X, Y, we put $Y[X] := {}^tYXY$ if it is defined.

For square matrices $A_1, \cdots, A_n$, $\mathrm{diag}(A_1, \cdots, A_n)$ means

$$\begin{pmatrix} A_1 & & \\ & \ddots & \\ & & A_n \end{pmatrix}$$

If the product $x \circ y$ is defined for $x \in X$, $y \in Y$, then $S \circ T$ denotes $\{s \circ t \mid s \in S, t \in T\}$ for $S \subset X, T \subset Y$.

Let F be a field and R a ring such that R generates F, and V be a vector space over F and M be a finitely generated submodule in V over R. A submodule N over R of M is called primitive if $N = FN \cap M$. A matrix $X \in M_{m,n}(R)$ is said to be primitive if X can be extended to a matrix in $GL_m(R)$ or $GL_n(R)$, that is every elementary divisor of X is in $R^\times$ if R is a principal ideal domain.

1

General theory of quadratic forms

Throughout this book, rings R are commutative and contain the unity 1, *and modules* M *over* R *are finitely generated with* $1m = m$ *for* $m \in M$.

In this chapter, we give some basics about quadratic forms which include the so-called Witt theorem, Clifford algebra and quaternion algebra. Besides them, the theory of quadratic forms over finite fields is outlined. It is useful for readers to get used to how to deal with quadratic forms and also to their applications.

1.1 Symmetric bilinear forms

Let M be a module over a ring R and b a mapping from $M \times M$ to R satisfying the conditions

(1) $b(x, y) = b(y, x)$ for $x, y \in M$
(2) $b(rx + sy, z) = rb(x, z) + sb(y, z)$ for $r, s \in R, \quad x, y, z \in M$.

We call b a *symmetric bilinear form* and the pair (M, b) or simply M a *symmetric bilinear module* over R. When R is a field, we often use "space" instead of "module".

If M is free and $\{v_i\}_{i=1}^n$ is a basis of M, then we write

$$M \cong \langle A \rangle$$

for $A = (b(v_i, v_j))$. For another basis $\{u_i\}_{i=1}^n$, there is a matrix $T = (t_{ij})$ such that $(u_1, \cdots, u_n) = (v_1, \cdots, v_n)T$, $t_{ij} \in R, \det T \in R^\times$, and

$(b(u_i,u_j)) = (b(\sum_k t_{ki}v_k, \sum_h t_{hj}v_h)) = (\sum_{k,h} t_{ki}b(v_k,v_h)t_{hj}) = {}^tTAT = A[T]$ holds and so we have $M \cong \langle A\rangle \cong \langle A[T]\rangle$. Thus $\det A(R^\times)^2$ is independent of the choice of a basis and is uniquely determined by M, and we denote it by

$$\mathrm{d}\,M$$

and call it the *discriminant*. (It is defined only for free modules.)

For a symmetric bilinear space U over a field F, we call U *regular* if $\mathrm{d}\,U \neq 0$, and it is easy to see that

U is regular
- $\Leftrightarrow$ if $b(x,U)=0$, then $x=0$
- $\Leftrightarrow$ $\mathrm{Hom}_F(U,F) = \{y \mapsto b(x,y) \mid x \in U\}$
- $\Leftrightarrow$ for a basis $\{u_i\}$ of U, there is a subset $\{v_i\}$ of U such that $b(u_i,v_j) = \delta_{i,j}$ (Kronecker's delta).

(We use "regular" for spaces only!)

For a subset S of a symmetric bilinear module M over a ring R, we put

$$S^\perp = \{x \in M \mid b(x,s) = 0 \quad \text{for} \quad s \in S\}.$$

For symmetric bilinear modules $M, M_1, \cdots, M_m$ over a ring R such that $M = M_1 \oplus \cdots \oplus M_m$, $b(M_i, M_j) = 0$ if $i \neq j$, we call M the *orthogonal sum* of $M_1, \cdots, M_m$ and write

$$M = M_1 \perp \cdots \perp M_m.$$

When M has a basis $\{v_i\}$ such that $M = Rv_1 \perp \cdots \perp Rv_n$, $\{v_i\}$ is called an *orthogonal basis* of M. By the above notation, $M \cong \perp_i \langle a_i\rangle$ with $a_i = b(v_i, v_i)$.

For a symmetric bilinear space $U = U_1 \perp \cdots \perp U_m$ over a field F, $\mathrm{d}\,U = \mathrm{d}\,U_1 \cdots \mathrm{d}\,U_m$ is clear and so U is regular if and only if every U_i is regular.

Proposition 1.1.1. *Let U be a symmetric bilinear space over a field F and V a subspace of U. If V is regular, then $U = V \perp V^\perp$ holds.*

Proof. Since V is regular, we have $V \cap V^\perp = \{x \in V \mid b(x,V) = 0\} = \{0\}$ and every linear mapping $f \in \mathrm{Hom}_F(V,F)$ is given by $x \mapsto b(x,y)$ for $y \in V$. For $u \in U$, $x \mapsto b(x,u)$ is a linear mapping from V to F. Therefore, there is an element $y \in V$ such that $b(x,u) = b(x,y)$ for $x \in V$, which implies $u - y \in V^\perp$. This means $U = V + V^\perp$ and then $U = V \oplus V^\perp = V \perp V^\perp$. □

Proposition 1.1.2. *Let U be a symmetric bilinear space over a field F and V a subspace of U. Then $V \cap U^\perp = \{0\}$ if and only if* $\mathrm{Hom}_F(V, F) = \{x \mapsto b(x,y) \mid y \in U\}$, *and then* $\dim V^\perp = \dim U - \dim V$.

Proof. Suppose $V \cap U^\perp = \{0\}$. Take a subspace $W(\supset V)$ such that $U = W \oplus U^\perp$. If $w \in W$ satisfies $b(w, W) = \{0\}$, then $b(w, U) = \{0\}$ and hence $w \in U^\perp \cap W = \{0\}$. Thus W is regular, and for $f \in \mathrm{Hom}_F(V, F)$ we extend it to $\mathrm{Hom}_F(W, F)$ by $f(W_0) = 0$ with $W = W_0 \oplus V$. Because of regularity of W, there is an element y in W so that $f(x) = b(x, y)$ for x in W and especially in V. If, conversely $V \cap U^\perp \neq \{0\}$, then we take a basis $\{v_i\}$ of V such that $v_1 \in V \cap U^\perp$. A linear mapping f defined by $f(v_1) = 1, f(v_i) = 0\ (i \geq 2)$ is not of form $f(x) = b(x, y)$, since $b(v_1, U) = 0$. Thus the former part has been proved.

Suppose $V \cap U^\perp = \{0\}$ and for a basis $\{v_i\}_{i=1}^m$ of V we can take a subset $\{u_i\}_{i=1}^m$ of U such that $b(v_i, u_j) = \delta_{ij}$ for $1 \leq i, j \leq m$. Then $\{u_i\}_{i=1}^m$ is linearly independent. We define a linear mapping

$$f : U \to U_0 := \oplus_{i=1}^m F u_i$$

by $f(u) = \sum_i b(u, v_i) u_i$. Then f is surjective by virtue of $f(u_i) = u_i$ and $\ker f = V^\perp$ is clear. Thus we have $\dim U - \dim V^\perp = \dim U_0 = \dim V$. □

The proof shows

Corollary 1.1.1. *In Proposition 1.1.2, $V \cap U^\perp = \{0\}$ if and only if V is contained in some regular subspace of U.*

1.2 Quadratic forms

Let M be a module over a ring R and q a mapping from M to R which satisfies the conditions

(i) $q(ax) = a^2 q(x)$ for $a \in R$, $x \in M$,
(ii) $b(x, y) := q(x + y) - q(x) - q(y)$ is a symmetric bilinear form.

We call the pair (M, q) or simply M a *quadratic module* over R, q a *quadratic form* and b the *associated symmetric bilinear form.* When R is a field, we often use "space" instead of "module".

Putting $x = y$, we have

$$b(x, x) = 2q(x) \quad \text{for } x \in M.$$

If 2 is in $R^\times$, then we can associate a symmetric bilinear form $\frac{1}{2} b(x, y) = B(x, y)$. Then $B(x, x) = q(x)$ and conversely for a given symmetric bilinear form $B(x, y)$, $q(x) := B(x, x)$ is a quadratic form and $q(x+y) - q(x) - q(y) =$

$2B(x,y)$. Thus "quadratic module" and "symmetric bilinear module" are equivalent if $2 \in R^\times$.

In this section and the next we will associate $b(x,y)$ with M, but after 1.4, the fields are of characteristic 0, and we will prefer the bilinear form B instead of b. The difference between them is minor and the choice depends on the individual's taste.

Considering a quadratic module M as a symmetric bilinear module with $b(x,y)$ ($B(x,y)$ after section 1.4), the notations and terminology $\perp, M \cong \langle A \rangle, \mathrm{d}\,M$ and regular remain meaningful.

If 2 is a unit, then $\perp$ and "regular" are independent of the choice of $B(x,y)$ or $b(x,y)$. But $\mathrm{d}\,M$ differs by 2^n with $n = \operatorname{rank} M$, since $M \cong \langle A \rangle$ or $\langle 2^{-1}A \rangle$ for $A = (b(x,y))$, according to $b(x,y)$ or $B(x,y)$, respectively.

For a quadratic module M, we put

$$\operatorname{Rad} M := \{x \in M^\perp \mid q(x) = 0\}.$$

This is a submodule of M and if $2 \in R^\times$, then $\operatorname{Rad} M = M^\perp$.

Theorem 1.2.1. *Let U be a quadratic space over a field F. If $\operatorname{ch} F \neq 2$, then we have*

$$U = U_1 \perp \cdots \perp U_r \perp U^\perp$$

where the U_i's are regular and 1-dimensional.

If $\operatorname{ch} F = 2$, then we have

$$U = V_1 \perp \cdots \perp V_s \perp W_1 \perp \cdots \perp W_t \perp \operatorname{Rad} U,$$

where the V_i's are regular and 2-dimensional, the W_i's are 1-dimensional and $0 \leq t \leq [F^\times : (F^\times)^2]$ and

$$U^\perp = W_1 \perp \cdots \perp W_t \perp \operatorname{Rad} U.$$

Proof. Suppose $\operatorname{ch} F \neq 2$. If there is an element $u_1 \in U$ such that $q(u_1) \neq 0$, then $b(u_1, u_1) = 2q(u_1) \neq 0$ and so $U_1 = Fu_1$ is regular. Proposition 1.1.1 implies $U = U_1 \perp U_1^\perp$. Repeating this, we have $U = U_1 \perp \cdots \perp U_r \perp U_{r+1}$ where $U_1, \cdots, U_r$ are regular and 1-dimensional and $q(x) = 0$ for all $x \in U_{r+1}$, which implies $b(U_{r+1}, U_{r+1}) = 0$ and hence $U_{r+1} \subset U^\perp$. Decomposing $x \in U^\perp$ along the above orthogonal decomposition of U, $U^\perp \subset U_{r+1}$ is easy to see.

Next, suppose $\operatorname{ch} F = 2$. If $x, y \in U$ satisfies $b(x,y) \neq 0$, then $V_1 = F[x,y]$ is regular with $\mathrm{d}\,V_1 = -b(x,y)^2$, noting $b(x,x) = b(y,y) = 0$. Hence we have $U = V_1 \perp V_1^\perp$. Repeating this, we have $U = V_1 \perp \cdots \perp V_s \perp U_0$

where $V_1, \cdots, V_s$ are regular and 2-dimensional and $b(U_0, U_0) = 0$. As above, $U_0 = U^\perp$ holds. We take a direct sum decomposition

$$U^\perp = W_1 \oplus \cdots \oplus W_t \oplus \operatorname{Rad} U$$

where the W_i's are 1-dimensional. Since $b(U^\perp, U^\perp) = 0$, $U^\perp = W_1 \perp \cdots \perp W_t \perp \operatorname{Rad} U$ holds. Put $W_i = Fw_i$. Since $w_i \in U^\perp$ but $w_i \notin \operatorname{Rad} U$, $q(w_i) \neq 0$ holds. If $q(w_i)(F^\times)^2 = q(w_j)(F^\times)^2$ for $i \neq j$, then $q(w_i) = a^2 q(w_j)$ for $a \in F^\times$ holds and this means $q(w_i - aw_j) = 0$. Hence we have a contradiction $w_i - aw_j \in \operatorname{Rad} U$. Thus the $q(w_i)$'s give distinct representatives of $F^\times/(F^\times)^2$. □

Let M, N be quadratic modules over a ring R. If a linear mapping σ from M to N satisfies that

$$\sigma \text{ is injective and}$$
$$q(\sigma(x)) = q(x) \text{ for } x \in M,$$

we call σ an *isometry* from M to N and say that M is *represented* by N and write

$$\sigma : M \hookrightarrow N.$$

When $\sigma(M) = N$, we write

$$\sigma : M \cong N,$$

and say that M and N are *isometric*. The group of all isometries from M to itself is denoted by

$$O(M).$$

Suppose that R is a subring of a field F and generates F. For an R-submodule M of a quadratic space V over F satisfying $FM = V$, we denote by

$$O^+(M) := \{\sigma \in O(M) \mid \det \sigma = 1\}$$

where $\det \sigma$ is defined by $\det T$ for a matrix T with

$$(\sigma(v_1), \cdots, \sigma(v_n)) = (v_1, \cdots, v_n)T$$

for a basis $\{v_i\}$ of V.

For an isometry σ from M to N, it is easy to see

$$b(\sigma(x), \sigma(y)) = b(x, y)$$

for $x, y \in M$.

Conversely, an injective linear mapping σ from M to N which satisfies $b(\sigma(x), \sigma(y)) = b(x, y)$ is an isometry if $2 \in R^\times$, since $q(x) = 2^{-1}b(x, x)$.

For quadratic modules M, N and a linear mapping σ from M to N with $q(\sigma(x)) = q(x)$ for $x \in M$, $\sigma(x) = 0$ yields $x \in M^\perp$ since $b(x, M) = b(\sigma(x), \sigma(M)) = \{0\}$. If, moreover M is a regular quadratic space, then σ is an isometry.

Now we give an important example of isometry.

Proposition 1.2.1. *Let M be a quadratic module over a ring R. For an element x in M with $q(x) \in R^\times$, we put*

$$\tau_x(y) := y - b(x,y)q(x)^{-1}x \quad \text{for } y \in M.$$

Then τ_x is an isometry from M to M and satisfies

$$\tau_x(x) = -x,\ \tau_x(y) = y \ \text{for } y \in x^\perp, \ \text{and } \tau_x^2 = \mathrm{id}\,.$$

Proof. $\tau_x(x) = -x$ and $\tau_x(y) = y$ for $y \in x^\perp$ are easy.

$$\tau_x^2(y) = \tau_x(y) - b(x,y)q(x)^{-1}\tau_x(x) = y$$

implies $\tau_x^2 = \mathrm{id}$ and the bijectivity of τ_x. Finally

$$q(\tau_x(y)) = q(y) + b(x,y)^2 q(x)^{-2} q(x) - b(x,y)q(x)^{-1}b(x,y) = q(y)$$

implies that τ_x is an isometry. □

τ_x is called a *symmetry.* If R is a field, then the determinant of a transformation τ_x is -1.

The following theorem of Witt type is due to Kneser.

Theorem 1.2.2. *Let $\tilde{R} \supset R$ be rings and P a proper ideal of R satisfying $R = R^\times \cup P$ and $R^\times \cap P = \emptyset$. Let L, M, N, H be R-submodules of a quadratic module U over $\tilde{R}$ such that they are finitely generated over R, M, N are free R-modules and $L \supset M, N, H$. Suppose that*

(1) $$q(H) \subset R,\ b(L,H) \subset R,$$

(2a) $$\mathrm{Hom}_R(M,R) = \{x \mapsto b(x,y) \mid y \in H\},$$

(2b) $$\mathrm{Hom}_R(N,R) = \{x \mapsto b(x,y) \mid y \in H\},$$

and $\sigma : M \cong N$ is an isometry such that

(3) $$\sigma(x) \equiv x \bmod H$$

for $x \in M$.

Then σ can be extended to an element of $O(L)$ which satisfies

(4) $$\sigma = \mathrm{id} \quad \text{on } H^\perp$$

and (3) *for every x in L.*

Proof. For z in H with $q(z) \in R^\times$, the symmetry

$$\tau_z(x) = x - b(x,z)q(z)^{-1}z$$

satisfies three properties: $\tau_z(L) = L$ because of the property (1), secondly (3) for $x \in L$ and (4) by definition of τ_z. For a submodule J of H, we denote by $S(J)$ a subgroup of $O(L)$ generated by τ_x ($x \in J, q(x) \in R^\times$), and then every element in $S(J)$ satisfies the condition (3) for $x \in L$ and the condition (4). Note that the quotient ring $\bar{R} := R/P$ is a field. First, we will prove a preparatory assertion:

Assertion 1. *We assume, moreover*

(5) $\sharp\bar{R} > 2 \Rightarrow q(x) \in R^\times$ *for some* $x \in H$,

(6) $\sharp\bar{R} = 2 \Rightarrow q(x) \in R^\times$ *and* $b(x, H) \subset P$ *for some* $x \in H$.

Then σ *is a restriction of an element of* $S(H)$.

We prove this by induction of the rank of M.

Suppose rank $M = 1$ and $M = Rm, N = Rn, n = \sigma(m)$, and put, by (3)

(7) $$h = n - m \in H.$$

Then $q(h) = 2q(n) - b(n, m) = 2q(m) - b(n, m)$ and (7), (1) imply

(8) $$q(h) = b(n, h) = -b(m, h) \in R.$$

If $q(h) \in R^\times$, then $\tau_h(m) = m - b(h, m)q(h)^{-1}h = n$ by (8), (7), and we complete the proof in the case of rank $M = 1$.

Suppose

(9) $$q(h) \in P.$$

First, we show that if there is an element f in H satisfying

(10) $$q(f), b(f, m), b(f, n) \in R^\times,$$

then the proof in the case of rank $M = 1$ is completed.

Supposing the existence of such f, we put $g := n - \tau_f(m)$; then (7), (10) imply

(11) $$g = h + b(f, m)q(f)^{-1}f \in H,$$

and

$$\begin{aligned} q(g) &= q(h) + b(f, m)^2 q(f)^{-2} q(f) + b(f, m)q(f)^{-1} b(h, f) \\ &= q(h) + b(f, m)b(f, n)q(f)^{-1} \in R^\times \end{aligned} \tag{12}$$

by (9), (10).

Moreover, we have

$$\begin{aligned} \tau_g(n) &= n - b(n, g)q(g)^{-1}g \\ &= n - (b(n, h) + b(f, m)q(f)^{-1}b(n, f))q(g)^{-1}g \quad \text{by (11)} \\ &= n - (b(n, h) + q(g) - q(h))q(g)^{-1}g \quad \text{by (12)} \\ &= n - g \quad \text{by (8)} \\ &= \tau_f(m) \quad \text{by definition of } g. \end{aligned}$$

Thus σ is a restriction of $\tau_g \tau_f \in S(H)$.

It remains to show the existence of f.

We denote a vector space H/PH over $\bar{R}$ by $\bar{H}$ and an element of $\bar{H}, \bar{R}$ represented by $x \in H, y \in R$ by $\bar{x}, \bar{y}$, respectively. By virtue of (1), $\bar{H}$ becomes a quadratic space over $\bar{R}$ by $\bar{q}(\bar{x}) := \overline{q(x)}$ for $x \in H$. This is well defined by (1). Put

$$\bar{m}^\perp = \{\bar{x} \in \bar{H} \mid \bar{b}(\bar{x}, \bar{m}) = 0\},$$
$$\bar{n}^\perp = \{\bar{x} \in \bar{H} \mid \bar{b}(\bar{x}, \bar{n}) = 0\}.$$

The condition (10) is equivalent to $\bar{q}(\bar{H} \setminus (\bar{m}^\perp \cup \bar{n}^\perp)) \neq 0$. From (8), (9) follows $\bar{m}^\perp \cap \bar{n}^\perp \ni \bar{h}$, and by virtue of (2) there exist $h_m, h_n \in H$ such that $b(m, h_m) = 1$ and $b(n, h_n) = 1$. Thus $\dim \bar{m}^\perp = \dim \bar{n}^\perp = \dim \bar{H} - 1$.

Suppose

$$\text{(13)} \qquad \bar{q}(\bar{H} \setminus (\bar{m}^\perp \cup \bar{n}^\perp)) = 0.$$

For $\bar{x} \in \bar{m}^\perp \cap \bar{n}^\perp$ and $\bar{y} \in \bar{H} \setminus (\bar{m}^\perp \cup \bar{n}^\perp)$, we have, by (13)

$$\text{(14)} \qquad \bar{q}(a\bar{x} + \bar{y}) = a^2 \bar{q}(\bar{x}) + a\bar{b}(\bar{x}, \bar{y}) + \bar{q}(\bar{y}) = 0 \quad \text{for } a \in \bar{R}.$$

If $\sharp \bar{R} > 2$, then (14) implies

$$\text{(15)} \qquad \bar{q}(\bar{x}) = \bar{b}(\bar{x}, \bar{y}) = \bar{q}(\bar{y}) = 0.$$

Note that $\bar{h} \in \bar{m}^\perp \cap \bar{n}^\perp$, and the vectors which are not in the union of two hyperplanes $\bar{m}^\perp$, $\bar{n}^\perp$ span the whole space if the number of elements of the coefficient field is greater than 2. Putting $\bar{x} := \bar{h}$ in (15), we have $\bar{b}(\bar{h}, \bar{H}) = 0$ and then $\bar{m}^\perp = \bar{n}^\perp$ by (7). From this with (15) it follows that $\bar{q}(\bar{m}^\perp) = \bar{q}(\bar{H} \setminus \bar{m}^\perp) = 0$ and so $\bar{q}(\bar{H}) = 0$. This contradicts the assumption (5).

Suppose $\sharp \bar{R} = 2$ and put

$$H_1 := \bar{m}^\perp \cap \bar{n}^\perp \cap \bar{H}^\perp \text{ and } H_2 := (\bar{H} \setminus (\bar{m}^\perp \cup \bar{n}^\perp)) \cap \bar{H}^\perp;$$

then (14) implies $\bar{q}(H_1) = \bar{q}(H_2) = 0$, noting $\bar{b}(\bar{H}_1, \bar{H}_2) = 0$. For $\bar{x} \in \bar{H}$, we have

$$\begin{aligned} \bar{x} \in \bar{m}^\perp \cap \bar{H}^\perp &\Leftrightarrow \bar{b}(\bar{x}, \bar{m}) = 0, \bar{b}(\bar{x}, \bar{H}) = 0 \\ &\Leftrightarrow \bar{b}(\bar{x}, \bar{n}) = 0, \bar{b}(\bar{x}, \bar{H}) = 0 \text{ by (7)} \\ &\Leftrightarrow \bar{x} \in \bar{n}^\perp \cap \bar{H}^\perp, \end{aligned}$$

and hence $H_1 = \bar{m}^{\perp} \cap \bar{H}^{\perp} = \bar{n}^{\perp} \cap \bar{H}^{\perp}$, and then

$$H_2 = (\bar{H} \setminus \bar{m}^{\perp}) \cap (\bar{H} \setminus \bar{n}^{\perp}) \cap \bar{H}^{\perp} = \bar{H}^{\perp} \setminus H_1.$$

This means $\bar{q}(\bar{H}^{\perp}) = \bar{q}(H_1) \cup \bar{q}(H_2) = 0$ which contradicts (6). Thus we have completed the proof in the case of rank $M = 1$.

Suppose $r = \operatorname{rank} M > 1$. Let $\{m_1, \cdots, m_r\}$ be a basis of M and we take $h_i \in H$ such that $b(m_i, h_j) = \delta_{ij}$ by virtue of the assumption (2). It is easy to see

$$(16) \qquad H = \oplus_i Rh_i \oplus K, K := M^{\perp} \cap H.$$

From the assumption (5), (6) there is an element $\bar{h} \in \bar{H}$ or $\bar{H}^{\perp}$ such that $\bar{q}(\bar{h}) \neq 0$ according to $\sharp \bar{R} > 2$ or $\sharp \bar{R} = 2$. Changing a basis if necessary, we may suppose

$$(17) \qquad \bar{h} \in \bar{R}\bar{h}_r + \bar{K}.$$

Applying the inductive assumption to $M_0 := \oplus_{i=1}^{r-1} Rm_i$, there is $\tau \in S(H)$ such that $\tau = \sigma$ on M_0. Taking $\tau^{-1}(N), \tau^{-1}\sigma$ instead of N, σ, they satisfy conditions (2), (3), because of $\tau(H) = H$ and a remark at the beginning of the proof. Hence we may suppose

$$(18) \qquad \sigma(m_i) = m_i \text{ for } 1 \leq i \leq r-1,$$

taking $\tau^{-1}\sigma$ as σ again. Then, for $x \in M$ and $1 \leq i \leq r-1$, we have $b(\sigma(x) - x, m_i) = b(\sigma(x), m_i) - b(x, m_i) = 0$ by (18) and hence by (3), (16)

$$(19) \qquad \sigma(x) - x \in H \cap M_0^{\perp} = Rh_r + K \text{ for } x \in M.$$

We show that conditions (1), $\cdots$, (6) are satisfied for $M_1 = Rm_r, H_0 = Rh_r + K$ instead of M, H respectively. (1) follows from $H_0 \subset H$. By definition of h_r, we have $b(m_r, h_r) = 1$ and so the condition (2a) for M_1. Since $\{m_1, \cdots, m_{r-1}, \sigma(m_r)\}$ is a basis of N, there is an element $h' \in H$ such that $b(m_i, h') = 0$ for $1 \leq i \leq r-1$ and $b(\sigma(m_r), h') = 1$ from the original assumption (2b). Hence h' is in $H \cap M_0^{\perp} = Rh_r + K = H_0$ and so (2b) holds for $\sigma(M_1)$. The condition (3) follows from (19), and (5), (6) do from (17). Hence, by the inductive assumption, there exists $\tau \in S(H_0)$ such that $\tau(m_r) = \sigma(m_r)$, moreover $m_1, \cdots, m_{r-1} \in H_0^{\perp}$ implies $\tau(m_i) = m_i = \sigma(m_i)$ for $1 \leq i \leq r-1$. Thus we have $\sigma = \tau$ on M and $\tau \in S(H_0) \subset S(H)$ completes the proof of the assertion.

Proof of Theorem 1.2.2. Let V be a binary quadratic module over $\tilde{R}$ with basis $\{v_1, v_2\}$ such that $q(a_1v_1 + a_2v_2) = a_1a_2$ for $a_i \in \tilde{R}$. Put

$$U' = U \perp V, L' = L \perp (Rv_1 + Rv_2), M' = M \perp Rv_1,$$
$$N' = N \perp Rv_1, H' = H \perp R(v_1 + v_2), \sigma' = \sigma \perp (\text{id on } Rv_1).$$

Then $q(v_1 + v_2) = 1$ and moreover $b(v_1 + v_2, H') \subset P$ if $\sharp\bar{R} = 2$. With M', N', H', σ' instead of M, N, H, σ, conditions (1), $\cdots$, (6) are satisfied and hence there exists $\tau \in S(H')$ such that $\tau = \sigma'$ on M'. Now $b(v_1 + v_2, v_1 - v_2) = 0$ implies $v_1 - v_2 \in H'^{\perp}$ and so $\tau(v_1 - v_2) = v_1 - v_2$, hence $\tau(v_i) = v_i$ for $i = 1, 2$ by $\tau(v_1) = \sigma'(v_1) = v_1$. Also $L = L' \cap \{v_1, v_2\}^{\perp}$ and $\tau(L') = L'$ yield $\tau \in O(L)$. Thus we have completed the proof. □

We note that the conditions (1) and (3) are absorbed in definitions in the case of $\tilde{R} = R$ and $U = L = H$ and that this case is also quite useful.

Corollary 1.2.1. *Let U be a quadratic space over a field F, and V, W subspaces satisfying $V \cap U^{\perp} = W \cap U^{\perp} = \{0\}$. Then every isometry $\sigma : V \cong W$ is extended to an isometry of U, and if $\operatorname{ch} F \neq 2$ and $q(U) \neq 0$ then it is a product of symmetries.*

Proof. In the theorem, we put $\tilde{R} = R := F$, $P := \{0\}$, $L = H := U$, $M := V$, $N := W$. Conditions (1), (3) are obviously satisfied and (2) is done by Proposition 1.1.2. The latter part follows from Assertion 1 with the condition (5). □

Corollary 1.2.2. *Let U be a regular quadratic space over a field F with $\operatorname{ch} F \neq 2$. Then $O(U)$ is generated by symmetries.*

Proof. In the previous corollary, we have only to put $U = V = W$. □

Corollary 1.2.3. *Let $V = V_1 \perp V_2$, $W = W_1 \perp W_2$ be quadratic spaces over a field F and suppose that $V \hookrightarrow W$, $V_1 \cong W_1$ and V_1 is regular. Then V_2 is represented by W_2. If, moreover $V \cong W$, then $V_2 \cong W_2$.*

Proof. Since V is represented by W, we may suppose that V is a subspace of W, and let σ be an isometry from V_1 to W_1. We can extend it to σ_1 in $O(W)$ by Corollary 1.2.1, since V_1 is regular. Then we have $\sigma_1(V_2) \subset \sigma_1(V_1^{\perp}) \subset \sigma_1(V_1)^{\perp} = W_1^{\perp} = W_2$. Hence V_2 is represented by W_2. If $V \cong W$, then we have $\dim V_2 = \dim W_2$ and hence $\sigma_1(V_2) = W_2$. □

Other applications are given later.

Let U be a quadratic space over a field F. For a non-zero vector x, we call x *anisotropic* if $q(x) \neq 0$, *isotropic* if $q(x) = 0$, respectively. If U contains an isotropic vector, then U is called *isotropic*, otherwise *anisotropic*, that is if $q(x) = 0$ for $x \in U$ implies $x = 0$, then U is anisotropic. If $q(U) = 0$, then U is called *totally isotropic.*

If H is a 2-dimensional quadratic space over a field F which has a basis $\{h_1, h_2\}$ such that $q(a_1h_1 + a_2h_2) = a_1a_2$ for $a_i \in F$, then we call H the *hyperbolic plane* and $\{h_1, h_2\}$ a *hyperbolic basis*, and then H is clearly regular, $\mathrm{d}\,H = -1$ and $q(H) = F$. An orthogonal sum of hyperbolic planes or $\{0\}$ is called a *hyperbolic space.*

Proposition 1.2.2. *Let U be a quadratic space over a field F and V a subspace with $V \cap U^\perp = \{0\}$. If V is totally isotropic, then there exists a totally isotropic subspace W of U such that $\dim W = \dim V$ and $V \oplus W$ is a hyperbolic space. In particular a regular 2-dimensional isotropic quadratic space is a hyperbolic plane.*

Proof. We use induction on $\dim V$. Let $\{v_1, \cdots, v_n\}$ be a basis of V. Then by virtue of Proposition 1.1.2, there is a $y \in U$ such that $b(v_1, y) = 1, b(v_i, y) = 0$ for $i \geq 2$, and $w_1 := y - q(y)v_1$ satisfies $q(w_1) = 0$ and $b(v_1, w_1) = 1$. This means that $W_1 := F[v_1, w_1]$ is a hyperbolic plane. Thus, in particular the case of $\dim V = n = 1$ has been completed. Suppose $n > 1$. Since V is totally isotropic, $b(V, V) = 0$ and $\{v_2, \cdots, v_n\} \subset W_1^\perp$ holds. Since

$$((W_1^\perp))^\perp := \{x \in W_1^\perp \mid B(x, W_1^\perp) = 0\} \subset U^\perp$$

because of $U = W_1 \perp W_1^\perp$, we have

$$F[v_2, \cdots, v_n] \cap ((W_1^\perp))^\perp \subset V \cap U^\perp = 0,$$

and hence by the assumption of the induction, there is a totally isotropic subspace W_2 in $W_1^\perp$ with $\dim W_2 = n - 1$ such that $F[v_2, \cdots, v_n] \oplus W_2$ is a hyperbolic space. So $W := Fw_1 \oplus w_2$ is what we want. □

From Corollary 1.2.1 it follows that all maximal hyperbolic subspaces in a quadratic space U have the same dimension. A half of the dimension of a maximal hyperbolic subspace is called the *Witt index* of U which we write $\mathrm{ind}\, U$. By virtue of the above proposition, $\mathrm{ind}\, U$ is equal to the dimension of a maximal totally isotropic subspace if U is regular.

Corollary 1.2.4. *Let U be a regular quadratic space over a field F and $V = Fv$ a 1-dimensional quadratic space with $q(v) = -a \in F^\times$. Then a is represented by U, i.e. $a \in q(U)$ if and only if $U \perp V$ is isotropic.*

Proof. Suppose $q(x) = a$ for $x \in U$; then $x + v \in U \perp V$ is isotropic. Conversely, suppose that $U \perp V$ is isotropic and $u + bv$ is an isotropic vector for $u \in U$, $b \in F$. If $b \neq 0$, then $q(b^{-1}u) = -q(v) = a$. If $b = 0$, then u is isotropic. Hence the regular space U contains a hyperbolic plane and then $q(U) = F \ni a$.

Proposition 1.2.3. *Let U be a 2-dimensional regular quadratic space over a field F with* $\mathrm{ch}\, F \neq 2$. *Then assertions (i), (ii), (iii) are equivalent;*

(i) *U is isotropic;*
(ii) *U is a hyperbolic plane;*
(iii) $\mathrm{d}U = -1$ *(in $F^\times/(F^\times)^2$).*

Proof. (i) ⇒ (ii) follows from the previous proposition. (ii) ⇒ (iii) is clear. Suppose that (iii) holds. Since $\mathrm{ch}\, F \neq 2$, there is an orthogonal basis $\{u_1, u_2\}$ of U and $q(u_1)q(u_2) = -a^2$ for some $a \in F^\times$. Hence $u_1 + q(u_1)a^{-1}u_2$ is an isotropic vector of U. □

Exercise 1. Prove Sylvester's law of inertia, that is for a quadratic space U over the field $\mathbb{R}$ of real numbers, there is a basis $\{v_i\}$ satisfying $q(\sum a_i v_i) = \sum_{i=1}^m a_i^2 - \sum_{i=1}^n a_{m+i}^2$, where m and n are dependent only on U.

Exercise 2. Let V be a quadratic space over a field F. If $V = V_1 \perp \mathrm{Rad}\, V = V_2 \perp \mathrm{Rad}\, V$, then V_1 is isometric to V_2. If $\mathrm{ch}\, F = 2$, then this does not hold for $V^\perp$ instead of $\mathrm{Rad}\, V$. Namely, suppose that $\mathrm{ch}\, F = 2$ and that H is a hyperbolic plane, V is a regular 2-dimensional quadratic space; then $V \perp Fu \cong H \perp Fu$ if $q(u) \in q(V)$. (Hint: Note that for a basis $\{v_1, v_2\}$ of V with $q(v_1) = q(u)$, $F[u + v_1, v_2]$ is regular and isotropic and hence hyperbolic.)

Exercise 3. Let W, H be quadratic modules over a ring R and suppose that H is a hyperbolic plane, that is H has a basis $\{h_1, h_2\}$ over R satisfying $q(a_1h_1 + a_2h_2) = a_1a_2$. For $w \in W$ and $v \in V := W \perp H$, we put

$$E_w(v) := v + b(v, h_1)w - b(v, w)h_1 - q(w)b(v, h_1)h_1.$$

E_w is an isometry of V such that $E_w(h_1) = h_1$ and $E_{w_1+w_2} = E_{w_1}E_{w_2}$. If an isometry σ of V satisfies $\sigma(h_1) = h_1$, then $\sigma^{-1}E_w = \mathrm{id}$ on H for some $w \in W$. This is called a *Siegel-Eichler transformation.*

1.3 Quadratic forms over finite fields

Throughout this section, let F be a finite field $\mathbb{F}_q$ with q elements.

Let $F' = F(\theta)$ be a unique quadratic extension of F; then the norm N and the trace tr from F' to F are surjective mappings. For $x, y \in F'$, $\mathrm{tr}(x\bar{y}) = \mathrm{N}(x + y) - \mathrm{N}(x) - \mathrm{N}(y)$ is clear, where $\bar{y}$ denotes the conjugate of y over F. If $\mathrm{N}(x) = 0$ for $x \in F'$, then $x = 0$ holds, and $\mathrm{tr}(x\overline{F'}) = 0$, for $x \in F'$, implies $x = 0$. Thus (F', N) is an anisotropic regular quadratic space over F.

Proposition 1.3.1. *Let U be a 2-dimensional anisotropic quadratic space over a finite field F. Then $U \cong (F', \mathrm{N})$ holds, and hence U is regular and $q(U) = F$.*

Proof. For a basis $\{u_1, u_2\}$ of U and $x \in F$, we have

$$q(xu_1 + u_2) = q(u_1)x^2 + b(u_1, u_2)x + q(u_2) \neq 0$$

since U is anisotropic. Hence there exists $\alpha \in F'$ such that

$$q(x_1u_1 + x_2u_2) = q(u_1)(x_1 - \alpha x_2)(x_1 - \bar{\alpha} x_2) = q(u_1)\,\mathrm{N}(x_1 - \alpha x_2)$$

for $x_1, x_2 \in F$. Taking $\beta \in F'$ with $\mathrm{N}(\beta) = q(u_1)$, a mapping $x_1u_1 + x_1u_2 \mapsto \beta(x_1 - \alpha x_2)$ gives an isometry from U onto (F', N). □

Corollary 1.3.1. *A regular 2-dimensional quadratic space over F is isometric to a hyperbolic plane or an anisotropic space (F', N).*

Proof. Unless a quadratic space is anisotropic, then it is isotropic. Hence Propositions 1.2.2 and 1.3.1 yield immediately the corollary. □

Proposition 1.3.2. *A quadratic space U over a finite field F with* $\dim U > 2$ *is isotropic.*

Proof. Suppose that U is anisotropic. Take a 2-dimensional subspace V; then V is regular and $q(V) = F$ by virtue of the previous proposition. For a non-zero vector $x \in V^\perp (\neq \{0\})$, there is a vector $v \in V \cong (F', \mathrm{N})$ such that $q(v) = -q(x) (\neq 0)$. Thus $v + x$ is isotropic. This contradicts our assumption on U. □

Corollary 1.3.2. *An orthogonal sum V of two isometric 2-dimensional regular quadratic spaces over F is a hyperbolic space.*

Proof. Suppose that an orthogonal sum V of two 2-dimensional anisotropic quadratic spaces is not a hyperbolic space. Since $\dim V = 4 > 2$, V is isotropic by Proposition 1.3.2. Then it is an orthogonal sum of a hyperbolic plane and an anisotropic 2-dimensional quadratic space (F', N). From Corollary 1.2.3 it follows that a hyperbolic plane is isometric to (F', N), which is a contradiction. □

Exercise 1. Suppose $\mathrm{ch}\, F \neq 2$. Since a regular 2-dimensional quadratic space represents 1, every regular quadratic space V is isometric to

$$\langle 1 \rangle \perp \cdots \perp \langle 1 \rangle \perp \langle \mathrm{d}\, V \rangle,$$

and hence it is classified by $\dim V$ and $\mathrm{d}\, V$ $(\in F^\times/(F^\times)^2)$.

Suppose $\operatorname{ch} F = 2$, and let $V = \perp_{i=1}^{n} F[e_i, f_i]$ be a regular quadratic space over F with $b(e_i, f_i) = 1$ for all i. Then

$$A(V) := \sum_{i=1}^{n} q(e_i)q(f_i) \bmod P$$

depends only on V, where $P := \{x^2 + x \mid x \in F\}$. We call $A(V)$ the *Arf invariant* of V. (It is not necessary to confine ourselves to finite fields.)

(Hint: First, verify the case of $n = 1$. Suppose

$$V \cong \perp_{i=1}^{m} H \perp (\perp_{i=m+1}^{n} (F', N))$$

where H is a hyperbolic plane; then $A(V) = (n - m)A((F', N))$ holds and hence, noting $A((F', N)) \neq 0$, we have $A(V) = 0$ if and only if V is a hyperbolic space.)

Theorem 1.3.1. *A regular quadratic space over a finite field F which is not isometric to a hyperbolic space is isometric to an orthogonal sum of a hyperbolic space and an anisotropic 2-dimensional space ($\cong (F', \mathrm{N})$), or an orthogonal sum of a hyperbolic space and a 1-dimensional regular quadratic space. Here the latter case can occur only when* $\operatorname{ch} F \neq 2$.

Proof. If a regular quadratic space U over F is of $\dim U > 2$, then Proposition 1.3.2 and Proposition 1.2.2 imply that U contains a hyperbolic plane H. Applying a similar argument to $H^\perp$ we have, finally, $U = V \perp W$ for a hyperbolic space V and W with $\dim W \leq 2$. By Proposition 1.2.2 and Proposition 1.3.1, W is a hyperbolic plane or isometric to (F', N) if $\dim W = 2$. If $\operatorname{ch} F = 2$, then $\dim U$ is even by Theorem 1.2.1 and hence $\dim W = 1$ does not happen. □

We put, for U with $\dim U$ even, $\chi(U) = 1$ if U is a hyperbolic space, and $\chi(U) = -1$, otherwise. For an odd-dimensional regular quadratic space U, we put $\chi(U) = 0$. Then it is easy to see that $\chi(U \perp H) = \chi(U)$ if H is a hyperbolic space, and $\chi(U \perp V) = -\chi(U)$ if V is an anisotropic 2-dimensional space and U is regular, since an orthogonal sum of two 2-dimensional anisotropic spaces is a hyperbolic space.

Let us evaluate the number $r(V, W)$ of isometries from V to W. For a quadratic space U over F and $t \in F$, we put

$$a(t, U) := \sharp\{x \in U \mid q(x) = t\}.$$

If Fu is a quadratic space with $q(u) = t \neq 0$, then $r(Fu, U) = a(t, U)$ is obvious.

Lemma 1.3.1. *For a regular quadratic space U over F with* $\dim U = u$ *we have*

$$a(0, U) = q^{u-1} - \chi(U)q^{u/2-1} + \chi(U)q^{u/2}.$$

Proof. We use induction on u. If $u = 1$, then $a(0, U) = 1$ is easy, since the regularity of U with $u = 1$ implies $\operatorname{ch} F \neq 2$, and $q(U) \neq \{0\}$. Suppose $u = 2$. If U is anisotropic, then $a(0, U) = 1$ and $\chi(U) = -1$ imply the assertion. If U is a hyperbolic plane, then

$$a(0, U) = \sharp\{x_1, x_2 \in F \mid x_1 x_2 = 0\} = 2q - 1$$

which is the assertion. Suppose $\dim U \geq 3$ and decompose U as $U = H \perp U_0$, where H is a hyperbolic plane, and we take a hyperbolic basis $\{h_1, h_2\}$. Writing $u = a_1 h_1 + a_2 h_2 + u_0$ for $u \in U, u_0 \in U_0$, we have

$$\begin{aligned} a(0, U) &= \sharp\{a_1, a_2 \in F, u_0 \in U_0 \mid a_1 a_2 + q(u_0) = 0\} \\ &= (2q - 1)a(0, U_0) + \sum_{t \in F^\times} (q - 1)a(t, U_0) \\ &= (2q - 1)a(0, U_0) + (q - 1)(q^{u-2} - a(0, U_0)) \\ &= (q - 1)q^{u-2} + qa(0, U_0). \end{aligned}$$

By the induction hypothesis, we complete the proof. □

Lemma 1.3.2. *Let U be a regular quadratic space over F with even dimension. For $t \in F^\times$, there is an automorphism f of U such that $q(f(u)) = tq(u)$ for $u \in U$, and $a(t, U) = a(1, U) = q^{u-1}(1 - \chi(U)q^{-u/2})$ holds with* $u := \dim U$.

Proof. Let H be a hyperbolic plane with hyperbolic basis $\{h_1, h_2\}$. For $t \in F^\times, f(h_1) = h_1, f(h_2) = th_2$ is an automorphism of H with $q(f(h)) = tq(h)$ for $h \in H$. If K is an anisotropic 2-dimensional quadratic space, then identifying K with F' at the beginning and taking an element $k \in F'$ such that $q(k) = t$, $f(x) = kx$ is an automorphism of K with $q(f(x)) = tq(x)$ for $x \in K$. By virtue of Theorem 1.3.1 we obtain the existence of an automorphism f of U. Then

$$\begin{aligned} a(t, U) = a(1, U) &= (q - 1)^{-1} \sum_{t \in F^\times} a(t, U) \\ &= (q - 1)^{-1}(q^u - a(0, U)) = q^{u-1} - \chi(U)q^{u/2-1} \end{aligned}$$

from the previous lemma. □

For a regular quadratic space (U, q), we denote a new quadratic space $(U, -q)$ by $U^{(-1)}$. Let U be a regular quadratic space such that $\dim U = 1$

if $\operatorname{ch} F \neq 2$, $\dim U = 2$ if $\operatorname{ch} F = 2$; then $U \perp U^{(-1)}$ is hyperbolic, by Proposition 1.2.3 if $\operatorname{ch} F \neq 2$, Corollary 1.3.1 and 1.3.2 if $\operatorname{ch} F = 2$. Thus, for a regular quadratic space U, $U \perp U^{(-1)}$ is hyperbolic and hence

$$\chi(U \perp U^{(-1)}) = 1.$$

Theorem 1.3.2. *Let V, W be regular quadratic spaces over a finite field $F(= F_q)$ and put $v := \dim V$, $w := \dim W$ and suppose $v \leq w$. Then the number $r(V, W)$ of isometries from V to W is equal to*

$$q^{vw-v(v+1)/2}(1 - \chi(W)q^{-w/2})(1 + \chi(V \perp W^{(-1)})q^{(v-w)/2}) \times \prod_{\substack{w-v+1 \leq e \leq w-1 \\ e:\text{even}}} (1 - q^{-e}).$$

Proof. Let $r'(V, W)$ be the number given by the above formula; we show $r(V, W) = r'(V, W)$ by induction on $v = \dim V$. We note first that for regular quadratic spaces U, U_1, U_2, we have

$$r(U \perp U_1, U \perp U_2) = \sum_{\sigma: U \hookrightarrow U \perp U_2} r(U_1, \sigma(U)^{\perp}) = r(U, U \perp U_2) r(U_1, U_2)$$

since $\sigma(U)^{\perp} \cong U_2$ follows from Corollary 1.2.3.

It is easy to verify $r'(U \perp U_1, U \perp U_2) = r'(U, U \perp U_2) r'(U_1, U_2)$ for $\dim U = 1$ if $\operatorname{ch} F \neq 2$, a hyperbolic plane U if $\operatorname{ch} F = 2$. To show $r(V, W) = r'(V, W)$, we have only to prove $r(V, W) = r'(V, W)$ for 1-dimensional spaces V if $\operatorname{ch} F \neq 2$, and for 2-dimensional spaces V if $\operatorname{ch} F = 2$, noting that a regular quadratic space is an orthogonal sum of 1-dimensional subspaces if $\operatorname{ch} F \neq 2$, or an orthogonal sum of hyperbolic planes and at most one anisotropic 2-dimensional subspace if $\operatorname{ch} F = 2$.

Claim. *If $\operatorname{ch} F \neq 2$, then $r(V, W) = r'(V, W)$ in the case of $v = \dim V = 1$.*

Proof. We have only to prove $a(t, W) = r'(V, W)$ if $V = Fv, q(v) = t (\neq 0)$.

If $w = \dim W$ is even, then $a(t, W) = q^{w-1}(1 - \chi(W)q^{-w/2})$ follows from Lemma 1.3.2 and it is equal to $r'(V, W)$.

Suppose $w = \dim W = 1$; then we know $\chi(V \perp W^{(-1)}) = 1 \Leftrightarrow dV = dW \Leftrightarrow V \cong W$, and $r(V, W) = 2$ or 0 depending on whether $V \cong W$ or $V \ncong W$ respectively. Hence $r(V, W) = r'(V, W)$ follows.

Suppose that w is odd and put $W = Fw_1 \perp W_0$. Then from

$$a(t, W) = \sharp\{x \in F, w_0 \in W_0 \mid x^2 q(w_1) + q(w_0) = t\}$$

it follows that $a(t, W) = qa(1, W_0)$ if $x^2 q(w_1) \neq t$ for all $x \in F$, and

$$a(t, W) = 2a(0, W_0) + (q - 2)a(1, W_0)$$

if $x^2q(w_1) = t$ for some $x \in F$. Here we note that $t \neq 0$. Suppose $x^2q(w_1) \neq t$ for all $x \in F$; then $Fw_1 \not\cong V$, and hence $V \perp (Fw_1)^{(-1)}$ is anisotropic. Thus

$$\chi(V \perp W^{(-1)}) = -\chi(W_0^{(-1)}) = -\chi(W_0)$$

and

$$a(t, W) = q^{w-1}(1 + \chi(V \perp W^{(-1)})q^{(1-w)/2}) = r'(V, W).$$

Suppose $x^2q(w_1) = t$ for some $x \in F$; then $Fw_1 \cong V$ and $V \perp (Fw_1)^{(-1)}$ is hyperbolic. Thus $\chi(V \perp W^{(-1)}) = \chi(W_0)$ follows. Finally

$$\begin{aligned} a(t, W) &= 2(q^{w-2} - \chi(W_0)q^{(w-1)/2-1} + \chi(W_0)q^{(w-1)/2}) \\ &\quad + (q-2)q^{w-2}(1 - \chi(W_0)q^{-(w-1)/2}) \\ &= q^{w-1}(1 + \chi(W_0)q^{(1-w)/2}) = r'(V, W). \end{aligned}$$

Thus we have completed the proof in the case of $\operatorname{ch} F \neq 2$. □

Claim. *If* $\operatorname{ch} F = 2$, *then* $r(V, W) = r'(V, W)$ *in the case of* $v = \dim V = 2$.

Proof. Let U be a hyperbolic plane and $\{u_1, u_2\}$ a hyperbolic basis. By virtue of the fact that {isotropic vectors in U} = $\{tu_1, tu_2 \mid t \in F^\times\}$, an isometry σ of U is $\sigma(u_1) = tu_1, \sigma(u_2) = t^{-1}u_2$ or $\sigma(u_1) = tu_2, \sigma(u_2) = t^{-1}u_1$. Hence we have $r(U, U) = 2(q-1) = r'(U, U)$.

Let K be an anisotropic 2-dimensional quadratic space identified with (F', N) as explained at the beginning of this section. If σ is an isometry of K with $\sigma(1) = 1$, then σ is a linear isomorphism of F' over F. Now $q(\sigma(x)) = x$ and $b(1, \sigma(x)) = b(1, x)$ imply $\mathrm{N}(\sigma(x)) = x$ and $\operatorname{tr}\sigma(x) = \operatorname{tr} x$, and hence the generator θ with $F' = F(\theta)$ is transformed by σ to a conjugate. Thus σ is either the identity or the conjugate mapping. Hence an isometry σ of K is $\sigma(x) = \alpha x$ or $\overline{\alpha x}$ for some $\alpha \in K$ with $\mathrm{N}(\alpha) = 1$, and $r(K, K) = 2a(1, K) = 2q(1 + q^{-1}) = r'(K, K)$.

If $v = w = 2$ and $V \not\cong W$, then $V \perp W^{(-1)}$ is not a hyperbolic space and hence $\chi(V \perp W^{(-1)}) = -1$ and $r(V, W) = 0 = r'(V, W)$ follows. Thus we have completed the proof of $w = 2$.

Suppose $w \geq 4$ and put

$$W = H \perp W_0$$

where H is a hyperbolic plane with hyperbolic basis $\{h_1, h_2\}$.

If V is a hyperbolic plane, then Corollary 1.2.1 shows that every isotropic vector in W is transformed to h_1 by an isometry of W. We have

$$r(V, W) = (a(0, W) - 1) \cdot \sharp\{w \in W \mid b(h_1, w) = 1,\ q(w) = 0\}.$$

By Lemma 1.3.1, we know

$$a(0,W)-1=q^{w-1}(1-\chi(W)q^{-w/2})(1+\chi(W)q^{1-w/2}).$$

On the other hand, we have

$$\begin{aligned}&\sharp\{w\in W \mid b(h_1,w)=1,\ q(w)=0\}\\&=\sharp\{w=x_1h_1+x_2h_2+w_0\in W \mid x_1,x_2\in F,\ w_0\in W_0,\ b(h_1,w)=1,\\&\qquad q(w)=0\}\\&=\sharp\{x_1\in F,\ w_0\in W_0 \mid x_1+q(w_0)=0\}\\&=q^{w-2},\end{aligned}$$

and hence

$$r(V,W)=q^{2w-3}(1-\chi(W)q^{-w/2})(1+\chi(W)q^{1-w/2})=r'(V,W).$$

Suppose that V is an anisotropic 2-dimensional quadratic space with $q(x_1v_1+x_2v_2)=ax_1^2+bx_1x_2+cx_2^2$ for $x_1,x_2\in F$; then $r(V,W)$ equals

$$\begin{aligned}&a(a,W)\sharp\{w\in W \mid b(h_1+ah_2,w)=b,\ q(w)=c\}\\&\quad=a(1,W)\sharp\{w=a_1h_1+a_2h_2+w_0 \mid a_1,a_2\in F,\ w_0\in W_0,\ a_2\\&\qquad+aa_1=b, a_1a_2+q(w_0)=c\}\\&\quad=a(1,W)\sharp\{a_1\in F,\ w_0\in W_0 \mid q(w_0)=aa_1^2+ba_1+c\}\\&\quad=a(1,W)a(1,W_0)q,\end{aligned}$$

since $aa_1^2+ba_1+c=q(a_1v_1+v_2)\neq 0$ for $a_1\in F$.

Thus we have

$$r(V,W)=q^{2w-3}(1-\chi(W)q^{-w/2})(1-\chi(W_0)q^{(2-w)/2})=r'(V,W),$$

noting that $\chi(W_0)=\chi(W^{(-1)})=-\chi(V\perp W^{(-1)})$, which completes the proof in the case of $\operatorname{ch}F=2$ and so the proof of Theorem 1.3.2. □

Proposition 1.3.3. *Let V,W be quadratic spaces over F and suppose that V is regular and represented by W. Put $W=W_0\perp W^\perp$ and $v=\dim V$, $w=\dim W_0$, $r=\dim \operatorname{Rad} W$. Then we have*

$$r(V,W)=\begin{cases} r(V,W_0)q^{vr} & \text{if } W^\perp=\operatorname{Rad}W,\\ q^{v\dim W-v(v+1)/2}\prod_{\substack{w-v+2\le e\le w\\ e:\text{even}}}(1-q^{-e}) & \text{if } W^\perp\neq\operatorname{Rad}W.\end{cases}$$

Proof. Let $\{v_i\}$ be a basis of V. Every isometry σ from V to W is given by $\sigma(v_i)=\eta(v_i)+w_i$ for an isometry η from V to W_1 and $w_i\in\operatorname{Rad}W$ where

W_1 is defined by $W = W_1 \perp \operatorname{Rad} W$. Hence we have $r(V,W) = r(V,W_1)q^{vr}$ and the first equation is clear. Note that $W^\perp = \operatorname{Rad} W$ if $\operatorname{ch} F \neq 2$.

Next we suppose $\operatorname{ch} F = 2$, $W^\perp \neq \operatorname{Rad} W$. From $(F^\times)^2 = F^\times$ follows $W^\perp = Fw_0 + \operatorname{Rad} W$, $q(w_0) = 1$ by virtue of Theorem 1.2.1. We have only to show

$$r(V,W_1) = q^{(w+1)v - v(v+1)/2} \prod_{\substack{w-v+2 \le e \le w \\ e:\text{even}}} (1 - q^{-e}).$$

We may suppose $W_1 = W_0 \perp Fw_0$. We use induction on v. Since V is regular, v is even. Suppose $v = 2$ and $q(v_i) = a_i$, $b(v_1, v_2) = 1$.

$$r(V,W_1) = \sharp\{(w_1,w_2) \in W_1 \times W_1 | q(w_i) = a_i,\ b(w_1,w_2) = 1\}$$

is clear. If $q(w_1) = q(w_1') = a_1$ and $b(w_1, W_1) \neq 0$, $b(w_1', W_1) \neq 0$ for $w_1, w_1' \in W_1$, then they are transformed by an isometry of W_1 by virtue of Corollary 1.2.1. Hence

$$\begin{aligned} r(V,W_1) = \sharp\{x \in W_1 \mid q(x) = a_1,\ b(x,W_1) \neq 0\} \\ \times \sharp\{x \in W_1 \mid q(x) = a_2,\ b(x,w_1) = 1\} \end{aligned}$$

where $w_1 \in W_1$ is any vector such that $q(w_1) = a_1$, $b(w_1, W_1) \neq 0$. Let us evaluate $\sharp\{x \in W_1 \mid q(x) = a_1, b(x,W_1) \neq 0\}$. Since $q(Fw_0) = F$, there is, for an element $y \in W_0$, a unique $f \in F$ such that $q(y + fw_0) = q(y) + f^2 = a_1$; moreover $b(y + fw_0, W_1) = b(y, W_0) \neq 0$ is equivalent to $y \neq 0$. Hence, writing $x = y + fw_0$ $(y \in W_0)$, it follows easily that $\sharp\{x \in W_1 \mid q(x) = a_1, b(x,W_1) \neq 0\} = q^w - 1$.

Next, let us evaluate

$$\sharp\{x \in W_1 \mid q(x) = a_2,\ b(x,w_1) = 1\}$$

and put $w_1 := y_0 + aw_0$, $y_0 \in W_0$, $a \in F$. Suppose $q(w_1) = a_1$, $b(w_1, W_1) \neq 0$. For $x = y + cw_0$ $(y \in W_0,\ c \in F)$, we have $b(x, w_1) = b(y_0, y)$, and $q(x) = q(y) + c^2 = a_2$ uniquely determines c by y. Thus we have

$$\sharp\{x \in W_1 \mid q(x) = a_2,\ b(x,w_1) = 1\} = \sharp\{y \in W_0 \mid b(y,y_0) = 1\} = q^{w-1}.$$

Hence $r(V,W_1) = (q^w - 1)q^{w-1}$ is what we desired for $v = 2$.

If $v > 2$, then we decompose V as $V = V_0 \perp V_1$ with $\dim V_0 = 2$ and use $r(V,W_1) = r(V_0,W_1)r(V_1, V_0^\perp \text{ in } W_1)$, (regarding $V_0 \subset W_1$) and the induction hypothesis while noting that $\dim(V_0^\perp \text{ in } W_1)^\perp = 1$. □

1.4 The Clifford algebra

Henceforth, unless mentioned otherwise, the characteristic of fields is different from 2 and a ring is embedded in a field. We replace a quadratic form q by Q, the associated bilinear form b by $B := \frac{1}{2}b$. Thus we have, for a quadratic form Q and the associated bilinear form B,

$$Q(x+y) - Q(x) - Q(y) = 2B(x,y),\ Q(x) = B(x,x).$$

A quadratic module (M, Q) is considered as a symmetric bilinear module with respect to B. So, $M \cong \langle A \rangle$ means $(B(v_i, v_j)) = A$ for a basis $\{v_i\}$ of M.

For a quadratic space V over a field F, we put

$$T(V) := \oplus_{m=0}^{\infty} \otimes^m V,\ \otimes^0 V := F,\ \otimes^1 V := V,$$

where $\otimes$ denotes the tensor product over F. For $u := u_1 \otimes \cdots \otimes u_m \in \otimes^m V, v := v_1 \otimes \cdots \otimes v_n \in \otimes^n V$ ($u_i, v_j \in V$), we define a product of u and v by

$$u \otimes v := u_1 \otimes \cdots \otimes u_m \otimes v_1 \otimes \cdots \otimes v_n \in \otimes^{m+n} V.$$

This is well defined and $T(V)$ is called the *tensor algebra* of V. Denoting by $I(V)$ the two-sided ideal of $T(V)$ generated by $v \otimes v - Q(v)$ ($v \in V$), we put

$$C(V) := T(V)/I(V),$$

which is called the *Clifford algebra*. For simplicity, an element of $C(V)$ which is represented by $v_1 \otimes \cdots \otimes v_m$ ($v_i \in V$) is denoted by $v_1 \cdots v_m$. By definition, we have, in $C(V)$

$$\text{(1)} \qquad \begin{cases} v \in V & \Rightarrow v^2 = Q(v) \\ u, v \in V & \Rightarrow uv + vu = 2B(u,v),\ \ uv = -vu \ \text{ if } B(u,v) = 0. \end{cases}$$

Proposition 1.4.1. *Let V be a quadratic space over F and A an algebra over F. If f is a linear mapping from V to A with $f(v)^2 = Q(v)$ for $v \in V$, then there is a homomorphism g from $C(V)$ to A such that $f(v) = g(v)$ for $v \in V$.*

Proof. We define an algebra homomorphism g' from $T(V)$ to A by

$$g'(v_1 \otimes \cdots \otimes v_m) = f(v_1) \cdots f(v_m) \text{ for } v_i \in V.$$

Then $\ker g'$ is a two-sided ideal of $T(V)$ containing $v \otimes v - Q(v)$ for $v \in V$ and hence $\ker g'$ contains $I(V)$. Thus g' induces a homomorphism g from $C(V) = T(V)/I(V)$ to A satisfying $g(v) = g'(v) = f(v)$ for $v \in V$. □

Corollary 1.4.1. *For isometric quadratic spaces V, W over F, Clifford algebras $C(V), C(W)$ are isomorphic.*

Proof. Let σ be an isometry from V on W; then σ induces a linear mapping from V to $C(W)$ by $x \mapsto \sigma(x) \in C(W)$. Hence there exists a homomorphism f from $C(V)$ to $C(W)$ satisfying $f(x) = \sigma(x)$ for $x \in V$. Similarly, there exists a homomorphism g from $C(W)$ to $C(V)$ satisfying $g(w) = \sigma^{-1}(w)$ for $w \in W$. Thus gf (resp. fg) is the identity mapping on elements in $C(V)$ (resp. $C(W)$) represented by elements in V (resp. W) and hence on $C(V)$ (resp. $C(W)$) respectively. □

Theorem 1.4.1. *Let V be a quadratic space over F with an orthogonal basis $\{v_i\}_{i=1}^n$. Then $\{v_1^{e_1} \cdots v_n^{e_n} \mid e_i = 0$ or $1\}$ is a basis of the Clifford algebra $C(V)$. In particular $\dim C(V) = 2^n$.*

Proof. By virtue of (1), the Clifford algebra is spanned by $v_1^{e_1} \cdots v_n^{e_n}$ ($e_1 = 0$ or 1) as a vector space and hence $\dim C(V) \leq 2^n$. We must show the converse inequality. Let F_2 be a prime field with 2 elements and $K = F_2^n$ an n-dimensional vector space over F_2, and moreover let A be a 2^n-dimensional vector space whose basis are elements of K. When we regard an element f in K as a basis vector of A, we will write $[f]$. So, note $[f_1 + f_2] \neq [f_1] + [f_2]$ for $f_1, f_2 \in K$. We introduce a product in A by

$$[f_1] \cdot [f_2] = \prod_{1 \leq i < j \leq n} (-1)^{a_j b_i} \prod_{1 \leq i \leq n} q_i^{a_i b_i} [f_1 + f_2],$$

where $f_1 = (a_1, \cdots, a_n)$, $f_2 = (b_1, \cdots, b_n)$, $q_i = Q(v_i)$. Here we put, for $x \in F$, $a \in F_2$

$$x^a := \begin{cases} x & \text{if} \quad a = 1, \\ 1 & \text{if} \quad a = 0, \end{cases}$$

hence in particular $0^0 = 1$ and

$$(2) \qquad (-1)^{a+b} = (-1)^a (-1)^b, \quad \text{and} \quad x^{ab} x^{(a+b)c} = x^{bc} x^{a(b+c)}$$
$$(x \in F, \quad a, b, c \in F_2)$$

are easy.

Let $e_i = (0, \cdots, 0, 1, 0, \cdots, 0) \in K$ with i-th component 1. We show that $[e_i]$'s generate A. To do it, we have only to prove that

$$[e_S][e_T] = c[e_{S \cup T}] \quad \text{for} \quad c \in F^\times$$

where S, T are disjoint subsets of $\{1, \cdots, n\}$, and $e_S = (a_1, \cdots, a_n)$ with $a_i = 1$ for $i \in S$, $a_i = 0$ for $i \notin S$, and so on. By definition, for S, T as above we have $c = \prod_{1 \le i<j \le n}(-1)^{a_j b_i} \prod_{1 \le i \le n} q_i^{a_i b_i}$, putting $e_T = (b_1, \cdots, b_n)$, and from $a_i b_i = 0$ follows $c = \prod_{1 \le i<j \le n}(-1)^{a_j b_i} = \pm 1 \ne 0$, which is what we want.

By definition, we have $[0][f] = [f][0] = [f]$ for $f \in K$. For $f_1 = (a_1, \cdots, a_n)$, $f_2 = (b_1, \cdots, b_n)$, $f_3 = (c_1, \cdots, c_n) \in K$, we have

$$
\begin{aligned}
&([f_1][f_2])[f_3] \\
&= \prod_{1 \le i<j \le n} (-1)^{a_j b_i} \prod_{1 \le i \le n} q_i^{a_i b_i} [f_1 + f_2][f_3] \\
&= \prod_{1 \le i<j \le n} (-1)^{a_j b_i} \prod_{1 \le i \le n} q_i^{a_i b_i} \prod_{1 \le i<j \le n} (-1)^{(a_j + b_j) c_i} \\
&\quad \times \prod_{1 \le i \le n} q_i^{(a_i + b_i) c_i} [f_1 + f_2 + f_3] \\
&= \prod_{1 \le i<j \le n} (-1)^{a_j b_i + a_j c_i + b_j c_i} \prod_{1 \le i \le n} q_i^{a_i (b_i + c_i)} \\
&\quad \times \prod_{1 \le i \le n} q_i^{b_i c_i} [f_1 + f_2 + f_3] \quad \text{by (2)} \\
&= [f_1]([f_2][f_3]).
\end{aligned}
$$

Hence A is an algebra over F and we define a linear mapping f from V to A by $f(v_i) = [e_i]$. Then $f(v_i)^2 = q_i[0]$ is clear and so, by Proposition 1.4.1 there exists a homomorphism g from $C(V)$ to A. Since A is generated by $[e_i]$, g is a surjective mapping and hence we have $\dim C(V) \ge \dim A = 2^n$. □

Corollary 1.4.2. *V is canonically embedded in $C(V)$.*

Proof. This follows immediately from Theorem 1.4.1. □

The mapping $v \mapsto -v$ from V to $C(V)$ induces an automorphism J of $C(V)$ by Proposition 1.4.1 which we call the *canonical involution.* We put

$$C_0(V) = \{c \in C(V) \mid Jc = c\},$$

which is called the *second Clifford algebra.*

Corollary 1.4.3. *Let V be a regular quadratic space over F with an orthogonal basis $\{v_i\}_{i=1}^n$. Then elements of $C_0(V)$ are linear combinations of products of even number of elements of V, and*

$$\text{the centre of } C(V) = \begin{cases} F & \text{if} \quad 2 \mid n, \\ F + F v_1 \cdots v_n & \text{if} \quad 2 \nmid n, \end{cases}$$

$$\textit{the centre of } C_0(V) = \begin{cases} F + Fv_1 \cdots v_n & \textit{if} \quad 2|n, \\ F & \textit{if} \quad 2 \not| n, \end{cases}$$

Proof. For $x_1, \cdots, x_m \in V$, $J(x_1 \cdots x_m) = (-1)^m x_1 \cdots x_m$ follows from the definition, and $\{v_{i_1} \cdots v_{i_m} \mid i_1 < \cdots < i_m\}$ is a basis of $C(V)$. Hence $C_0(V)$ is spanned over F by $v_{i_1} \cdots v_{i_m}$ (m : even), and generated by $v_i v_j$ as an algebra.

For a subset $S = \{i_1, \cdots, i_m \mid i_1 < \cdots < i_m\}$, we put $v(S) = v_{i_1} \cdots v_{i_m} \in C(V)$. Then we see, by virtue of (1)

$$(3) \qquad v_i v(S) = \begin{cases} (-1)^m v(S) v_i & \text{if} \quad i \notin S, \\ (-1)^{m-1} v(S) v_i & \text{if} \quad i \in S. \end{cases}$$

Suppose that $x = \sum_{S \subset \{1, \cdots, n\}} x_S v(S)$ ($x_S \in F$) is in the centre of $C(V)$. Then, comparing coefficients of $v(T)$, $T = \{i_1 < \cdots < i_m\}$ in the equation $v_i x = x v_i$, we have

$$(-1)^{m-1} x_{T \setminus \{i\}} = x_{T \setminus \{i\}} \quad \text{if} \quad i \in T,$$
$$(-1)^m x_{T \cup \{i\}} = x_{T \cup \{i\}} \quad \text{if} \quad i \notin T.$$

Here we note that $v_i^2 \neq 0$ holds since V is regular. Hence, if $i \in T$ and $\sharp T \equiv 0 \bmod 2$, then $x_{T \setminus \{i\}} = 0$, and if $i \notin T$ and $m \equiv 1 \bmod 2$, then $x_{T \cup \{i\}} = 0$. Therefore, for a proper non-empty subset $S \subset \{1, \cdots, n\}$, applying this to $T := S \cup \{i\}$ for $i \notin S$ if $\sharp S \equiv 1 \bmod 2$, and $T := S \setminus \{i\}$ for $i \in S$ if $\sharp S \equiv 0 \bmod 2$, we have $x_S = 0$. Hence it yields $x = a_1 + a_2 v_1 \cdots v_n$, $a_i \in F$. From (3) it follows that $v_1 \cdots v_n$ is in the centre of $C(V)$ if and only if n is odd. Thus the centre of $C(V)$ is what we desired.

Next suppose $x = \sum x_S v(S)$ is in the centre of $C_0(V)$ and then $v_i v_j x = x v_i v_j$ holds. We have only to prove $x = a_1 + a_2 v_1 \cdots v_n$ ($a_1, a_2 \in F$). Comparing the terms of $v(T)$ in the equation $v_i v_j x = x v_i v_j$ for $i \neq j$, we have

$$v_i v_j x_{(T \setminus \{i\}) \cup \{j\}} \cdot v((T \setminus \{i\}) \cup \{j\}) = x_{(T \setminus \{i\}) \cup \{j\}} \cdot v((T \setminus \{i\}) \cup \{j\}) v_i v_j$$

if $i \in T$ and $j \notin T$. Since $x \in C_0(V)$, $x_S = 0$ if $\sharp S \equiv 1 \bmod 2$. Suppose that S is a non-empty proper subset of $\{1, \cdots, n\}$. Putting $T := (S \cup \{i\}) \setminus \{j\}$ for $i \notin S$, $j \in S$ in the above, we have

$$v_i v_j x_S v(S) = x_S v(S) v_i v_j.$$

By virtue of (3), $v_i v_j v(S) = -v(S) v_i v_j$ holds, and hence we have $2 x_S v(S) v_i v_j = 0$. This means $x_S v(T) = 0$ and hence $x_S = 0$. Thus we have $x = a_1 + a_2 v_1 \cdots v_n$. □

Remark.

Since $(v_1 \cdots v_n)^2 = (-1)^{(n-1)n/2} Q(v_1) \cdots Q(v_n)$, we have

$$F + Fv_1 \cdots v_n \cong F[X]/(X^2 - a),$$

where we put $a = (-1)^{(n-1)n/2} Q(v_1) \cdots Q(v_n) = (-1)^{(n-1)n/2} \mathrm{d}V \bmod (F^\times)^2$.

1.5 Quaternion algebras

Let C be an algebra which contains a field F in the centre of C. It is called a *quaternion algebra* over F if it has a basis $\{1, x_1, x_2, x_3\}$ over F such that

$$x_1^2 = \alpha,\ x_2^2 = \beta,\ x_3 = x_1 x_2 = -x_2 x_1\ (\alpha, \beta \in F^\times)$$

and we denote it by

$$(\alpha, \beta).$$

Then it is easy to see

$$x_3^2 = -\alpha\beta,\ x_2 x_3 = -x_3 x_2 = -\beta x_1,\ x_1 x_3 = -x_3 x_1 = \alpha x_2,$$

from which follows that F is the centre of C.

Proposition 1.5.1. *Put $C^0 := Fx_1 + Fx_2 + Fx_3$ for a quaternion algebra C. Then we have*

$$C^0 = \{x \in C \mid x^2 \in F,\ x \notin F\} \cup \{0\}.$$

Proof. Put $x = a_0 + \sum_{i=1}^{3} a_i x_i$, $a_i \in F$. Then, putting $x_0 := 1$, we have

$$\begin{aligned} x^2 &= \sum_{i=0}^{3} a_i^2 x_i^2 + \sum_{i<j} a_i a_j (x_i x_j + x_j x_i) \\ &= \sum_{i=0}^{3} a_i^2 x_i^2 + 2a_0 \sum_{j=1}^{3} a_j x_j. \end{aligned}$$

Hence, $x \in C^0$ implies $x^2 \in F$ and $x \notin F$ if $x \neq 0$. Conversely, suppose that $x^2 \in F$, $x \notin F$; then $a_0 a_j = 0$ for $j = 1, 2, 3$. Since $x \notin F$ implies $a_j \neq 0$ for some $j \geq 1$, $a_0 = 0$ follows, that is $x \in C^0$. □

We call an element of C^0 *pure*.

For $x = a_0 + \sum_{i=1}^3 a_i x_i \in C$, we put

$$\bar{x} := a_0 - \sum_{i=1}^{3} a_i x_i.$$

Then it is easy to see, for $x, y \in C$,

$$(1) \qquad \overline{x+y} = \bar{x} + \bar{y}, \quad \overline{xy} = \bar{y}\bar{x},$$

since $\overline{x_1 x_2} = \bar{x}_3 = -x_3 = x_2 x_1 = \bar{x}_2 \bar{x}_1$ and so on.

We put, for $x = a_0 + \sum_{i=1}^3 a_i x_i$,

$$\mathrm{T}(x) := x + \bar{x} = 2a_0 \in F, \ \ \mathrm{N}(x) := x\bar{x} = a_0^2 - a_1^2\alpha - a_2^2\beta + a_3^2\alpha\beta \in F.$$

T and N are called the trace and norm of the quaternion algebra respectively. It is easy to see from (1) that

$$\mathrm{T}(x) = \mathrm{T}(\bar{x}),\ \mathrm{N}(x) = \mathrm{N}(\bar{x}),\ \mathrm{T}(x+y) = \mathrm{T}(x) + \mathrm{T}(y),\ \mathrm{N}(xy) = \mathrm{N}(x)\,\mathrm{N}(y),$$

and the determinant of the homomorphism $y \mapsto yx$ is $(\mathrm{N}(x))^2$. Moreover for $x, y \in C$,

$$\mathrm{N}(x+y) - \mathrm{N}(x) - \mathrm{N}(y) = x\bar{y} + y\bar{x} = \mathrm{T}(x\bar{y})$$

is a symmetric bilinear form on C.

Thus (C, N) is a quadratic space over F with $\mathrm{d}(C, \mathrm{N}) = 1$ and

$$(C, \mathrm{N}) \cong \langle 1\rangle \perp \langle -\alpha\rangle \perp \langle -\beta\rangle \perp \langle \alpha\beta\rangle,$$
$$(C, \mathrm{N})^0 \cong \langle -\alpha\rangle \perp \langle -\beta\rangle \perp \langle \alpha\beta\rangle.$$

Proposition 1.5.2. *Let C, D be quaternion algebras over F. Then the following are equivalent:*

(i) *C and D are isomorphic as algebras;*
(ii) *(C, N) is isometric to (D, N);*
(iii) *$(C, \mathrm{N})^0$ is isometric to $(D, \mathrm{N})^0$.*

Proof. (i) $\Rightarrow$ (ii). Let φ be an isomorphism from C to D. By virtue of the previous proposition, we have $\varphi(C^0) = D^0$. Hence $\varphi(\bar{x}) = \overline{\varphi(x)}$ holds for $x \in C$ and then $\varphi(\mathrm{N}(x)) = \mathrm{N}(\varphi(x))$.

(ii) $\Rightarrow$ (iii). Let σ be an isometry from (C, N) to (D, N). Since $\mathrm{N}(\sigma(1)) = \mathrm{N}(1) = 1$, $\sigma(1)$ is transformed to 1 by an isometry of (D, N), and hence we may suppose $\sigma(1) = 1$. Then we have $\sigma(C^0) = D^0$, taking orthogonal complements of 1.

(iii) $\Rightarrow$ (i). Let $V = Fv_1 \perp Fv_2 \perp Fv_3$ be a regular quadratic space with $Q(v_1) = -\alpha$, $Q(v_2) = -\beta$, $Q(v_3) = \alpha\beta$; then we consider the second Clifford algebra $C_0(V) = F[1, v_1v_2, v_2v_3, v_1v_3] = F[1, \beta^{-1}v_2v_3, \alpha^{-1}v_1v_3, v_1v_2]$. Since $(\beta^{-1}v_2v_3)^2 = -\beta^{-2}v_2^2v_3^2 = \alpha$, $(\alpha^{-1}v_1v_3)^2 = -\alpha^{-2}v_1^2v_3^2 = \beta$, $(\beta^{-1}v_2v_3)(\alpha^{-1}v_1v_3) = v_1v_2$, it follows that $C_0(V)$ is isomorphic to the quaternion algebra (α, β). Therefore if V is isometric to C^0, then $C_0(V)$ is isomorphic to the original quaternion algebra C. This completes the proof. □

Proposition 1.5.3. *For $a, b \in F^\times$, the following are equivalent.*

(i) $(a, b) \cong (1, -1)$ *as algebras.*

(ii) (a, b) *is not a division algebra, that is there is a non-zero element x in (a, b) with $\mathrm{N}(x) = 0$.*

(iii) (a, b) *is isotropic as a quadratic space with respect to* N.

(iv) $(a, b)^0$ *is isotropic as a quadratic space with respect to* N.

(v) $\langle a\rangle \perp \langle b\rangle$ *represents* 1.

(vi) $a \in \mathrm{N}_{F'/F}(F')$ *for* $F' := F(\sqrt{b})$.

Proof. (i) $\Rightarrow$ (ii). Take a basis $\{1, x_1, x_2, x_3\}$ of $(1, -1)$ such that $x_1^2 = 1$, $x_2^2 = -1, x_3 = x_1x_2$; then $\mathrm{N}(1 + x_1) = (1 + x_1)(1 - x_1) = 0$ holds.

(ii) $\Rightarrow$ (iii). Obvious.

(iii) $\Rightarrow$ (iv). As a quadratic space, (a, b) is isometric to $\langle 1\rangle \perp \langle -a\rangle \perp \langle -b\rangle \perp \langle ab\rangle$ and so $(a, b)^0$ ($\cong \langle -a\rangle \perp \langle -b\rangle \perp \langle ab\rangle$) represents -1 by virtue of Corollary 1.2.4, since (a, b) is isotropic. Hence we can write $(a, b)^0 \cong \langle -1\rangle \perp U$ for a 2-dimensional subspace U. Comparing the discriminants, we have $\mathrm{d}U = -1$. From Proposition 1.2.3 it follows that U is isotropic and hence $(a, b)^0$ is isotropic.

(iv) $\Rightarrow$ (v). Since $(a, b)^0 \cong \langle -a\rangle \perp \langle -b\rangle \perp \langle ab\rangle$ is isotropic, $\langle -a\rangle \perp \langle -b\rangle$ represents $-ab$. Hence $-ax^2 - by^2 = -ab$ for some x, $y \in F$. Thus $a(a^{-1}y)^2 + b(b^{-1}x)^2 = 1$ means (v).

(v) $\Rightarrow$ (vi). Suppose $ax^2 + by^2 = 1$ for x, $y \in F$; then, in the case of $x \neq 0$ we have $a = \{x^{-1}(1 - \sqrt{b}y)\}(x^{-1}(1 + \sqrt{b}y)) \in \mathrm{N}_{F'/F}(F')$. In the case of $x = 0$, we have $b = y^{-2}$ and $F' = F$, and hence $a \in \mathrm{N}_{F'/F}(F')$.

(vi) $\Rightarrow$ (i). We show first that (vi) implies $\langle a\rangle \perp \langle b\rangle \cong \langle 1\rangle \perp \langle ab\rangle$. If $\sqrt{b} \in F$, then it is clear. If $\sqrt{b} \notin F$, then $a = x^2 - by^2$ is soluble for $x, y \in F$. $x^2 = a + by^2$ means that $\langle a\rangle \perp \langle b\rangle$ represents 1. Comparing discriminants, we have $\langle a\rangle \perp \langle b\rangle \cong \langle 1\rangle \perp \langle ab\rangle$. Hence we have

$$\begin{aligned}(a, b)^0 &\cong \langle -a\rangle \perp \langle -b\rangle \perp \langle ab\rangle \cong \langle -1\rangle \perp \langle -ab\rangle \perp \langle ab\rangle \\ &\cong \langle -1\rangle \perp \langle 1\rangle \perp \langle -1\rangle \cong (1, -1)^0.\end{aligned}$$

By the previous proposition, we have $(a,b) \cong (1,-1)$. $\square$

Exercise. Take a basis $\{1, x_1, x_2, x_3\}$ of $(1,-1)$ such that $x_1^2 = 1, x_2^2 = -1, x_3 = x_1x_2$; show the mapping $1 \mapsto 1_2$, $x_1 \mapsto \begin{pmatrix} 1 & 0 \\ 0 & -1 \end{pmatrix}$, $x_2 \mapsto \begin{pmatrix} 0 & -1 \\ 1 & 0 \end{pmatrix}, x_3 \mapsto \begin{pmatrix} 0 & -1 \\ -1 & 0 \end{pmatrix}$, gives $(1,-1) \cong M_2(F)$ and $\overline{\begin{pmatrix} a & b \\ c & d \end{pmatrix}} = \begin{pmatrix} d & -b \\ -c & a \end{pmatrix}$, $\mathrm{N}\begin{pmatrix} a & b \\ c & d \end{pmatrix} = ad - bc$, $\mathrm{T}\begin{pmatrix} a & b \\ c & d \end{pmatrix} = a + d$, and $\begin{pmatrix} a & b \\ c & d \end{pmatrix}$ corresponds to $2^{-1}(a+d) + 2^{-1}(a-d)x_1 + 2^{-1}(c-b)x_2 - 2^{-1}(b+c)x_3$.

Proposition 1.5.4. *For $a, b \in F^\times$, the following are true:*

(i) $O^+((a,b)) = \{x \mapsto \alpha x \beta \mid \alpha, \beta \in (a,b),\ \mathrm{N}(\alpha\beta) = 1\}$;
(ii) $x \mapsto \bar{x}$ *is an isometry of* (a,b), *but not in* $O^+((a,b))$;
(iii) *For an element* u *in* $(a,b)^0$ *with* $\mathrm{N}(u) \neq 0$, *we have*

$$\tau_u(x) = -uxu^{-1} \quad \text{for} \quad x \in (a,b)^0.$$

Proof. For an element u in (a,b) with $\mathrm{N}(u) \neq 0$, we have

$$\begin{aligned} \tau_u(x) &= x - \frac{\mathrm{T}(x\bar{u})}{\mathrm{N}(u)} u \\ &= x - (x\bar{u} + u\bar{x})(\bar{u}^{-1}u^{-1})u \\ &= -u\bar{x}\bar{u}^{-1}. \end{aligned}$$

This implies (iii). (ii) follows from $\bar{x} = -\tau_1(x)$ and $\det \tau_1 = -1$. Since $\det \tau_u = -1$, $O^+((a,b))$ is the set of all products of even number of symmetries. For $u_1, u_2 \in (a,b)$, with $\mathrm{N}(u_1u_2) \neq 0$, we have

$$\tau_{u_1}\tau_{u_2}(x) = -u_1\overline{(-u_2\bar{x}\bar{u}_2^{-1})}\bar{u}_1^{-1} = u_1u_2^{-1}x\bar{u}_2\bar{u}_1^{-1}.$$

Hence the left-hand side in (i) is contained in the right-hand side.

Conversely, let $\sigma(x) = \alpha x \beta$, $\alpha, \beta \in (a,b)$, $\mathrm{N}(\alpha\beta) = 1$; then σ is an isometry. Since $\tau_1\sigma(x) = -\overline{\alpha x \beta} = -\bar{\beta}\bar{x}\beta^{-1}\beta\bar{\alpha} = \tau_{\bar{\beta}}(x)\beta\bar{\alpha}$, we have $\det \tau_1\sigma = -\mathrm{N}(\beta\bar{\alpha})^2 = -1$. This implies $\tau_1\sigma \notin O^+$ and thus $\sigma \in O^+$. $\square$

Now we define the tensor product of quadratic spaces. Let U, V be quadratic spaces over a field F. We introduce a bilinear form B on $U \otimes V$ by $B(u_1 \otimes v_1, u_2 \otimes v_2) := B(u_1, u_2)B(v_1, v_2)$ for $u_1, u_2 \in U$, $v_1, v_2 \in V$, and define a quadratic form Q on $U \otimes V$ by $Q(x) := B(x,x)$. This quadratic space is called a *tensor product* of quadratic spaces U and V. By definition, $Q(u \otimes v) = Q(u)Q(v)$ for $u \in U$, $v \in V$. For a basis $\{u_i\}$ of U and a basis $\{v_i\}$ of V, we have $(B(u_i \otimes v_j, u_h \otimes v_k)) = (B(u_i, u_h)) \otimes (B(v_j, v_k))$ (Kronecker product of matrices). Suppose $\dim U = 1$ and $U = Fu$; then any element x of $U \otimes V$ is of form $x = u \otimes v$, $v \in V$ and hence $Q(x) = Q(u)Q(v)$. Thus $U \otimes V$ is isometric to (V, aQ) with $a = Q(u)$, which we write as $V^{(a)}$ and call it a *scaling* of V by a. If $U \cong \perp_i \langle a_i \rangle$, $V \cong \perp_j \langle b_j \rangle$, that is U, V have an orthogonal basis $\{u_i\}$, $\{v_i\}$ such that $Q(u_i) = B(u_i, u_i) = a_i$, $Q(v_j) = B(v_j, v_j) = b_j$, then $U \otimes V \cong \perp_{i,j} \langle a_i b_j \rangle$ holds, taking a basis $\{u_i \otimes v_j\}$ of $U \otimes V$. Hence, U, V are regular if and only if $U \otimes V$ is regular, and $\mathrm{d}(U \otimes V) = (\mathrm{d}\, U)^{\dim V} (\mathrm{d}\, V)^{\dim U}$ (in $F^\times/(F^\times)^2$).

Let us examine Clifford algebras for low dimensional quadratic spaces V.

(i) $\dim V = 1$. Let $V = Fv \cong \langle a \rangle$; then

$$C(V) = F + Fv \cong F[x]/(x^2 - a), \; C_0(V) = F.$$

(ii) $\dim V = 2$. Suppose $V = Fv_1 \perp Fv_2$, $Q(v_i) = q_i \in F^\times$; then

$$C(V) = F + Fv_1 + Fv_2 + Fv_1v_2 \cong (q_1, q_2).$$

and

$$C_0(V) = F + Fv_1v_2 \cong F[x]/(x^2 + q_1q_2) \cong F[x]/(x^2 + \mathrm{d}\, V).$$

(iii) $\dim V = 3$. Suppose $V = Fv_1 \perp Fv_2 \perp Fv_3$, $Q(v_i) = q_i \in F^\times$; then $C_0(V) = F + Fv_1v_2 + Fv_2v_3 + Fv_1v_3$ and

$$(v_1v_2)^2 = -q_1q_2, \; (v_2v_3)^2 = -q_2q_3, \; (v_1v_2)(v_2v_3) = q_2(v_1v_3).$$

Hence we have $C_0(V) \cong (-q_1q_2, -q_2q_3)$, and

$$\begin{aligned}(-q_1q_2, -q_2q_3)^0 &\cong \langle q_1q_2 \rangle \perp \langle q_2q_3 \rangle \perp \langle q_1q_2^2q_3 \rangle \\ &\cong \langle q_1q_2q_3 \rangle \otimes (\langle q_1 \rangle \perp \langle q_2 \rangle \perp \langle q_3 \rangle) \cong \langle \mathrm{d}\, V \rangle \otimes V.\end{aligned}$$

Thus, if $\mathrm{d}\, V = 1$, then we may consider V as a subspace of pure elements of the quaternion algebra $(-q_1q_2, -q_2q_3)$.

(iv) $\dim V = 4$. Suppose $V = Fv_1 \perp Fv_2 \perp Fv_3 \perp Fv_4$, $Q(v_i) = q_i \in F^\times$. Then, putting $Z :=$ centre of $C_0(V) = F + Fv_1v_2v_3v_4$, we have

$$C_0(V) = Z + Zv_1v_2 + Zv_2v_3 + Zv_1v_3$$

and

$$(v_1v_2)^2 = -q_1q_2, \; (v_2v_3)^2 = -q_2q_3, \; (v_1v_2)(v_2v_3) = q_2v_1v_3.$$

If $\mathrm{d}\,V \notin (F^\times)^2$, then $Z(\cong F[x]/(x^2 - \mathrm{d}\,V))$ is a quadratic extension of F, and $C_0(V)$ is isomorphic to a quaternion algebra $(-q_1q_2, -q_2q_3)$ over Z, and moreover $C_0(V)$ is isometric over Z to

$$\langle 1\rangle \perp \langle q_1q_2\rangle \perp \langle q_2q_3\rangle \perp \langle q_1q_2^2q_3\rangle \cong \langle q_1\rangle \otimes (\langle q_1\rangle \perp \langle q_2\rangle \perp \langle q_3\rangle \perp \langle q_4\rangle),$$

since $q_1q_4 = q_2q_3$ in $Z^\times/(Z^\times)^2$.

Thus if $Q(V) \ni 1$ and $\mathrm{d}\,V \notin (F^\times)^2$, then taking $v_1 \in V$ such that $Q(v_1) = 1$, we have $ZV \cong C_0(V)$ as quadratic spaces over Z.

1.6 The spinor norm

Let V be a quadratic space over F. Since the mapping $v_1 \otimes \cdots \otimes v_n \mapsto v_n \otimes \cdots \otimes v_1$ of the tensor algebra $T(V)$ transforms $I(V)$ to itself, it induces an involution ι of the Clifford algebra $C(V)$.

Proposition 1.6.1. *Let V be a quadratic space over F and $v_1, \cdots, v_m \in V$ be anisotropic. If the product of the symmetries $\tau_{v_1}, \cdots, \tau_{v_m}$ is the identity, then $Q(v_1)\cdots Q(v_m) \in (F^\times)^2$ holds.*

Proof. For an anisotropic vector $v \in V$, we have, in $C(V)$

$$\begin{aligned}\tau_v(x) &= x - \frac{2B(x,v)}{Q(v)}v \\ &= x - Q(v)^{-1}(xv + vx)v \quad \text{(because of } Q(x+v) = (x+v)^2) \\ &= -Q(v)^{-1}vxv \\ &= -v^{-1}xv,\end{aligned}$$

and hence $\tau_{v_1}\cdots\tau_{v_m}x = (-1)^m v_1^{-1}\cdots v_m^{-m} x v_m \cdots v_1$. If $\tau_{v_1}\cdots\tau_{v_m} = 1$, then, comparing determinants we have $m \equiv 0 \bmod 2$ and $xv_m\cdots v_1 = v_m\cdots v_1 x$ for $x \in V$. Thus $v_m\cdots v_1$ is in the centre of $C(V)$ and so $v_m\cdots v_1 = a + bx_1\cdots x_n$, where $\{x_i\}$ is an orthogonal basis of V and $a, b \in F$ by Corollary 1.4.3. Applying the automorphism J, we have $v_m\cdots v_1 = a + b(-1)^n x_1\cdots x_n$. If $n := \dim V$ is odd, then b is equal to 0. If n is even, then the centre of $C(V)$ is F and so b is also 0. Thus $v_m\cdots v_1 = a \in F$ follows and then $Q(v_1)\cdots Q(v_m) = v_m\cdots v_1v_1\cdots v_m = a\iota(a) = a^2 \in (F^\times)^2$. □

Corollary 1.6.1. *Let V be a regular quadratic space over F. For $\sigma = \tau_{v_1}\cdots\tau_{v_n} \in O(V)$,*

$$\theta(\sigma) := Q(v_1)\cdots Q(v_n) \in F^\times/(F^\times)^2$$

is uniquely determined by σ.

Proof. Since V is regular and $\mathrm{ch}\,F \neq 2$ is assumed, $O(V)$ is generated by symmetries. Suppose $\sigma = \tau_{v_1}\cdots\tau_{v_n} = \tau_{u_1}\cdots\tau_{u_m}$; then $\tau_{v_1}\cdots\tau_{v_n}\tau_{u_m}\cdots\tau_{u_1}$

$= 1$ implies $(Q(v_1)\cdots Q(v_n))(Q(u_m)\cdots Q(u_1)) \in (F^\times)^2$. Hence we have $Q(v_1)\cdots Q(v_n) = Q(u_m)\cdots Q(u_1)$ in $F^\times/(F^\times)^2$. □

The mapping θ from $O(V)$ to $F^\times/(F^\times)^2$ is called the *spinor norm* and it is obviously a homomorphism. We put

$$O'(V) = O^+(V) \cap \ker\theta.$$

We often identify the image of θ in $F^\times/(F^\times)^2$ with the inverse image by $F^\times \to F^\times/(F^\times)^2$ as far as there is no confusion.

Proposition 1.6.2. *Let $U = V \perp W$ be a regular quadratic space over F and suppose that $\sigma \in O(U)$ satisfies $\sigma(V) = V, \sigma(W) = W$. Then we have $\theta(\sigma) = \theta(\sigma|_V)\theta(\sigma|_W)$.*

Proof. Expressing $\sigma|_V = \tau_{v_1}\cdots\tau_{v_m}$ $(v_i \in V)$, $\sigma|_W = \tau_{w_1}\cdots\tau_{w_n}$ $(w_i \in W)$, we have $\sigma = \tau_{v_1}\cdots\tau_{v_m}\tau_{w_1}\cdots\tau_{w_n}$ and hence $\theta(\sigma) = Q(v_1)\cdots Q(v_m)Q(w_1)\cdots Q(w_n) = \theta(\sigma|_V)\theta(\sigma|_W)$. □

Corollary 1.6.2. *If V is a regular isotropic quadratic space, then $\theta(O^+(V)) = F^\times/(F^\times)^2$.*

Proof. Decompose V as $V = H \perp W$, where H is a hyperbolic plane. $Q(H) = F$ implies $\theta(O^+(H)) = F^\times$ and hence $\theta(O^+(V)) = F^\times$. □

Exercise 1. If V is regular, show $\theta(-1) = \mathrm{d}V$. (Hint: For an orthogonal basis $\{v_i\}$ of V, $\tau_{v_1}\cdots\tau_{v_n} = -1$ holds.)

Exercise 2. Let V be a regular quadratic space over F and denote by G a set of all elements $g \in C_0(V)$ such that there exists the inverse g^{-1} and $gVg^{-1} \subset V$; prove $\sigma(v) = gvg^{-1}$ is an isometry of V and $g \mapsto \sigma$ induces an injective homomorphism $G/F^\times \to O(V)$ with image $= O^+(V)$ or $O(V)$. (Hint: $Q(\sigma(v)) = (\sigma(v))^2 = (gvg^{-1})^2 = gv^2g^{-1} = Q(v)$ is obvious. If $\sigma = 1$, then $g \in C_0(V)$ is in the centre of $C(V)$. This implies $g \in F$. If $\sigma = \tau_{v_1}\cdots\tau_{v_m} \in O^+(V)$, then $\sigma(x) = (v_m\cdots v_1)^{-1}x(v_m\cdots v_1) = gxg^{-1}$ with $g = (v_m\cdots v_1)^{-1}$.)

Proposition 1.6.3. *Let V be a regular isotropic quadratic space over a field F. Then $O'(V)$ is generated by $\tau_x\tau_y$ for $x, y \in V$ satisfying $Q(x) = Q(y) \neq 0$.*

Proof. Let Ω be the subgroup of $O(V)$ which is generated by $\tau_x\tau_y$ for $x, y \in V$ satisfying $Q(x) = Q(y) \neq 0$. Noting $\sigma\tau_x\sigma^{-1} = \tau_{\sigma(x)}$ for $\sigma \in O(V)$, Ω is a normal subgroup of $O'(V)$. Let $V = H \perp W$ where H is a hyperbolic plane with a basis $\{e_1, e_2\}$ such that $Q(e_1) = Q(e_2) = 0$, $B(e_1, e_2) = 1$. Let $\sigma = \tau_{x_1}\cdots\tau_{x_n} \in O'(V)$. We take $y_i \in H$ such that $Q(x_i) = Q(y_i)$ for $i = 1, \cdots, n$. Then

(1) $$\sigma^{-1}\tau_{y_1}\cdots\tau_{y_n} \in \Omega$$

follows from $\tau_{x_i}\tau_{y_i} \in \Omega$ and the normality of Ω in $O'(V)$. Put $\eta := \tau_{y_1} \cdots \tau_{y_n}$; then $\eta|_W = \mathrm{id}$ since $y_i \in H$, and $\eta' := \eta|_H \in O'(H)$ holds by $\theta(\eta) = \theta(\eta|_W)\theta(\eta|_H) = \theta(\eta')$ from Proposition 1.6.2. Put $\eta'(e_1) = a_1e_1 + a_2e_2$, $\eta'(e_2) = b_1e_1 + b_2e_2$; then $a_1a_2 = b_1b_2 = 0$ and $\eta' \in O^+(H)$ imply $\eta'(e_1) = a_1e_1$, $\eta'(e_2) = a_1^{-1}e_2$ and hence $\eta' = \tau_{e_1-e_2}\tau_{e_1-a_1e_2}$ is easily seen. Therefore $\eta = \tau_{e_1-e_2}\tau_{e_1-a_1e_2}$ on V since they are equal on H and W, and $\theta(\eta') = a_1$ and $\theta(\eta') \in (F^\times)^2$ implies $a_1 = a^2$ for some $a \in F^\times$. Usng this, we have $\eta = \tau_{a(e_1-e_2)}\tau_{e_1-a_1e_2}$ and $Q(a(e_1 - e_2)) = Q(e_1 - a_1e_2) = -2a_1$. Thus η is in Ω and by (1) σ is in Ω. This means $\Omega = O'(V)$. □

Proposition 1.6.4. *Let $a, b \in F^\times$. Then an element η in $O^+((a,b)^0)$ is written as $\eta(x) = vxv^{-1}$ for $v \in (a,b)$ and we have $\theta(\eta) = \mathrm{N}(v)$.*

Proof. Put $\eta = \tau_{u_1} \cdots \tau_{u_m}$ for $u_i \in (a,b)^0$ with $\mathrm{N}(u_i) \neq 0$; then by virtue of Proposition 1.5.4 we have $\eta(x) = u_1 \cdots u_m x u_m^{-1} \cdots u_1^{-1}$. Hence $\eta(x) = vxv^{-1}$ holds for $v = u_1 \cdots u_m$ and then $\theta(\eta) = \mathrm{N}(u_1) \cdots \mathrm{N}(u_m) = \mathrm{N}(v)$. □

Let F be a topological field, that is F is a field with topology, and mappings $(x, y) \to x \pm y$ and xy are continuous from $F \times F$ to F and $x \mapsto x^{-1}$ is also continuous from $F^\times$ to $F^\times$. For a vector space V over F with $\dim V = n$, we introduce a topology on V, identifying $V = F^n$. The topology is independent of the identification.

Proposition 1.6.5. *Let F be a topological field such that $(F^\times)^2$ is open in $F^\times$. For a regular quadratic space V over F, the spinor norm $\theta : O(V) \to F^\times/(F^\times)^2$ is continuous.*

Proof. We have only to show that if an isometry σ is sufficiently close to the identity mapping, then $\theta(\sigma) = 1$ in $F^\times/(F^\times)^2$. We use induction on $\dim V$. If $\dim V = 1$, then $\sigma = \pm 1$ and so $\sigma = 1$ if σ is close to 1. Hence θ is continuous. Suppose that $\dim V \geq 2$ and $\{v_i\}$ is an orthogonal basis of V. Let σ be an isometry of V close to 1, and put $\sigma(v_i) = v_i + \sum_j a_{ij}v_j$; then the a_{ij} are close to 0, and hence

$$Q(\sigma(v_1) + v_1) = (2 + a_{11})^2 Q(v_1) + \sum_{j \neq 1} a_{1j}^2 Q(v_j)$$

is close to $4Q(v_1)(\neq 0)$, since $Q(v_1) \neq 0$ and $2 + a_{11}$ is close to 2. Hence we have $Q(\sigma(v_1) + v_1) = 4Q(v_1)a^2$ for $a \in F^\times$, since $(F^\times)^2$ is open. Putting $\tau := \tau_{\sigma(v_1)+v_1}$, we have

$$\begin{aligned} \tau(v_1) &= v_1 - 2B(v_1, \sigma(v_1) + v_1)Q(\sigma(v_1) + v_1)^{-1}(\sigma(v_1) + v_1) \\ &= -\sigma(v_1), \\ \theta(\tau) &= Q(\sigma(v_1) + v_1) = Q(v_1) \text{ in } F^\times/(F^\times)^2, \end{aligned}$$

and

$$\tau(v_i) = v_i - 2B(v_i, \sigma(v_1) + v_1)Q(\sigma(v_1) + v_1)^{-1}(\sigma(v_1) + v_1)$$

is close to v_i for $i \geq 2$ since $\sigma(v_1) + v_1$ is close to $2v_1$. Hence $\tau\sigma(v_i)$ and v_i are close for $i \geq 2$ and so $\tau\sigma(v_1) = -v_1$ yields that $\tau\sigma|_{v_1^\perp}$ is also close to 1 on $v_1^\perp$. From the induction hypothesis, it follows that $\theta(\tau\sigma|_{v_1^\perp}) = 1$ and then we have

$$\begin{aligned}\theta(\sigma) &= \theta(\tau^2\sigma) = \theta(\tau)\theta(\tau\sigma)\\ &= Q(v_1)\theta(\tau\sigma) = Q(v_1)\theta(-1 \text{ on } Fv_1)\theta(\tau\sigma|_{v_1^\perp})\\ &= Q(v_1)^2 = 1 \text{ in } F^\times/(F^\times)^2.\end{aligned}$$

□

1.7 Scalar extensions

Let $R' \supset R$ be commutative rings and (M, B) be a symmetric bilinear module over R. Denote $R' \otimes_R M$ by $R'M$; then we can define a symmetric bilinear form B' by $B'(rx, sy) = rsB(x, y)$ for $r, s \in R'$ and $x, y \in M$. The new symmetric bilinear module $(R'M, B')$ is called the *scalar extension* of (M, B) and usually we use the same letter B instead of B'. For a quadratic module (M, Q), we associate the symmetric bilinear module (M, B) with $B(x, x) = Q(x)$ and take the scalar extension $(R'M, B)$ to define the *scalar extension* of (M, Q) as $(R'M, Q)$ which is associated with $(R'M, B)$. Hence for $a_i \in R', x_i \in M$, we have

$$Q\Big(\sum a_i x_i\Big) = B\Big(\sum a_i x_i, \sum a_i x_i\Big) = \sum_{i,j} a_i a_j B(x_i, x_j).$$

If M has a basis over R, then M is identified with R^n for some n by using a basis of M over R, and $R'M$ is identified with R'^n through the identification of M and R^n; thus the scalar extension is simply extending the domain R of variables to R', keeping the corresponding matrix.

2

Positive definite quadratic forms over $\mathbb{R}$

The aim of this chapter is to give a reduction theory of quadratic forms over $\mathbb{R}$. Let P be the set of all positive definite matrices of degree m and denote $GL_m(\mathbb{Z})$ by Γ. Then Γ acts discontinuously on P by $p \to p[g]$ ($p \in P, g \in \Gamma$). An important problem is to determine the fundamental domain P/Γ, that is to give a "canonical" matrix in the orbit $p[g]$ ($g \in \Gamma$). This was done by Minkowski for general m and the set of canonical matrices is called Minkowski's domain. Unless there is a need for an exact fundamental domain, Siegel's domain is more useful, especially for analysis on P. So we introduce Siegel's domain instead of Minkowski's. As an application, we prove the finiteness of the class number of (not necessarily positive definite) quadratic forms. In Section 2.2, we give a better estimate due to Blichfeldt of Hermite's constant introduced in Section 2.1. The proof will be an introduction to the theory of geometry of numbers. We need it in the last chapter.

2.1 Reduction theory

In this section, we describe a reduction theory of positive definite symmetric matrices. Let m be an integer larger than 1 and put

$$\begin{aligned} G = G(m) :=& GL_m(\mathbb{R}), \\ A = A(m) :=& \{\mathrm{diag}(a_1, \cdots, a_m) \mid a_i > 0 \text{ for } i = 1, \cdots, m\}, \end{aligned}$$

$$N = N(m) := \text{the subgroup of all upper triangular matrices with all diagonals being 1 in } G,$$
$$K = K(m) := \text{ the subgroup of all orthogonal matrices in } G\,.$$

For positive numbers t, u, we define subsets A_t, N_u as follows:

$$A_t := \{\mathrm{diag}(a_1, \cdots, a_m) \in A \,|\, a_i/a_{i+1} \le t \text{ for } i = 1, \cdots, m-1\},$$
$$N_u := \{n = (n_{ij}) \in N \,|\, |n_{ij}| \le u \text{ for } 1 \le i < j \le m\}.$$

N_u is obviously compact, but A_t is not; being rather a cone; that is for $a \in A_t, 0 < r \in \mathbb{R}$, ra is also in A_t. Finally we put

$$\Gamma = \Gamma(m) := GL_m(\mathbb{Z}),$$
$$\tilde{S}_{t,u} = \tilde{S}_{t,u}(m) := KA_tN_u = \{kan \,|\, k \in K, a \in A_t, n \in N_u\},$$

which is called a *Siegel domain* for G.

For $g \in G$, ${}^tgg$ is a positive definite symmetric matrix and for it, there exists an upper triangular matrix p whose diagonals are positive such that ${}^tgg = {}^tpp$. This implies $gp^{-1} \in K$, and hence we have

$$g \in KAN,$$

decomposing p as a product of elements of A, N. The decomposition of $g = kan$ ($k \in K, a \in A, n \in N$) is called the *Iwasawa decomposition* of g. The first step is to prove

$$g\Gamma \cap \tilde{S}_{2/\sqrt{3},1/2} \neq \emptyset \quad \text{for every} \quad g \in G,$$

which is done in Lemma 2.1.3.

Lemma 2.1.1. *Put $N_{\mathbb{Z}} = N \cap \Gamma$; then we have $N = N_{1/2} \cdot N_{\mathbb{Z}}$.*

Proof. Let $x = (x_{ij}), y = (y_{ij}) \in N$; then by definition, $x_{ij} = y_{ij} = 0$ for $i > j$, $x_{ii} = y_{ii} = 1$ for all i. Putting $z = (z_{ij}) := xy$, we have

$$z_{ij} = y_{ij} + \sum_{i<k<j} x_{ik}y_{kj} + x_{ij}.$$

Making use of this, we can choose $y_{ij} \in \mathbb{Z}$ for $i < j$ such that $|z_{ij}| \le 1/2$ in order of $(m-1)$th row , $\cdots$, the first row. Hence, $(z_{ij}) \in N_{1/2}$ and so $x = zy^{-1} \in N_{1/2} \cdot N_{\mathbb{Z}}$ follows. □

We denote by $|x|$ the ordinary distance $\sqrt{(\sum_{i=1}^m x_i^2)}$ for $x = {}^t(x_1, \cdots, x_m) \in \mathbb{R}^m$, and define the function ϕ on G by

$$\phi(g) := |g\mathbf{e}| \text{ for } \mathbf{e} = {}^t(1, 0, \cdots, 0).$$

Then it is clear for $g \in G, k \in K, a \in A, n \in N$ that

$$\phi(kg) = \phi(g) = \phi(gn), \qquad (1)$$
$$\phi(a) = \text{ the (1,1) entry of } a.$$

Hence if $g = kan$, then $\phi(g) = \phi(a) =$ the (1,1) entry of a. Since $g\Gamma\mathbf{e} \subset g(\mathbb{Z}^m \setminus \{0\})$ for $g \in G$, the minimum $\min_{\gamma\in\Gamma} \phi(g\gamma)$ exists and is positive.

Lemma 2.1.2. *Let $g = kan$ ($g \in G$, $k \in K$, $a = \mathrm{diag}(a_1, a_2, \cdots) \in A$, $n \in N$) and assume that $\phi(g\gamma) \geq \phi(g)$ for all $\gamma \in \Gamma$; then $a_1/a_2 \leq 2/\sqrt{3}$.*

Proof. By the previous lemma, there exists $n' \in N_{\mathbf{Z}} \subset \Gamma$ such that $h = (h_{ij}) := nn' \in N_{1/2}$ and so in particular $|h_{12}| \leq 1/2$. For

$$\gamma = \mathrm{diag}(\begin{pmatrix} 0 & 1 \\ 1 & 0 \end{pmatrix}, 1_{m-2}),$$

we have

$$gn'\gamma\mathbf{e} = gn'^t(0, 1, \cdots, 0) = ka^t(h_{12}, 1, 0, \cdots, 0).$$

Applying the assumption to $n'\gamma \in \Gamma$ instead of γ, we have

$$a_1 = \phi(g) \leq \phi(gn'\gamma) = \sqrt{(a_1 h_{12})^2 + a_2^2}$$

and hence

$$a_2^2 \geq a_1^2(1 - h_{12}^2) \geq (3/4)a_1^2$$

follows. □

Lemma 2.1.3. *Let $g \in G$; then there exists an element $\gamma_0 \in \Gamma$ such that $\phi(g\gamma_0) = \min_{\gamma \in \Gamma} \phi(g\gamma)$ and $g\gamma_0 \in \tilde{S}_{2/\sqrt{3},1/2}$.*

Proof. We use induction on m. We can choose $\gamma' \in \Gamma$ such that $\phi(g\gamma') = \min_{\gamma \in \Gamma} \phi(g\gamma)$, and write

$$g\gamma' = kan$$

where $k \in K$, $a = \mathrm{diag}(a_1, a_2, \cdots) \in A$, $n \in N$, and moreover we may assume $n \in N_{1/2}$ by virtue of Lemma 2.1.1 and (1). First suppose $m = 2$; by the previous lemma, we have $a_1/a_2 \leq 2/\sqrt{3}$. Putting $\gamma_0 = \gamma'$, we have $g\gamma_0 = kan \in \tilde{S}_{2/\sqrt{3},1/2}$, and $\phi(g\gamma_0) = \phi(g\gamma')$ is the minimum. This completes the case of $m = 2$. Suppose $m \geq 3$. $k^{-1}g\gamma' = an$ is of the form

$$\begin{pmatrix} a_1 & * \cdots * \\ 0 & \\ \vdots & b \\ 0 & \end{pmatrix}, \quad b \in G(m-1).$$

Applying the assumption of the induction, we write

$$bx = k'a'n'$$

where $x \in \Gamma(m-1), k' \in K(m-1), a' = \mathrm{diag}(a'_1, \cdots, a'_{m-1}) \in A_{2/\sqrt{3}}(m-1), n' \in N_{1/2}(m-1)$. Then

$$\tilde{g} := k^{-1} g \gamma' \begin{pmatrix} 1 & \\ & x \end{pmatrix} = an \begin{pmatrix} 1 & \\ & x \end{pmatrix} = \begin{pmatrix} a_1 & * \cdots * \\ 0 & \\ \vdots & bx \\ 0 & \end{pmatrix}$$

$$(2) \qquad = \begin{pmatrix} 1 & 0 \cdots 0 \\ 0 & \\ \vdots & k' \\ 0 & \end{pmatrix} \begin{pmatrix} a_1 & 0 \cdots 0 \\ 0 & \\ \vdots & a' \\ 0 & \end{pmatrix} \begin{pmatrix} 1 & * \cdots * \\ 0 & \\ \vdots & n' \\ 0 & \end{pmatrix}.$$

Noting $\phi(\tilde{g}) = \phi(g\gamma')$ by (1) and the minimality of it, Lemma 2.1.2 implies $a_1/a'_1 \le 2/\sqrt{3}$, which yields $\mathrm{diag}(a_1, a') \in A_{2/\sqrt{3}}$ with $a' \in A_{2/\sqrt{3}}(m-1)$. Now applying Lemma 2.1.1 to the final matrix in (2) we can take $\tilde{n} \in N_{\mathbb{Z}}$ such that $\tilde{g}\tilde{n} \in \tilde{S}_{2/\sqrt{3},1/2}$. Thus, putting $\gamma_0 = \gamma' \begin{pmatrix} 1 & 0 \\ 0 & x \end{pmatrix} \tilde{n}$ we have $g\gamma_0 = k\tilde{g}\tilde{n} \in \tilde{S}_{2/\sqrt{3},1/2}$, and γ_0 is a required matrix in Γ, since $\phi(g\gamma_0) = \phi(\tilde{g}) = \phi(g\gamma')$ by (1). □

Now we apply the reduction theory of $GL_m(\mathbb{R})$ to the reduction theory of quadratic forms. Denote by $P = P(m)$ the set of all positive definite symmetric matrices of degree m and put

$$S_{t,u} = S_{t,u}(m) := \{a[n] \mid a \in A_t, n \in N_u\},$$

which is called *Siegel's domain.*

Theorem 2.1.1.

$$P = \cup_{\gamma \in \Gamma} S_{4/3,1/2}[\gamma] \quad \textit{and} \quad \mu_m := \sup_{p \in P} \frac{\min p}{(\det p)^{1/m}} \le (4/3)^{(m-1)/2},$$

where $\min p$ *denotes the minimum of* $p[x]$ $(0 \ne x \in \mathbb{Z}^m)$.

Proof. For $p \in P$, write $p = {}^t gg$ $(g \in G)$, and making use of the previous lemma, we can write $g\gamma_0 = kan$ $(\gamma_0 \in \Gamma, k \in K, a \in A_{2/\sqrt{3}}, n \in N_{1/2})$. Moreover we may assume $\phi(g\gamma_0) \le \phi(g\gamma)$ for all $\gamma \in \Gamma$. Then $p[\gamma_0] = a^2[n] \in S_{4/3,1/2}$ is clear. This implies the first assertion. To prove the inequality, we show first,

$$\min_{x \in \mathbb{Z}^m \setminus \{0\}} |gx| \le (4/3)^{(m-1)/4} |\det g|^{1/m}.$$

Since the left-hand side is attained by the primitive vector x, it is equal to

$$\min_{\gamma\in\Gamma}|g\gamma\mathbf{e}| = \min_{\gamma\in\Gamma}\phi(g\gamma) = \phi(g\gamma_0) = a_1,$$

where we put $a = \mathrm{diag}(a_1, \cdots, a_m)$. Now

$$a_1/a_i = a_1/a_2 \cdots a_{i-1}/a_i \leq (2/\sqrt{3})^{i-1}$$

implies

$$a_1^m / \det a \leq \prod_{i=2}^{m}(2/\sqrt{3})^{i-1} = (2/\sqrt{3})^{m(m-1)/2}.$$

This is what we want, noting $\det g = \pm\det a$. Now we can complete the proof of the theorem as follows:

$$\begin{aligned}\min p =& \min_{x\in\mathbb{Z}^m\setminus\{0\}} {}^t(gx)gx = \min_{x\in\mathbb{Z}^m\setminus\{0\}} |gx|^2 \\ \leq& ((4/3)^{(m-1)/4}|\det g|^{1/m})^2 = (4/3)^{(m-1)/2}(\det p)^{1/m}.\end{aligned}$$

□

We call μ_m *Hermite's constant.* The following two theorems are useful for analysis on the space of positive definite matrices.

Theorem 2.1.2. *Let $p \in S_{t,u}$ and write $p = a[n]$ $(a \in A_t, n \in N_u)$. Then we have*

$$c_1 \leq p[x]/a[x] \leq c_2 \text{ for all non-zero } x \in \mathbb{R}^m,$$

where c_1, c_2 are dependent only on t, u, m. In particular

$$c_1 \leq p_{ii}/a_i \leq c_2 \text{ for } i = 1, \cdots, m$$

where we put $p = (p_{ij}), a = \mathrm{diag}(a_1, \cdots, a_m)$.

Proof. Put $n = (n_{ij})$ $(n_{ii} = 1, n_{ij} = 0$ for $i > j)$, and for $x = {}^t(x_1, \cdots, x_m)$ we put $y = {}^t(y_1, \cdots, y_m) := nx$; then we have

$$\begin{aligned}a_i y_i^2 =& a_i(x_i + \sum_{j>i} n_{ij}x_j)^2 \\ \leq& a_i(|x_i| + \sum_{j>i}|n_{ij}||x_j|)^2 \\ =& (|\sqrt{a_i}x_i| + \sum_{j>i}|n_{ij}||\sqrt{a_i}x_j|)^2.\end{aligned}$$

Noting $|\sqrt{a_i}x_i| \le \sqrt{a[x]}$ and

$$\begin{aligned}|\sqrt{a_i}x_j| =& \sqrt{a_i/a_{i+1}} \cdots \sqrt{a_{j-1}/a_j}\sqrt{a_j}|x_j| \\ \le & t^{(j-i)/2}\sqrt{a[x]} \quad \text{for } i < j,\end{aligned}$$

we have

$$\begin{aligned}a_i y_i^2 \le & (\sqrt{a[x]} + \sum_{j>i} ut^{(j-i)/2}\sqrt{a[x]})^2 \\ =& c_{2,i}a[x],\end{aligned}$$

where $c_{2,i} = (1 + \sum_{j>i} ut^{(j-i)/2})^2$; then it follows that

$$p[x] = a[y] = \sum a_i y_i^2 \le a[x] \sum_i c_{2,i}.$$

We can take $\sum_i c_{2,i}$ as c_2. From the formula of the inverse matrix it follows that $n^{-1} \in N_{u'}$ for all $n \in N_u$ for an appropriate $u' > 0$. Hence $a[n^{-1}] \in S_{t,u'}$ and applying the above for $S_{t,u'}$ instead of $S_{t,u}$, there is a constant c_1 such that

$$a[n^{-1}][x] \le c_1^{-1} a[x] \text{ for all } x \in \mathbb{R}^m.$$

Replacing x by nx, we have $a[x] \le c_1^{-1} p[x]$, which yields the first inequality. For $f = {}^t(0, \cdots, 0, 1, 0, \cdots, 0)$, we have $p[f] = p_{ii}, a[f] = a_i$ and hence the latter inequality in the assertion. □

Theorem 2.1.3. *Put $S = S_{t,u}$ for $t, u > 0$ and let d be a positive number. Then there is only a finite number of $m \times m$ integral matrices x such that $0 < |\det x| < d$ and $S \cap S[x] \ne \emptyset$.*

Proof. We use induction on m. Suppose that for $x \in M_m(\mathbb{Z})$ with $0 < |\det x| < d$ there is a matrix p such that

$$p = a[n],\ p[x] = a'[n']\ (a, a' \in A_t, n, n' \in N_u).$$

(i) We assume first that we can write

$$x = \begin{pmatrix} x_1 & x_3 \\ 0 & x_2 \end{pmatrix} (x_1 \in M_{m_1}(\mathbb{Z}), x_2 \in M_{m_2}(\mathbb{Z}), x_3 \in M_{m_1,m_2}(\mathbb{Z}));$$

then putting

$$\begin{aligned}&a = \operatorname{diag}(a_1, a_2), (a_1 \in A_t(m_1), a_2 \in A_t(m_2)), \\ &n = \begin{pmatrix} n_1 & n_3 \\ 0 & n_2 \end{pmatrix} (n_1 \in N_u(m_1), n_2 \in N_u(m_2), n_3 \in M_{m_1,m_2}(\mathbb{R})),\end{aligned}$$

we have

$$p = a[n] = \begin{pmatrix} a_1 & 0 \\ 0 & a_2 \end{pmatrix} \left[\begin{pmatrix} n_1 & 0 \\ 0 & n_2 \end{pmatrix} \begin{pmatrix} 1 & n_1^{-1} n_3 \\ 0 & 1 \end{pmatrix} \right]$$

(3)

$$= \begin{pmatrix} a_1[n_1] & 0 \\ 0 & a_2[n_2] \end{pmatrix} \left[\begin{pmatrix} 1 & n_1^{-1} n_3 \\ 0 & 1 \end{pmatrix} \right],$$

and

$$p[x] = \begin{pmatrix} (a_1[n_1][x_1] & 0 \\ 0 & a_2[n_2][x_2] \end{pmatrix} \left[\begin{pmatrix} 1 & (n_1 x_1)^{-1}(n_1 x_3 + n_3 x_2) \\ 0 & 1 \end{pmatrix} \right].$$

We write similarly

$$a' = \mathrm{diag}(a_1', a_2'), n' = \begin{pmatrix} n_1' & n_3' \\ 0 & n_2' \end{pmatrix}$$

where a_i', n_i' have the same degree as a_i, n_i, and $a'[n']$ has the similar decomposition to (3). Noting the matrix identity

$$\begin{pmatrix} b & 0 \\ 0 & c \end{pmatrix} \left[\begin{pmatrix} 1 & f \\ 0 & 1 \end{pmatrix} \right] = \begin{pmatrix} b & bf \\ {}^t(bf) & b[f] + c \end{pmatrix},$$

we have from $p[x] = a'[n']$ that

(4)

$$(a_1[n_1])[x_1] = a_1'[n_1'], a_2[n_2][x_2] = a_2'[n_2']$$
$$(n_1 x_1)^{-1}(n_1 x_3 + n_3 x_2) = n_1'^{-1} n_3'.$$

The last equality gives

(5)

$$x_3 = x_1 n_1'^{-1} n_3' - n_1^{-1} n_3 x_2$$

If $m = 2$, then we have $m_1 = m_2 = 1, n_1 = n'_1 = n_2 = n'_2 = 1, |n_3|, |n'_3| \leq u$ and $x_3 = x_1 n_3' - n_3 x_2$. Since $0 < |x_1 x_2| < d$, the number of possible values of x_1, x_2 and $x_3 = x_1 n'_3 - n_3 x_2$ is finite. Thus the case of $m = 2$ has been proved. Coming back to the general case, (4) means

$$S_{t,u}(m_i)[x_i] \cap S_{t,u}(m_i) \neq \emptyset \ (i = 1, 2).$$

Now $0 < |\det x| = |\det x_1||\det x_2| < d$ yields $0 < |\det x_i| < d$ $(i = 1, 2)$, the induction hypothesis implies the finiteness of possibilities of x_1 and x_2. Since by virtue of $n, n' \in N_u$, all entries of $n_1'^{-1} n_3', n_1^{-1} n_3$ are bounded by a

constant dependent on u and m, (5) implies the finiteness of the possibilities of x_3. Thus the first case has been proved.
(ii) Suppose that x does not have a decomposition as in (i). Put $x = (x_{ij}) = (x^{(1)}, \cdots, x^{(m)})$ ($x^{(i)}$: i-th column). By the assumption, every left lower submatrix $(x_{h,k})_{\substack{h>i \\ k\le i}}$ is not a zero matrix for $i = 1, \cdots, m-1$. Hence some entry $x_{h(i),k(i)}$ is not zero for $k(i) \le i < h(i)$. By Theorem 2.1.2, there exist constants c_1, c_2 such that

$$c_1 a[y] \le p[y] \le c_2 a[y], \ c_1 a'[y] \le p'[y] \le c_2 a'[y] \ \text{ for } y \in \mathbb{R}^m,$$

where $p' := p[x] = a'[n']$. For simplicity, we write $e \succ f$ if $e > cf$ for a positive constant c dependent only on m, t, u, d. By definition,

$$a_1 \prec \cdots \prec a_m, \ a'_1 \prec \cdots \prec a'_m, \tag{6}$$

Noting that $a'_i \succ p'_{ii} = p[x^{(i)}] \succ a[x^{(i)}] = \sum_j a_j x_{ji}^2$, we have

$$a'_i \succ a_j x_{ji}^2 \tag{7}$$

and for $i \le m-1$,

$$\begin{aligned} a'_i &\succ a'_{k(i)} && \text{by } i \ge k(i) \text{ and (6)} \\ &\succ a_{h(i)} x_{h(i),k(i)}^2 && \text{by (7)} \\ &> a_{h(i)} && \text{by } x_{h(i),k(i)} \ne 0 \\ &\succ a_i && \text{by } h(i) > i. \end{aligned} \tag{8}$$

Thus we have

$$a'_i \succ a_i. \tag{9}$$

For $d_1 := \det x$, and $p' = p[x]$ implies $p'[d_1 x^{-1}] = d_1^2 p$, but $p', d_1^2 p$ are also in $S_{t,u}$, and it is easy to see that $d_1 x^{-1}$ does not have a decomposition as in (i). By a similar argument, we have for $i \le m-1$,

$$a_i \succ a'_i. \tag{10}$$

Since $\det p \succ \det p'$ and $\det p' \succ \det p$, $\det p = \prod a_i$ and $\det p' = \prod a'_i$ imply the validity of (9), (10) for $i = m$. Then, for $i < m$ (8), (10) and (6) with $h(i) > i$ imply

$$a'_i \succ a_{h(i)} \succ a'_{h(i)} \succ a'_{i+1}. \tag{11}$$

(6), (9), (10) and (11) show that $a_1, \cdots, a_m, a'_1, \cdots, a'_m$ have the same magnitude up to constants dependent only on m, t, u, d and then (7) implies that $|x_{ji}| \prec 1$, which shows the finite possibility of x. □

Let q be a not necessarily positive definite symmetric regular matrix of degree m over $\mathbb{R}$. If a positive definite symmetric matrix p satisfies

$$q^{-1}[p] = q,$$

then p is called a *majorant* of q. The existence of a majorant is shown as follows: write $q = d[g]$ for $d = \mathrm{diag}(1_m, -1_n)$, $g \in GL_m(\mathbb{R})$ and put $p = {}^t gg$; then $q^{-1}[p] = (g^{-1}d^{-1}{}^t g^{-1})[{}^t gg] = d^{-1}[g] = q$. It is easy to see that if p is a majorant of q, then $p[u]$ is a majorant of $q[u]$ for every $u \in GL_m(\mathbb{R})$.

Corollary 2.1.1 (finiteness of class number). *Let d be a non-zero integer and put $Q_d = \{q \in M_m(\mathbb{Z}) \mid {}^t q = q, \det q = d\}$. We define an equivalence relation $q_1 \sim q_2$ by $q_1 = q_2[u]$ for $u \in GL_m(\mathbb{Z})$. Then the number of equivalence classes in Q_d is finite.*

Proof. Put $J = (x_{ij})$ where $x_{ij} = 1$ if $i + j = m + 1$ and $x_{ij} = 0$ if $i + j \neq m + 1$. J^2 is the identity matrix. We show

$$S_{t,u}^{-1} \subset JS_{t,u'}J \quad \text{for some } u'.$$

Let $a = \mathrm{diag}(a_1, \cdots, a_m) \in A_t, n \in N_u$; then

$$J(a[n])^{-1}J = {}^t(J{}^t nJ)^{-1}(Ja^{-1}J)(J{}^t nJ)^{-1},$$

and

$$Ja^{-1}J = \mathrm{diag}(a_m^{-1}, \cdots, a_1^{-1}) \in A_t,$$

and $J{}^t nJ \in N_u$ implies $(J{}^t nJ)^{-1} \in N_{u'}$ for some u'. Thus $Ja[n]^{-1}J \in S_{t,u'}$ follows, which is what we want. Let $q \in Q_d$ and p be a majorant of q. Choose $g \in GL_m(\mathbb{Z})$ such that $p[g] \in S_{4/3,1/2}$ by Theorem 2.1.1, and a positive number $u > 1/2$ such that $S_{4/3,1/2}^{-1} \subset JS_{4/3,u}J$. Then $q^{-1}[p] = q$ implies

$$\begin{aligned} p[g] &= q[g](p[g])^{-1}q[g] \\ &\in S_{4/3,1/2} \cap q[g]S_{4/3,1/2}^{-1}q[g] \subset S_{4/3,u} \cap (q[g]J)S_{4/3,u}(Jq[g]). \end{aligned}$$

By the previous theorem, the number of possible integral matrices x with $|\det x| = d$ such that $S_{4/3,u} \cap S_{4/3,u}[x] \neq \emptyset$ is finite. It follows that $Jq[g]$, and hence also $q[g]$, is in a finite set. □

There is an easier method to get the corollary. The following is enough. For given natural numbers m and d, there is a constant c satisfying that for any integral symmetric regular $m \times m$ matrix A with $|\det A| \le d$, there exists a non-zero vector $x \in \mathbb{Z}^m$ such that $|A[x]| \le c$. Indeed, we know the following:

Exercise (Hermite).

$$\min\{|A[x]| \mid x \in \mathbb{Z}^m, x \neq 0\} \le (4/3)^{(m-1)/2} |\det A|^{1/m}.$$

The above results are better derived as a byproduct of the analytic theory of quadratic forms.

2.2 An estimate of Hermite's constant

We give a better estimate due to Blichfeldt for Hermite's constant μ_m.

Theorem 2.2.1. *$\mu_m \le 2\pi^{-1}\Gamma(2+m/2)^{2/m}$ holds where Γ is the standard gamma function.*

We need several lemmas to prove this. Denote the unit ball in $\mathbb{R}^m$ by K_m, i.e. $K_m = \{x \in \mathbb{R}^m \mid |x| \le 1\}$; then the volume of K_m $\mathrm{vol}(K_m) = \pi^{m/2}\Gamma(1+m/2)^{-1}$ is well known, where vol means the volume by the standard Euclidean measure dx on $\mathbb{R}^m$. Let Λ be a lattice on $\mathbb{R}^m$, that is $\Lambda = \{gx \mid x \in \mathbb{Z}^m\}$, for some $g \in GL_m(\mathbb{R})$. If $(K_m + x) \cap (K_m + y) = \emptyset$ for $x, y \in \Lambda$ with $x \neq y$, then we say Λ gives a *packing* of $\mathbb{R}^m$ (by K_m). It is clear that Λ gives a packing if and only if $|x| > 2$ for all non-zero $x \in \Lambda$. $|\det g| = \mathrm{vol}(\mathbb{R}^m/\Lambda)$ is clear.

Lemma 2.2.1. *Suppose that a lattice Λ on $\mathbb{R}^m$ gives a packing of $\mathbb{R}^m$. Denoting $n_t(\Lambda) = \sharp\{x \in \Lambda \mid K_m + x \subset tW\}$, where $W := \{x = {}^t(x_1, \cdots, x_m) \in \mathbb{R}^m \mid |x_i| \le 1\}$, we have*

$$\lim_{t\to\infty} \frac{n_t(\Lambda)\,\mathrm{vol}(K_m)}{\mathrm{vol}(tW)} = \frac{\mathrm{vol}(K_m)}{\mathrm{vol}(\mathbb{R}^m/\Lambda)}.$$

Proof. Let $\Lambda = \{gx \mid x \in \mathbb{Z}^m\}$, and $\mathbf{e}_i = {}^t(0, \cdots, 0, 1, 0, \cdots, 0)$ (1 in the i-th coordinate). Define the fundamental parallelotope

$$F := \{\sum_i t_i g\mathbf{e}_i \mid 0 \le t_i < 1\};$$

then $\mathbb{R}^m = \cup_{x\in\Lambda}(F + x)$ is a disjoint union, and put

$$W_t := \cup_x (F + x) \quad \text{for } x \in \Lambda \text{ with } F + x \subset tW.$$

By definition, $\mathrm{vol}(\mathbb{R}^m/\Lambda) = \mathrm{vol}(F)$, and $W_t \subset tW$ is clear. Put $f = \sup_{x\in F}|x|$; then $F \subset fW$ is clear. We show

$$(1) \qquad (t-2f)W \subset W_t \subset tW \quad \text{for } t > 2f.$$

Suppose $(t-2f)w \in F + x$ for $w \in W, x \in \Lambda$; then $(t-2f)w = y + x$ for $y \in F$. Comparing coordinates, |the coordinate of $x| \le t - 2f +$ |the coordinate of $y| \le t - f$. Hence $x \in (t-f)W$ follows, and we have $F + x \subset fW + (t-f)W = tW$, which yields $(t-2f)w \in F + x \subset W_t$, by definition. Thus (1) has been shown. Put $V_t := \cup_{x\in\Lambda}\{(K_m + x) \cap W_t\}$; then

$$\begin{aligned}
\mathrm{vol}(V_t) &= \sum_{\substack{y\in\Lambda \\ F+y\subset tW}} \mathrm{vol}(\cup_{x\in\Lambda}((K_m + x) \cap (F+y))) \\
&= \sum_{\substack{y\in\Lambda \\ F+y\subset tW}} \mathrm{vol}(K_m) \\
(2) \qquad &= \mathrm{vol}(K_m)\,\mathrm{vol}(W_t)/\,\mathrm{vol}(F).
\end{aligned}$$

If $y \in V_t$, then $y \in (K_m + x) \cap W_t \subset tW$ for some $x \in \Lambda$ by (1), which implies, noting $K_m \subset W$, that $x \in y + K_m \subset (t+1)W$ and then $y \in K_m + x \subset (t+2)W$. Thus we have

$$V_t \subset \bigcup_{\substack{x\in\Lambda \\ K_m+x\subset(t+2)W}} (K_m + x)$$

and so

$$(3) \qquad V_{t-2} \subset \bigcup_{\substack{x\in\Lambda \\ K_m+x\subset tW}} (K_m + x) \subset V_{t+2f},$$

since $tW \subset W_{t+2f}$ by (1). Since

$$\mathrm{vol}\Big(\bigcup_{\substack{x\in\Lambda \\ K_m+x\subset tW}} (K_m + x)\Big) = n_t(\Lambda)\,\mathrm{vol}(K_m),$$

(3) implies

$$\begin{aligned}
\mathrm{vol}(V_{t-2})/\,\mathrm{vol}(tW) &\le n_t(\Lambda)\,\mathrm{vol}(K_m)/\,\mathrm{vol}(tW) \\
(4) \qquad &\le \mathrm{vol}(V_{t+2f})/\,\mathrm{vol}(tW).
\end{aligned}$$

(1) implies $\lim_{t\to\infty} \mathrm{vol}(W_t)/\,\mathrm{vol}(tW) = 1$ because of $\mathrm{vol}(tW) = (2t)^m$, and therefore (2) implies $\lim_{t\to\infty} \mathrm{vol}(V_t)/\,\mathrm{vol}(tW) = \mathrm{vol}(K_m)/\,\mathrm{vol}(F)$. Noting $\lim_{t\to\infty} \mathrm{vol}((t+a)W)/\,\mathrm{vol}(tW) = 1$ for any fixed number a, (4) completes the proof. □

Lemma 2.2.2. *For $x, x^{(1)}, \cdots, x^{(n)} \in \mathbb{R}^m$, we have*

$$\sum_{i,j} |x^{(i)} - x^{(j)}|^2 \leq 2n \sum_k |x - x^{(k)}|^2.$$

Proof. For $a_1, \cdots, a_n \in \mathbb{R}$, it is easy to see

$$\begin{aligned}\sum_{1\leq i,j \leq n} (a_i - a_j)^2 =& \sum_{i,j} ({a_i}^2 + {a_j}^2 - 2a_i a_j) \\ =& \sum_{1 \leq i,j \leq n} ({a_i}^2 + {a_j}^2) - 2(\sum_k a_k)^2 \\ \leq& \sum_{1 \leq i,j \leq n} ({a_i}^2 + {a_j}^2) \\ =& 2n \sum {a_i}^2.\end{aligned}$$

If, for a_j we substitute the i-th coordinate of $x - x^{(j)}$ and sum over i, then we get the required inequality. □

Lemma 2.2.3. *Suppose that a lattice Λ on $\mathbb{R}^m$ gives a packing of $\mathbb{R}^m$; then*

$$\operatorname{vol}(\mathbb{R}^m/\Lambda) \geq (2\pi)^{m/2} \Gamma(2 + m/2)^{-1}.$$

Proof. For $x \neq y \in \Lambda$, $(K_m + x) \cap (K_m + y) = \emptyset$ by definition and so

$$(5) \qquad |x - y| > 2.$$

We define the function δ by $\delta(\rho) = \max(0, 1 - \rho^2/2)$; then $\delta(\rho) = 0$ if $\rho \geq \sqrt{2}$ and for $x \in \mathbb{R}^m$, putting $S := \{y \in \Lambda \mid |x - y| \leq \sqrt{2}\}$

$$\begin{aligned}&\sum_{y\in S} \delta(|x - y|) = \sum_{y \in S} (1 - |x - y|^2/2) \\ &\leq \sharp S - (4\sharp S)^{-1} \sum_{y,z \in S} |y - z|^2 \qquad \text{by Lemma 2.2.2} \\ &\leq \sharp S - (4\sharp S)^{-1} (4\sharp S)(\sharp S - 1) \qquad \text{by (5)} \\ &= 1.\end{aligned}$$

Thus we have

$$(6) \qquad \sum_{y \in \Lambda} \delta(|x - y|) \leq 1 \text{ for } x \in \mathbb{R}^m.$$

Suppose that $y + K_m \subset tW$; then if $|x - y| < \sqrt{2}$, then

$$|\text{each coordinate of } x| < |\text{that of } y| + \sqrt{2}$$

and so $x \in (t+\sqrt{2})W$ since $y \in tW$. Hence $y+K_m \subset tW$ and $x \notin (t+\sqrt{2})W$ imply $|x - y| \geq \sqrt{2}$ and this means

$$\text{(7)} \qquad \int_{(t+\sqrt{2})W} \delta(|x-y|)dx = \int_{\mathbb{R}^m} \delta(|x-y|)dx \qquad \text{for } y + K_m \subset tW.$$

Thus we have

$$\begin{aligned}
\operatorname{vol}((t+\sqrt{2})W) &\geq \int_{(t+\sqrt{2})W} \sum_{\substack{y\in\Lambda \\ y+K_m\subset tW}} \delta(|x-y|)dx \quad \text{by (6)} \\
&= \sum_{\substack{y\in\Lambda \\ y+K_m\subset tW}} \int_{(t+\sqrt{2})W} \delta(|x-y|)dx \\
&= \sum_{\substack{y\in\Lambda \\ y+K_m\subset tW}} \int_{\mathbb{R}^m} \delta(|x|)dx \quad \text{by (7)} \\
&= n_t(\Lambda) \int_{0\leq|x|<\sqrt{2}} (1-|x|^2/2)dx \\
&= n_t(\Lambda)(m+2)^{-1}2^{1+m/2}\operatorname{vol}(K_m);
\end{aligned}$$

to see the final equality consider:

$$f(y) := \int_{0\leq|x|\leq y} (1-|x|^2/2)\,\mathrm{d}\,x,$$

$$f(y+h) - f(y) = \int_{y\leq|x|\leq y+h} (1-|x|^2/2)\,\mathrm{d}\,x = (1-y_0^2/2)\int_{y\leq|x|\leq y+h} \mathrm{d}\,x$$

for some $y_0 \in [y, y+h]$, and hence $f'(y) = m(1-y^2/2)y^{m-1}\operatorname{vol}(K_m)$. Thus we have

$$\frac{n_t(\Lambda)\operatorname{vol}(K_m)}{\operatorname{vol}((t+\sqrt{2})W)} \leq (m+2)2^{-1-m/2}.$$

Since $\operatorname{vol}((t+\sqrt{2})W) = (\frac{t+\sqrt{2}}{t})^m \operatorname{vol}(tW)$, Lemma 2.2.1 implies

$$\operatorname{vol}(K_m)\operatorname{vol}(\mathbb{R}^m/\Lambda)^{-1} \leq (m+2)2^{-1-m/2}$$

and hence

$$\operatorname{vol}(\mathbb{R}^m/\Lambda) \geq (m+2)^{-1}2^{1+m/2}\operatorname{vol}(K_m) = (2\pi)^{m/2}\Gamma(2+m/2)^{-1}.$$

□

Proof of Theorem 2.2.1.

First we note that $\min p/(\det p)^{1/m}$ for $p \in P(m)$ is invariant under multiplication of p by a positive number. Hence we have

$$\begin{aligned}\mu_m =& \sup_{p\in P(m)} \min p/(\det p)^{1/m} \\ =& \sup_{\substack{p\in P(m)\\ \min p=4}} 4(\det p)^{-1/m} \\ =& 4(\inf_{\substack{p\in P(m)\\ \min p=4}} \det p)^{-1/m},\end{aligned}$$

and so

$$(8) \qquad \inf_{\substack{p\in P(m)\\ \min p>4}} \det p = \inf_{\substack{p\in P(m)\\ \min p=4}} \det p = (4/\mu_m)^m.$$

For $g \in GL_m(\mathbb{R})$, the lattice $\Lambda = g\mathbb{Z}^m$ gives a packing of $\mathbb{R}^m$ if and only if $|x| > 2$ for all non-zero $x \in \Lambda$, which means that for $p := {}^t gg$, $p[y] > 4$ for all non-zero $y \in \mathbb{Z}^m$, putting $x = gy$. Moreover $\det p = (\det g)^2 = (\mathrm{vol}(\mathbb{R}^m/\Lambda))^2$ holds. Hence

$$\begin{aligned}\inf_{\substack{p\in P(m)\\ \min p>4}} \det p =& \inf\{(\mathrm{d}\,\Lambda)^2 \mid \Lambda \text{ gives a packing of } \mathbb{R}^m\} \\ \geq& (2\pi)^m \Gamma(2+m/2)^{-2} \text{ by Lemma 2.2.3.}\end{aligned}$$

By (8), we have

$$\mu_m \leq 4(2\pi)^{-1}\Gamma(2+m/2)^{2/m} = 2\pi^{-1}\Gamma(2+m/2)^{2/m}.$$

□

3

Quadratic forms over local fields

It is known that a locally compact non-discrete topological field which contains the rational number field $\mathbb{Q}$ as a dense subset is either the real number field $\mathbb{R}$ or a p-adic number field $\mathbb{Q}_p$ which, we will define in 3.1. These are called *local fields*. The theory of quadratic forms over $\mathbb{R}$ was developed in the previous chapter. The aim in this chapter is the classification of regular quadratic spaces over $\mathbb{Q}_p$. In 3.1, we define p-adic numbers, and introduce the quadratic residue symbol and the Hilbert symbol (of degree 2) in 3.2 and 3.3, respectively. After the definition of the Hasse invariant in 3.4, we classify regular quadratic spaces over $\mathbb{Q}_p$ in 3.5.

The classification theory can be generalized to regular quadratic spaces over a finite extension of $\mathbb{Q}_p$. But the generalization of results in 3.2 and 3.3 is not easy. It involves the theory of quadratic extension (the special case of class field theory).

3.1 p-adic numbers

Throughout this section, p denotes any fixed prime number.

We consider a sequence of integers

$$\{x_n\} = \{x_0, x_1, \cdots, x_n, \cdots\}$$

satisfying

(1) $$x_{n+1} \equiv x_n \bmod p^{n+1} \ (n = 0, 1, \cdots).$$

We denote by S the set of all such sequences. We introduce an equivalence relation $\sim$ in S by

$$\{x_n\} \sim \{y_n\} \Leftrightarrow x_n \equiv y_n \bmod p^{n+1} \ {\scriptstyle (n=0,1,\cdots)}.$$

We denote $S/\sim$ by $\mathbb{Z}_p$ which is called the ring of *p-adic integers*; an element of $\mathbb{Z}_p$ is called a *p-adic integer*. For $x \in \mathbb{Z}$, we put $x_n = x$ and then $\{x_n\} \in S$ is clear so by this correspondence $\mathbb{Z}$ is embedded in $\mathbb{Z}_p$.

It is easy to see that $\{x_n\}+\{y_n\} := \{x_n+y_n\}$, $\{x_n\}\{y_n\} := \{x_n y_n\}$ are well-defined and induce addition and multiplication in $\mathbb{Z}_p$. Then $0, 1 \in \mathbb{Z}$ are the zero and the unity, respectively, and $\mathbb{Z}_p$ is a commutative ring.

For $\{x_n\} \in S$, it is easy to see that there exists a unique sequence $\{y_n\}$ such that $\{x_n\} \sim \{y_n\}$, with $y_n = a_0 + a_1 p + \cdots + a_n p^n$ where each $a_i \in \mathbb{Z}$, $0 \le a_i < p$.

If there is no confusion then we denote elements of $\mathbb{Z}_p$ by elements of S.

Theorem 3.1.1. *For $x = \{x_n\} \in \mathbb{Z}_p$, $x \in \mathbb{Z}_p^\times$ holds if and only if $x_0 \not\equiv 0 \bmod p$.*

Proof. If $x \in \mathbb{Z}_p^\times$, then putting $x^{-1} = \{y_n\}$ we have $x_n y_n \equiv 1 \bmod p^{n+1}$. Thus we have $x_0 \not\equiv 0 \bmod p$. Conversely suppose $x_0 \not\equiv 0 \bmod p$. We have only to take integers y_n such that $y_{n+1} \equiv y_n \bmod p^{n+1}$ and $x_n y_n \equiv 1 \bmod p^{n+1}$ for all $n \ge 0$. The existence of y_0 follows from $x_0 \not\equiv 0 \bmod p$. Suppose that y_k are constructed for $k \le n$. Write $x_{n+1} = x_n + a p^{n+1}$, and put $y_{n+1} = y_n + b p^{n+1}$ where b is settled later; then $x_{n+1} y_{n+1} = x_n y_n + (a y_n + b x_n) p^{n+1} + ab p^{2n+2}$ holds. Writing $x_n y_n = 1 + c p^{n+1}$, we have

$$x_{n+1} y_{n+1} \equiv 1 + (a y_n + b x_n + c) p^{n+1} \bmod p^{n+2}.$$

Now $x_0 \not\equiv 0 \bmod p$ implies $x_k \not\equiv 0 \bmod p$ for all k, in particular $x_n \not\equiv 0 \bmod p$. Thus we can choose $b \in \mathbb{Z}$ such that $a y_n + b x_n + c \equiv 0 \bmod p$ and therefore $x_{n+1} y_{n+1} \equiv 1 \bmod p^{n+2}$. □

Corollary 3.1.1. $\mathbb{Z}_p^\times + p\mathbb{Z}_p \subset \mathbb{Z}_p^\times$.

Proof. This follows immediately from the theorem. □

Theorem 3.1.2. *If x is a non-zero p-adic integer, then we can write $x = p^m \epsilon$ for a non-negative integer m and $\epsilon \in \mathbb{Z}_p^\times$.*

Proof. Let $x = \{x_n\} \in \mathbb{Z}_p^\times$, $x_n = a_0 + a_1 p + \cdots + a_n p^n$, $0 \le a_i < p$. Since $x \ne 0$, there exists a non-negative integer m such that $x_m \not\equiv 0 \bmod p^{m+1}$, and assume m is minimal; then for $k < m$ we have $x_k \equiv 0 \bmod p^{k+1}$ and $x_m \not\equiv 0 \bmod p^{m+1}$, so by the choice of x_n we have $a_0 = \cdots = a_{m-1} = 0$ and $0 < a_m < p$. Putting $y_n = a_m + \cdots + a_{m+n} p^n$,

$$y_{n+1} \equiv y_n \bmod p^{n+1}$$

and

$$x_n - p^m y_n = \sum_{i=0}^{n} a_i p^i - \sum_{j=m}^{m+n} a_j p^j = - \sum_{n<j\leq m+n} a_j p^j \equiv 0 \bmod p^{n+1}$$

hold. Thus we have $x = p^m\{y_n\}$ and $\{y_n\} \in \mathbb{Z}_p^\times$. □

Corollary 3.1.2. *$\mathbb{Z}_p$ does not have a zero-divisor.*

Proof. Suppose $\alpha\beta = 0$ for non-zero α, $\beta \in \mathbb{Z}_p$ and put $\alpha = p^m\epsilon$ and $\beta = p^n\eta$ ($m, n \geq 0$, $\epsilon, \eta \in \mathbb{Z}_p^\times$). Multiplying $\alpha\beta = 0$ by $(\epsilon\eta)^{-1}$, we have $p^{m+n} = 0$ in $\mathbb{Z}_p$. Since $\mathbb{Z}$ is injectively embedded in $\mathbb{Z}_p$, this is a contradiction. □

By the corollary, we can define the quotient field $\mathbb{Q}_p$ of $\mathbb{Z}_p$ which is called the field of *p-adic numbers*. Thus a non-zero p-adic number x can be written uniquely as $p^m\epsilon$ ($m \in \mathbb{Z}, \epsilon \in \mathbb{Z}_p^\times$) and we put

$$\operatorname{ord}_p x := m \quad \text{and} \quad |x|_p := p^{-m}.$$

We set

$$|0|_p := 0$$

by definition. Then it is easy to see that $|x + y|_p \leq \max(|x|_p, |y|_p)$ with equality holding if $|x|_p \neq |y|_p$ and

$$\begin{cases} x \in \mathbb{Z}_p & \text{if and only if} \quad |x|_p \leq 1, \\ x \in p\mathbb{Z}_p & \text{if and only if} \quad |x|_p < 1, \\ x \in \mathbb{Z}_p^\times & \text{if and only if} \quad |x|_p = 1. \end{cases}$$

Now define a distance d in $\mathbb{Q}_p$ by $\mathrm{d}(x, y) = |x - y|_p$. Then it is easy to see that for $x, y, z \in \mathbb{Q}_p$

$$\begin{cases} \mathrm{d}(x, y) = \mathrm{d}(y, x) \\ \mathrm{d}(x, y) = 0 \Leftrightarrow x = y \\ \mathrm{d}(x, y) \leq \mathrm{d}(x, z) + \mathrm{d}(z, y). \end{cases}$$

Theorem 3.1.3. *$\mathbb{Q}_p$ is complete and $\mathbb{Z}_p$ is compact with respect to the distance* d.

Proof. Suppose that $\{x_n\}_{n=1}^{\infty}$ ($x_n \in \mathbb{Q}_p$) is a Cauchy sequence, that is for any $\epsilon > 0$ there exists a number $c(\epsilon)$ such that $\mathrm{d}(x_n, x_m) < \epsilon$ for $n, m > c(\epsilon)$. If $x_n = 0$ for infinitely many n, then it is easy to see that x_n converges to 0. Hence we may suppose that $x_n = 0$ for finitely many n.

Removing those n with $x_n = 0$, we may assume $x_n \neq 0$ for $n \geq 1$. Put $x_n = p^{a_n}\epsilon_n$ ($a_n \in \mathbb{Z}, \epsilon_n \in \mathbb{Z}_p^\times$). If there is a subsequence a_{n_m} such that $a_{n_1} > \cdots > a_{n_m} > \cdots \to -\infty$, then putting $a_{n_m} = b_m, \epsilon_{n_m} = \eta_m$,

$$\mathrm{d}(x_{n_m}, x_{n_{m+k}}) = |p^{b_m}\eta_m - p^{b_{m+k}}\eta_{m+k}| = p^{-b_{m+k}} \to \infty$$

which contradicts the property of the Cauchy sequence. Thus there exists an integer a such that $a_n \geq a$ and we have only to prove that $x_n p^{-a} (\in \mathbb{Z}_p)$ converges. Hence we may suppose $x_n \in \mathbb{Z}_p$. We write

$$x_n = \{x_{n,0}, \cdots, \sum_{i=0}^{m} x_{n,i}p^i, \cdots\}$$

with $0 \leq x_{n,i} < p$ as an element of S. For $m > 0$, we take a number $c(p^{-m})$ such that $\mathrm{d}(x_k, x_h) < p^{-m}$ if $k, h > c(p^{-m})$. Then $\mathrm{d}(x_k, x_h) < p^{-m}$ means $x_k - x_h \equiv 0 \bmod p^m$ and hence $x_{k,i} = x_{h,i}$ for $i < m$. So $x_{k,i}$ depends only on i if k is sufficiently large, $k > c(p^{-m})$ suffices. We denote it by b_i and put

$$y = \{b_0, \cdots, \sum_{i=0}^{m} b_i p^i, \cdots\} \in S.$$

Since

$$y_m - x_{k,m} = \sum_{i=0}^{m} b_i p^i - \sum_{i=0}^{m} x_{k,i}p^i = 0$$

for a sufficiently large k, that is $y - x_k \in p^m\mathbb{Z}_p$, $\{x_n\}$ converges to y.

Next we show that $\mathbb{Z}_p$ is compact. Let

$$y_n = \{y_{n,0}, \cdots, \sum_{i=0}^{m} y_{n,i}p^i, \cdots\} \in \mathbb{Z}_p \ (0 \leq y_{n,i} < p).$$

There is an infinite subsequence of y_n such that $y_{n,0} = a_0$ for some a_0 ($0 \leq a_0 < p$), and there is an infinite subsequence in this subsequence such that $y_{n,1} = a_1$ for some a_1. Repeating this, there exists a sequence $\{a_i\}_{i=0}^{\infty}$ ($0 \leq a_i < p$) such that there exist infinitely many y_n satisfying $y_{n,i} = a_i$ for $0 \leq i \leq$ any given integer. Hence $\{a_0, \cdots, \sum_{i=0}^{m} a_i p^i, \cdots\}$ is an accumulation point in $\mathbb{Z}_p$. □

Since $|\sum_{i=1}^{n} x_i|_p \leq \max_{1\leq i\leq n} |x_i|_p$ for $x_i \in \mathbb{Q}_p$, $\{\sum_{i=1}^{n} a_i\}_{n=1}^{\infty}$ ($a_i \in \mathbb{Q}_p$) is a Cauchy sequence if $|a_n|_p \to 0$. Hence $\lim_{n\to\infty} \sum_{i=1}^{n} a_i$ converges and we denote it by $\sum_{i=1}^{\infty} a_i$. As usual, we write $\alpha \equiv \beta \bmod p^n$ if $\alpha - \beta \in p^n\mathbb{Z}_p$.

Corollary 3.1.3. *$\mathbb{Q}_p$ is a locally compact non-discrete topological field.*

Proof. First, we show that the addition, product and division are continuous. It is easy to see for $x, x', y, y' \in \mathbb{Q}_p$, $z, z' \in \mathbb{Q}_p^\times$,

$$\begin{cases} \mathrm{d}(x+y, x'+y') = |x - x' + y - y'|_p \le \max(\mathrm{d}(x,x'), \mathrm{d}(y,y')) \\ \mathrm{d}(xy, x'y') = |xy - x'y + x'y - x'y'|_p \le \max(|y|_p\, \mathrm{d}(x,x'), |x'|_p\, \mathrm{d}(y,y')), \\ \mathrm{d}(z^{-1}, z'^{-1}) = |z - z'|_p |z^{-1}|_p |z'^{-1}|_p = {|z|_p}^{-1} {|z'|_p}^{-1}\, \mathrm{d}(z,z'). \end{cases}$$

Here we note that $|x'|_p = |x' - x + x|_p \le \max(\mathrm{d}(x', x), |x|_p)$ and if $\mathrm{d}(z, z') = |z - z'|_p < |z|_p$, then $|z'|_p = |z' - z + z|_p = |z|_p$. So, if x', y', z' are close to x, y, z respectively, then $x' + y', x'y', (z')^{-1}$ are close to $x + y, xy, z^{-1}$ respectively, which implies the continuity of the addition, product and division. Since $\mathbb{Z}_p$ is compact, $p^n\mathbb{Z}_p = \{x \in \mathbb{Q}_p \mid |x|_p < p^{-(n-1)}\}$ is a compact open neighbourhood of the zero. Hence $\mathbb{Q}_p$ is locally compact. The non-discrete property is clear. □

Let V be an n-dimensional vector space over $\mathbb{Q}_p$. Then, taking a basis $\{v_i\}$ of V, we can identify V and $\mathbb{Q}_p^n$ and introduce the topology on V induced by the product topology of $\mathbb{Q}_p^n$. Let $\{u_i\}$ be another basis and $(v_1, \cdots, v_n) = (u_1, \cdots, u_n)A$ for $A \in GL_n(\mathbb{Q}_p)$. Since $v = \sum a_i v_i$ has the expression

$$v = (u_1, \cdots, u_n)A^t(a_1, \cdots, a_n) = (u_1, \cdots, u_n)^t b$$

for $b := (b_1, \cdots, b_n) = A^t(a_1, \cdots, a_n)$, we have

$$c_1 \max_i(|a_i|_p) \le \max_i(|b_i|_p) \le c_2 \max_i(|a_i|_p),$$

where c_1, c_2 are positive numbers dependent on A. Hence the neighbourhood of zero can be defined independently of the choice of a basis of V. Thus the topology of V is uniquely determined.

Theorem 3.1.4. *For $a \in \mathbb{Z}_p^\times$, $a = x^2$ for $x \in \mathbb{Q}_p^\times$ if and only if $a \equiv 1 \bmod 8$ when $p = 2$, and $a \equiv y^2 \bmod p$ for $y \in \mathbb{Z}$ with $(y,p) = 1$ when $p \ne 2$.*

Proof. Suppose $a = x^2$ and put $x = \sum_{n=0}^{\infty} x_n p^n$ $(0 \le x_n < p)$; then $a \in \mathbb{Z}_p^\times$ yields $x \in \mathbb{Z}_p^\times$ by Theorem 3.1.2, and $a \equiv y^2 \bmod p$ holds for $y = x_0$, and moreover since $x_0 = 1$

$$\begin{aligned} a &= (x_0 + 2x_1)^2 + 2(x_0 + 2x_1)\sum_{n=2}^{\infty} x_n 2^n + (\sum_{n=2}^{\infty} x_n 2^n)^2 \\ &\equiv (x_0 + 2x_1)^2 \bmod 8 \equiv 1 \bmod 8 \text{ for } p = 2. \end{aligned}$$

Conversely, suppose $a \equiv y_0^2 \bmod q$ for $y_0 \in \mathbb{Z}$ where $(y_0, p) = 1$ and $q = p$ if $p \neq 2$ or $y_0 = 1$ and $q = 8$ if $p = 2$, and suppose $a \equiv y_k^2 \bmod qp^k$ for $0 \leq k \leq n$. Putting $y_{n+1} = y_n + bqp^n/2$ for $b \in \mathbb{Z}$, gives

$$a - y_{n+1}^2 = qp^n\{(a - y_n^2)/(qp^n) - by_n\} - (b^2q^2p^{2n}/4).$$

Now $a \in \mathbb{Z}_p^\times$ implies $(y_n, p) = 1$ and hence taking an integer b such that

$$(a - y_n^2)/(qp^n) \equiv by_n \bmod p,$$

$q^2p^{2n}/4 \in qp^{n+1}\mathbb{Z}_p$ implies $a \equiv y_{n+1}^2 \bmod qp^{n+1}$. Thus we have proved the existence of $y_n \in \mathbb{Z}$ such that $a \equiv y_n^2 \bmod qp^n$ and $y_{n+1} \equiv y_n \bmod p^{n+1}$ for every $n \geq 0$. On the other hand, $y_{n+1} - y_n \equiv 0 \bmod p^n$ implies that $y_{n+1} = y_0 + \sum_{i=0}^n (y_{i+1} - y_i)$ converges to some point x, and $a \equiv y_{n+1}^2 \bmod qp^{n+1}$ yields $a = x^2$. □

Corollary 3.1.4.

$$\mathbb{Q}_p^\times/(\mathbb{Q}_p^\times)^2 = \begin{cases} 1, \epsilon, p, \epsilon p \text{ if } p \neq 2, \\ 1, 3, 5, 7, 2\cdot 1, 2\cdot 3, 2\cdot 5, 2\cdot 7 \text{ if } p = 2, \end{cases}$$

where ϵ is any fixed element in $\mathbb{Z}_p^\times$ such that $\epsilon \not\equiv y^2 \bmod p$ for $y \in \mathbb{Z}$.

Proof. Let $x \in \mathbb{Q}_p^\times$; then $x(\mathbb{Q}_p^\times)^2 \ni \epsilon$ or $p\epsilon$ for some $\epsilon \in \mathbb{Z}_p^\times$. Hence the assertion is clear if $p = 2$. If $p \neq 2$, then $(\mathbb{Z}/p\mathbb{Z})^\times$ is a cylic group of even order $p - 1$ and so if $y, z \in \mathbb{Z}$ are not squares mod p, then $y \equiv w^2 z \bmod p$ for some $w \in \mathbb{Z}$. This completes the case of $p \neq 2$. □

Corollary 3.1.5. *$(\mathbb{Q}_p^\times)^2$ is open in $\mathbb{Q}_p$.*

Proof. $(\mathbb{Q}_p^\times)^2 = \cup_{n\in\mathbb{Z}}\, p^{2n}\{x^2 \mid x \in \mathbb{Z}_p^\times\}$, and

$$\{x^2 \mid x \in \mathbb{Z}_p^\times\} = \begin{cases} \cup_{\substack{y \bmod p \\ (y,p)=1}} \{z \in \mathbb{Q}_p \mid |z - y^2|_p < 1\} & \text{if } p \neq 2, \\ \{z \in \mathbb{Q}_p \mid |z - 1|_2 < 2^{-2}\} & \text{if } p = 2. \end{cases}$$

Hence $\{x^2 \mid x \in \mathbb{Z}_p^\times\}$ is open and so $(\mathbb{Q}_p^\times)^2$ is open. □

3.2 The quadratic residue symbol

Let p be an odd prime, and denote the field $\mathbb{Z}/p\mathbb{Z}$ by F_p. Since $F_p^\times$ is a cyclic group of order $p - 1$, there exists a unique subgroup N with $[F_p^\times : N] = 2$, i.e. $N = \{x^2 \mid x \in F_p^\times\}$. For $x \in F_p^\times$, we put $\chi(x) = 1$ if $x \in N$, and -1 if $x \notin N$. χ is a surjective homomorphism from $F_p^\times$ to $\{\pm 1\}$. For $x \in \mathbb{Z}$ with $(x, p) = 1$, we put $\left(\frac{x}{p}\right) := \chi(x)$ regarding $x \in F_p^\times$. It is called the *Legendre symbol* or *quadratic residue symbol.*

Theorem 3.2.1. *Let $p \neq q$ be odd prime numbers. Then the following hold.*

(1) *For $x, y \in \mathbb{Z}$ with $(xy, p) = 1$, $(\frac{xy}{p}) = (\frac{x}{p})(\frac{y}{p})$.*

(2) $(\frac{-1}{p}) = (-1)^{(p-1)/2}$.

(3) $(\frac{2}{p}) = (-1)^{(p^2-1)/8}$.

(4) $(\frac{q}{p})(\frac{p}{q}) = (-1)^{(p-1)/2 \cdot (q-1)/2}$ *(reciprocity law)*

Proof.

(1) is obvious from the definition.

(2): Let g be a generator of $F_p^\times$; then $g^{(p-1)/2} = -1$ is clear and so $-1 \in (F_p^\times)^2$ if and only if $(p-1)/2$ is even. This is the assertion (2).

(3): Let x be a primitive 8-th root of unity in the algebraic closure of F_p: then $x^4 = -1$ and $x^2 + x^{-2} = 0$. For $y = x + x^{-1}$, $y^2 = 2$ holds, and $(\frac{2}{p}) = 1 \Leftrightarrow y \in F_p \Leftrightarrow y^p = y$. If $p \equiv \pm 1 \bmod 8$, then $y^p = x^p + x^{-p} = y$ holds. If $p \equiv \pm 3 \bmod 8$, then $y^p = x^3 + x^{-3} = -(x^{-1} + x) = -y \neq y$. Thus $(\frac{2}{p}) = 1 \Leftrightarrow p \equiv \pm 1 \bmod 8 \Leftrightarrow (-1)^{(p^2-1)/8} = 1$ holds.

(4): Let x be a primitive p-th root of unity in the algebraic closure $\bar{F}_q$ of F_q, and put $\chi(a) := (\frac{a}{p})$ for $a \in F_p^\times$, and 0 for $a = 0$. Since x is a p-th root of unity, we can consider x^a canonically for $a \in F_p = \mathbb{Z}/p\mathbb{Z}$. Put $g := \sum_{a \in F_p} \chi(a) x^a \in \bar{F}_q$; then

$$g^2 = \sum_{a,b \in F_p} \chi(ab) x^{a+b} = \sum_{c \in F_p} x^c \sum_{a \in F_p} \chi(a(c-a)).$$

If $c \neq 0$, then we have

$$\begin{aligned}\sum_{a \in F_p} \chi(a(c-a)) &= \chi(-1) \sum_{a \in F_p^\times} \chi(a^2 - ac) \\ &= \chi(-1) \sum_{a \in F_p^\times} \chi(1 - a^{-1}c) = \chi(-1) \sum_{a \in F_p^\times} \chi(1-a) \\ &= \chi(-1)\{\sum_{b \in F_p} \chi(b) - \chi(1)\} = -\chi(-1).\end{aligned}$$

Thus splitting the expression for g^2 into two sums $c = 0$ and $c \neq 0$, we have

$$g^2 = \sum_{a \in F_p} \chi(-a^2) - \chi(-1) \sum_{c \in F_p^\times} x^c = \chi(-1)(p-1) + \chi(-1) = \chi(-1)p$$

since $\sum_{c \in F_p} x^c = 0$. On the other hand, we have

$$g^q = \sum_{a \in F_p} \chi(a) x^{qa} = \chi(q) \sum_{a \in F_p} \chi(qa) x^{qa} = \chi(q) g$$

and so $g^{q-1} = \chi(q)$. Thus we have

$$g^{q-1} = (g^2)^{(q-1)/2} = \chi(-1)^{(q-1)/2} p^{(q-1)/2} = \chi(q),$$

i.e.

$$(-1)^{(p-1)/2 \cdot (q-1)/2} p^{(q-1)/2} = (\frac{q}{p}) \text{ in } F_q.$$

Here $(p^{(q-1)/2})^2 = 1$ in F_q and so $p^{(q-1)/2} = \pm 1$ in F_q. It is easy to see that $p^{(q-1)/2} = 1$ in $F_q \Leftrightarrow p \in (F_q^\times)^2 \Leftrightarrow (\frac{p}{q}) = 1$ and so $p^{(q-1)/2} = (\frac{p}{q})$ in F_q. Thus we have $(\frac{q}{p})(\frac{p}{q}) = (-1)^{(p-1)/2 \cdot (q-1)/2}$ in F_q and hence in $\mathbb{Z}$ since they are ± 1. □

3.3 The Hilbert symbol

Let F be a local field, i.e. the real number field $\mathbb{R}$ or the p-adic number field $\mathbb{Q}_p$. For $a, b \in F^\times$, we define the *Hilbert symbol* $(a, b) = (a, b)_F$ by

$$(a, b) = \begin{cases} 1 & \text{if } \langle a \rangle \perp \langle b \rangle \perp \langle -1 \rangle \text{ is isotropic over } F, \\ -1 & \text{otherwise.} \end{cases}$$

Theorem 3.3.1. *For the Hilbert symbol, the following hold.*

(1) $(a, b) = (b, a), (a, b^2) = 1$ *for* $a, b \in F^\times$.
(2) $(a, -a) = (a, 1 - a) = 1$ *for* $a \in F$ *with* $a \neq 0, 1$.
(3) $(ab, c) = (a, c)(b, c)$ *for* $a, b, c \in F^\times$.
(4) $(a, b) \in F^\times \times F^\times \mapsto (a, b)_F \in F$ *is continuous.*
(5) $(a, b) = -1$ *if and only if* $a < 0$ *and* $b < 0$ *when* $F = \mathbb{R}$.
(6) *For* $F = \mathbb{Q}_p$ $(p \neq 2)$, *we have*

$$\begin{cases} (a, b) = 1 & \text{for } a, b \in \mathbb{Z}_p^\times, \\ (a, p) = (\frac{a_0}{p}) & \text{for } a \in \mathbb{Z}_p^\times, \end{cases}$$

where $a_0 \in \mathbb{Z}$ *with* $a \equiv a_0 \bmod p\mathbb{Z}_p$.

(7) *For* $F = \mathbb{Q}_2$, *we have for* $a, b \in \mathbb{Z}_2^\times$

$$\begin{cases} (a, b) = (-1)^{(a-1)/2 \cdot (b-1)/2} & (= 1 \text{ if and only if } a \text{ or } b \equiv 1 \bmod 4), \\ (a, 2) = (-1)^{(a^2-1)/8} & (= 1 \text{ if and only if } a \equiv \pm 1 \bmod 8). \end{cases}$$

(8) *For* $a \in F^\times$, $(a, F^\times) = 1$ *implies* $a \in (F^\times)^2$.
(9) *For* $a, b \in \mathbb{Q}^\times$, $\prod_p (a,b)_{\mathbb{Q}_p} \cdot (a,b)_{\mathbb{R}} = 1$ *holds.*

Proof. Since $(a,b)_F = 1$ means that $ax^2 + by^2 = z^2$ has a non-trivial solution in F, (1), (2), (5) are obvious. (4) follows from the fact that $(F^\times)^2$ is open.

Now let us prove (3). If $F = \mathbb{R}$, then (3) follows immediately from (5) and so we assume $F = \mathbb{Q}_p$. If $c \in (F^\times)^2$, then (3) is obvious by (1). Suppose $c \notin (F^\times)^2$; then $(d, c) = 1$ for $d \in F^\times$ if and only if $dx^2 + cy^2 = z^2$ has a non-trivial solution $x, y, z \in F$, where $x \neq 0$. $dx^2 = (z + \sqrt{c}y)(z - \sqrt{c}y)$ means that $d \in N := \mathrm{N}_{E/F}(E)$ for a quadratic extension $E = F(\sqrt{c})$. Thus $(d, c) = 1$ if and only if $d \in N$. Therefore (3) is valid if a or $b \in N$, since N is a group. It remains to show that $ab \in N$ for $a, b \notin N$. To do it, we have only to prove $[F^\times : N] \leq 2$. Since $[F^\times : (F^\times)^2] = 4$ or 8 according to whether $p > 2$ or $p = 2$ by Corollary 3.1.3, we have only to show $[N : (F^\times)^2] \geq 2$ or 4 according to $p > 2$ or $p = 2$, respectively.

Suppose $p > 2$; then $-c \in N$ is clear. If $\mathrm{ord}_p\, c$ is odd, then $-c \notin F^2$ and hence $[N : (F^\times)^2] \geq 2$. If $\mathrm{ord}_p\, c$ is even, then we may assume $c \in \mathbb{Z}_p^\times$, multiplying by an even power of p, and then by virtue of Proposition 1.3.2, $c^{-1}x^2 - y^2 = 1$ has a non-trivial solution in $\mathbb{Z}/p\mathbb{Z}$, regarding $c^{-1} \in \mathbb{Z}/p\mathbb{Z}$ by reduction modulo p. Hence for some $x, y \in \mathbb{Z}_p$, $c^{-1}x^2 - y^2 \equiv 1 \bmod p$ holds. Then by virtue of Theorem 3.1.4, $c^{-1}x^2 - y^2 = z^2$ holds for some $z \in \mathbb{Z}_p^\times$, which yields $cz^2 = x^2 - cy^2 \in N$, and hence $c \in N$, i.e. $[N : (F^\times)^2] \geq 2$.

Suppose $p = 2$; we have only to show the existence of $X, Y \in N$ with $X, Y, XY \notin (F^\times)^2$ to verify $[N : (F^\times)^2] \geq 4$. Noting $\mathrm{N}(x + \sqrt{c}y) = x^2 - cy^2$, we can take

$$\begin{array}{ll}
X = 2(= 1^2 - c \cdot 1^2), Y = 5(= 1^2 - c \cdot 2^2) & \text{if } c = -1; \\
X = 5(\equiv 1^2 - c \cdot 2^2 \bmod 8), Y = 2^2 \cdot 7(= 1^2 - c \cdot 3^2) & \text{if } c = -3; \\
X = 5(\equiv 0^2 - c \cdot 1^2 \bmod 8), Y = -2(= 1^2 - c \cdot 1^2) & \text{if } c = 3; \\
X = -2(= 0^2 - c \cdot 1^2), Y = -1(= 1^2 - c \cdot 1^2) & \text{if } c = 2; \\
X = 2(= 0^2 - c \cdot 1^2), Y = 3(= 1^2 - c \cdot 1^2) & \text{if } c = -2; \\
X = -3 \cdot 2(= 0^2 - c \cdot 1^2), Y = -5(= 1^2 - c \cdot 1^2) & \text{if } c = 3 \cdot 2; \\
X = 3 \cdot 2(= 0^2 - c \cdot 1^2), Y = 7(= 1^2 - c \cdot 1^2) & \text{if } c = -3 \cdot 2,
\end{array}$$

which exhaust all possibilities of c. Thus (3) has been proved.

Let us prove (6). Let $a, b \in \mathbb{Z}_p^\times$; then $ax^2 + by^2 = z^2$ has a non-trivial solution in $\mathbb{Z}/p\mathbb{Z}$, regarding $a, b \in \mathbb{Z}/p\mathbb{Z}$ by reduction modulo p. If $z \neq 0$, then we regard $x, y \in \mathbb{Z} \subset \mathbb{Z}_p$ and applying Theorem 3.1.4, $ax^2 + by^2 = z_0^2$ for some $z_0 \in \mathbb{Z}_p^\times$ which means $(a,b)_F = 1$. If $z = 0$, then $-ab^{-1}$ is a square in $(\mathbb{Z}/p\mathbb{Z})^2$ and applying the same theorem, $-ab^{-1} = w^2$ holds for some $w \in \mathbb{Z}_p^\times$ and then $a \cdot 1^2 + b \cdot w^2 = 0$ holds, which implies $(a,b)_F = 1$. If, for $a \in \mathbb{Z}_p^\times$ $(a,p)_F = 1$, that is $ax^2 + py^2 = z^2$ has a non-trivial solution, then multiplying by a power of p, we may assume $x, y, z \in \mathbb{Z}_p$ and one of them is in $\mathbb{Z}_p^\times$. Then $z \in \mathbb{Z}_p^\times$ follows necessarily. Hence $(a,p)_F = 1$ for $a \in \mathbb{Z}_p^\times$ implies that a is a square in $\mathbb{Z}/p\mathbb{Z}$ by reduction modulo p, and so $(\frac{a}{p}) = 1$. Conversely $(\frac{a}{p}) = 1$ implies $a \equiv z^2 \bmod p$ and then by Theorem 3.1.4, $a \in F^2$ and so $(a,p)_F = 1$.

Let us prove (7). Let $a \in \mathbb{Z}_2^\times$. Suppose $(a,2)_F = 1$; then $ax^2 + 2y^2 = z^2$ has a non-trivial solution in $\mathbb{Q}_2$ and we may assume $x, y, z \in \mathbb{Z}_2$ and one of them is in $\mathbb{Z}_2^\times$. Then $x, z \in \mathbb{Z}_2^\times$ holds, and $a + 2y^2 \equiv 1 \bmod 8$ since $x^2 \equiv z^2 \equiv 1 \bmod 8$. Hence $a \equiv 1 \bmod 8$ if $y \in 2\mathbb{Z}_2$, and $\equiv -1 \bmod 8$ if $y \in \mathbb{Z}_2^\times$. Thus $(2,a)_F = 1$ implies $a \equiv \pm 1 \bmod 8$. Conversely if $a \equiv \pm 1 \bmod 8$, then $\pm a \equiv 1 \bmod 8$ and so $\pm a = x^2$ for $x \in \mathbb{Z}_2^\times$. Hence $a \cdot 1^2 + 2 \cdot 0 = x^2$ if $a = x^2$, and $a \cdot 1^2 + 2 \cdot x^2 = x^2$ if $a = -x^2\cdot$ which mean $(a,2)_F = 1$. Thus $(2,a)_F = 1$ if and only if $a \equiv \pm 1 \bmod 8$.

Next, for $a, b \in \mathbb{Z}_2^\times$, we show $(a,b)_F = 1$ if and only if a or $b \equiv 1 \bmod 4$. Noting a unit in $\mathbb{Z}_2$ ($\equiv 1 \bmod 8$) is a square in $\mathbb{Z}_2$, we may assume $a, b = \pm 1, \pm 3$. First we show $(a,b)_F = 1$ if a or $b \equiv 1 \bmod 4$. We may assume $a \equiv 1 \bmod 4$, i.e. $a = 1$ or -3. $(1,b) = 1$ is clear and so we may assume $a = -3$. Noting $-7, -15 \in (\mathbb{Z}_2^\times)^2$ from $-7 \equiv -15 \equiv 1 \bmod 8$, we have $-3 \cdot 1^2 - 2^2 = \sqrt{-7}^2$, $-3 \cdot 1^2 + 3 \cdot 1^2 = 0$, $-3 \cdot 1^2 - 3 \cdot 2^2 = \sqrt{-15}^2$, which yield $(-3,-1)_F = (-3,3)_F = (-3,-3)_F = 1$. Hence a or $b \equiv 1 \bmod 4$ implies $(a,b)_F = 1$.

Conversely suppose $(a,b)_F = 1$ for $a, b \in \mathbb{Z}_2^\times$. Then $ax^2 + by^2 = z^2$ has a non-trivial solution, where $x, y, z \in \mathbb{Z}_2$ and one of them is in $\mathbb{Z}_2^\times$. If $z \in 2\mathbb{Z}_2$, then $ax^2 + by^2 \equiv 0 \bmod 4$ implies $x, y \in \mathbb{Z}_2^\times$, and so $a + b \equiv 0 \bmod 4$. Thus $a \equiv 1$ or $b \equiv 1 \bmod 4$. If $z \in \mathbb{Z}_2^\times$, then $ax^2 + by^2 \equiv 1 \bmod 8$, which implies that only one of x, y is in $\mathbb{Z}_2^\times$. If, for example $x \in \mathbb{Z}_2^\times$, then $y \in 2\mathbb{Z}_2$ and we have $a \equiv 1 \bmod 4$. Thus we have proved that $(a,b)_F = 1$ yields a or $b \equiv 1 \bmod 4$.

(8) easily follows from (5), (6), (7) using (3).

To prove (9), we may assume, by using (3) and the fact that $(a,a) = (a,-1)$, that either a, b are different primes, or else a may be a prime or -1 and $b = -1$. Now (9) follows from (6), (7) and (2), (3), (4) in Theorem 3.2.1 □

3.4 The Hasse invariant

Let the field F be a local field $\mathbb{R}$ or $\mathbb{Q}_p$ as in the previous section. We define an important invariant $S(V)$ for a regular quadratic space V over

F, which is called the *Hasse invariant.*

Theorem 3.4.1. *Let $V \cong \langle a_1 \rangle \perp \cdots \perp \langle a_n \rangle$ be a regular quadratic space over F. Then*

$$S(V) := \prod_{1 \le i \le j \le n} (a_i, a_j)_F = \prod_{i=1}^{n} (a_i, \prod_{i \le j \le n} a_j)_F$$

depends only on V.

Proof. Suppose $n = 1$ and $V \cong \langle a \rangle$; then $(a, a) = (a, -1) = (\mathrm{d}\,V, -1)$ depends only on V.

Suppose $n = 2$; then

$$\prod_{i \le j} (a_i, a_j) = (a_1, a_1 a_2)(a_2, a_2) = (a_1, -a_2)(a_2, -1)$$
$$= (a_1 a_2, -1)(a_1, a_2) = (\mathrm{d}\,V, -1)(a_1, a_2).$$

Now $(a_1, a_2) = 1$ if and only if $\langle a_1 \rangle \perp \langle a_2 \rangle \perp \langle -1 \rangle \cong V \perp \langle -1 \rangle$ is isotropic by definition. Thus $\prod_{i \le j} (a_i, a_j)$ is independent of the choice of an orthogonal basis.

When $n \ge 3$, we use the following.

Lemma. *We introduce the relation $\sim$ between orthogonal bases of V by writing $\{v_i\} \sim \{w_i\}$ if $Fv_i = Fw_j$ for some i, j. We say that $\{v_i\}, \{w_i\}$ are* linked *if there exist orthogonal bases $\{z_i^j\}$ of V such that $\{v_i\} = \{z_i^1\} \sim \{z_i^2\} \sim \cdots \sim \{z_i^m\} = \{w_i\}$. If $\dim V \ge 3$ and $\{v_i\}, \{w_i\}$ are orthogonal bases of V, then they are linked.*

Let us first show how the theorem follows from the lemma. Using the lemma, we only need that if

$$V \cong \langle a_1 \rangle \perp \cdots \perp \langle a_n \rangle \cong \langle b_1 \rangle \perp \cdots \perp \langle b_n \rangle$$

and $a_h = b_k$ for some h, k, then $\prod_{i \le j} (a_i, a_j) = \prod_{i \le j} (b_i, b_j)$. To show this, it is sufficient to consider two cases:

(i) the interchange of a_1 and a_k for $k \ne 1$ and
(ii) $a_1 = b_1$.

We note that for a fixed h $(1 \le h \le n)$,

$$\prod_{i \le j} (a_i, a_j) = \prod_{h \le j} (a_h, a_j) \prod_{i < h} (a_i, a_h) \prod_{\substack{i \le j \\ i, j \ne h}} (a_i, a_j)$$
$$= (a_h, \mathrm{d}\,V) \prod_{\substack{i \le j \\ i, j \ne h}} (a_i, a_j),$$

since $\prod_i a_i = \mathrm{d}\, V$ in $F^\times/(F^\times)^2$. Let us consider the first case;

If $a_1 = b_k, a_k = b_1, a_i = b_i$ for $i \neq 1, k$ $(k \neq 1)$, then applying the note to $h = 1$ and k,

$$\begin{aligned}\prod_{i \le j}(a_i, a_j) &= (a_1, \mathrm{d}\, V) \prod_{1<i\le j} (a_i, a_j) \\ &= (a_1, \mathrm{d}\, V)(a_k, a_1 \mathrm{d}\, V) \prod_{\substack{i \le j \\ i,j \neq 1,k}} (a_i, a_j) \\ &= (a_1 a_k, \mathrm{d}\, V)(a_k, a_1) \prod_{\substack{i \le j \\ i,j \neq 1,k}} (a_i, a_j) \\ &= (b_1 b_k, \mathrm{d}\, V)(b_1, b_k) \prod_{\substack{i \le j \\ i,j \neq 1,k}} (b_i, b_j) \\ &= \prod_{i \le j}(b_i, b_j)\end{aligned}$$

which completes the first case.

Next, if $a_1 = b_1$, then

$$\prod_{i \le j}(a_i, a_j) = (a_1, \mathrm{d}\, V) \prod_{1<i\le j} (a_i, a_j) = (a_1, \mathrm{d}\, V) S(\langle a_2 \rangle \perp \cdots \perp \langle a_n \rangle)$$

and the similar equation for $V \cong \langle b_1 \rangle \perp \cdots \perp \langle b_n \rangle$ completes the second case by the induction hypothesis on $\dim V$. □

Proof of Lemma.

Let $\{u_i\}$ be an orthogonal basis of V linked to $\{w_i\}$, and write $v_1 = a_1 u_1 + \cdots + a_m u_m$, $a_i \in F$, $a_m \neq 0$. We put $m = \nu(\{u_i\})$. Since $m = 1$ means that $\{v_i\}$ and $\{u_i\}$ are linked, we have only to show that if $m > 1$, then there exists an orthogonal basis $\{u_i'\}$ such that $\{u_i'\} \sim \{u_i\}$, $\nu(\{u_i'\}) < m$. If $a_j = 0$ for $j < m$, then putting $u_i' = u_i$ for $i < j$, $= u_{i+1}$ for $i < n$, and $= u_j$ for $i = n$, we have $\nu(\{u_i'\}) = m - 1$ and $\{u_i'\} \sim \{u_i\}$. Thus we may suppose $\prod_{i=1}^m a_i \neq 0$. If $m \geq 3$, then we may suppose

$$Q(a_1 u_1 + a_2 u_2) = a_1^2 Q(u_1) + a_2^2 Q(u_2) \neq 0,$$

since one of

$$Q(a_1 u_1 + a_2 u_2), Q(a_2 u_2 + a_3 u_3), Q(a_1 u_1 + a_3 u_3)$$

is not equal to 0. Then we can extend

$$\{a_1 u_1 + a_2 u_2, u_3, u_4, \cdots, u_m\}$$

to an orthogonal basis of V and for this basis the value of $\nu = m - 1$ and it is linked to $\{u_i\}$. This finishes the case of $m \geq 3$. Suppose $m = 2$; then $\{a_1 u_1 + a_2 u_2, a_2 Q(u_2)u_1 - a_1 Q(u_1)u_2, u_3, \cdots\}$ is an orthogonal basis whose value of ν is equal to 1, and hence it is linked to $\{u_i\}$. Thus we have completed the proof of the lemma. □

It should be remarked that the definition of Hasse's invariant is sometimes defined by replacing $\leq$ by $<$.

Theorem 3.4.2. *Let U, V be regular quadratic spaces over F and $\dim U = u, \dim V = v$. Then we have*

$$S(U \perp V) = S(U)S(V)(\mathrm{d}\,U, \mathrm{d}\,V)_F,$$
$$S(U \otimes V) = S(U)^v S(V)^u (\mathrm{d}\,U, \mathrm{d}\,V)_F^{uv+1} (\mathrm{d}\,U, -1)_F^{v(v-1)/2} (\mathrm{d}\,V, -1)_F^{u(u-1)/2},$$
$$S(\langle a \rangle \otimes V) = (a, -1)^{v(v+1)/2} (a, \mathrm{d}\,V)^{v+1} S(V).$$

Proof. Let $U \cong \langle a_1 \rangle \perp \cdots \perp \langle a_u \rangle, V \cong \langle b_1 \rangle \perp \cdots \perp \langle b_v \rangle$; then

$$\begin{aligned} S(U \perp V) &= \prod_{i \leq j} (a_i, a_j) \prod_{i,j} (a_i, b_j) \prod_{i \leq j} (b_i, b_j) \\ &= S(U)(\mathrm{d}\,U, \mathrm{d}\,V) S(V) \end{aligned}$$

is easy. We use induction on u to prove the formula for $S(U \otimes V)$. If $u = 1$, then

$$\begin{aligned} S(U \otimes V) &= \prod_{i \leq j} (a_1 b_i, a_1 b_j) = (a_1, a_1)^{v(v+1)/2} (a_1, \prod_{i \leq j} b_i b_j) \prod_{i \leq j} (b_i, b_j) \\ &= S(U)^{v(v+1)/2} (\mathrm{d}\,U, \mathrm{d}\,V)^{v+1} S(V) \end{aligned}$$

is what we want, noting $S(U) = (\mathrm{d}\,U, -1)$. Suppose $u \geq 2$ and put $U_0 = \perp_{i \geq 2} \langle a_i \rangle$; then we have

$$\begin{aligned} S(U \otimes V) &= S(\langle a_1 \rangle \otimes V \perp U_0 \otimes V) \\ &= S(\langle a_1 \rangle \otimes V) S(U_0 \otimes V)(\mathrm{d}(\langle a_1 \rangle \otimes V), \mathrm{d}(U_0 \otimes V)) \\ &= \{(a_1, -1)^{v(v+1)/2} (a_1, \mathrm{d}\,V)^{v+1} S(V)\} \\ &\quad \times \{S(U_0)^v S(V)^{u-1} (\mathrm{d}\,U_0, \mathrm{d}\,V)^{(u-1)v+1} \\ &\quad \times (a_1 \,\mathrm{d}\,U, -1)^{v(v-1)/2} (\mathrm{d}\,V, -1)^{(u-1)(u-2)/2} \\ &\quad \times (a_1^v \,\mathrm{d}\,V, (\mathrm{d}\,U_0)^v (\mathrm{d}\,V)^{u-1})\}. \end{aligned}$$

Substituting $S(U_0) = S(U)(a_1, -1)(a_1, a_1 \,\mathrm{d}\,U) = S(U)(a_1, \mathrm{d}\,U)$ and $\mathrm{d}\,U_0 = a_1 \,\mathrm{d}\,U$, we obtain the formula for $S(U \otimes V)$. The last follows immediately from the second. □

Exercise.

Let H_m be the hyperbolic space of dim $= 2m$ over F. Show that

$$S(H_m) = (-1,-1)_F^{m(m+1)/2}, \quad S(\langle\begin{pmatrix} 2 & 1 \\ 1 & 2 \end{pmatrix}\rangle) = 1.$$

3.5 Classification of quadratic spaces over p-adic number fields

In this section, we classify regular quadratic spaces over $\mathbb{Q}_p$. First we show

Theorem 3.5.1. *Let U be a regular quadratic space over $\mathbb{Q}_p$ with $\dim U = m$. Then U is isotropic if and only if $m \geq 5$ or*

$$\begin{cases} m = 2 & \Rightarrow \quad \mathrm{d}\,U = -1, \\ m = 3 & \Rightarrow \quad S(U) = (-1,-1), \\ m = 4 & \Rightarrow \quad \text{either } \mathrm{d}\,U = 1 \text{ and } S(U) = (-1,-1) \text{ or } \mathrm{d}\,U \neq 1. \end{cases}$$

Proof. Since U is regular, it is not isotropic if $m = 1$. Let

$$U \cong \langle a_1 \rangle \perp \cdots \perp \langle a_m \rangle.$$

If $m = 2$, then the assertion follows from Proposition 1.2.3.

Suppose $m = 3$; then U is isotropic if and only if

$$\langle -a_3 \rangle \otimes U \cong \langle -a_1 a_3 \rangle \perp \langle -a_2 a_3 \rangle \perp \langle -1 \rangle$$

is isotropic, equivalently $1 = (-a_1 a_3, -a_2 a_3)_F$ by definition of the Hilbert symbol, which is equal to

$$\begin{aligned} &(-1,-1)(-1, a_2 a_3)(a_1 a_3, -1)(a_1 a_3, a_2 a_3) \\ &\qquad = (-1,-1)(a_1, a_1 a_2 a_3)(a_2, a_2 a_3)(a_3, a_3) = (-1,-1)S(U). \end{aligned}$$

Thus we have finished this case.

Suppose $m = 4$. If U is isotropic and $\mathrm{d}\,U = 1$, then $U = H \perp V$ holds where H is a hyperbolic plane and $\mathrm{d}\,V = -1$, and so V is also a hyperbolic plane and therefore $S(U) = (-1,-1)$. Thus the " only if "-part has been proved. Before proving the converse, that is for an anisotropic quaternary quadratic space U, both $\mathrm{d}\,U = 1$ and $S(U) = -(-1,-1)$ hold, we claim first that for an anisotropic quadratic space

$$\langle 1 \rangle \perp \langle b_1 \rangle \perp \langle b_2 \rangle \perp \langle b_3 \rangle,$$

$(x, -b_1) = 1$ implies $(x, -b_2 b_3) = 1$ for $x \in F^\times$.

Proof of the claim.

It is easy to see
$(x, -b_1) = 1$

$\Rightarrow$ $\langle x\rangle \perp \langle -b_1\rangle \perp \langle -1\rangle$ is isotropic.
$\Rightarrow$ x is represented by $\langle b_1\rangle \perp \langle 1\rangle$.
$\Rightarrow$ x is not represented by $\langle -b_2\rangle \perp \langle -b_3\rangle$.
$\Rightarrow$ $\langle x\rangle \perp \langle b_2\rangle \perp \langle b_3\rangle$ is anisotropic.
$\Rightarrow$ $(x, xb_2b_3)(b_2, b_2b_3)(b_3, b_3) = -(-1, -1)$.
$\Rightarrow$ $(x, -b_2b_3) = 1$, using the fact that $\langle 1\rangle \perp \langle b_2\rangle \perp \langle b_3\rangle$ is anisotropic gives $(b_2, b_2b_3)(b_3, b_3) = -(-1, -1)$.

Thus the claim has been proved.

Suppose that U is anisotropic; then

$$\langle a_1\rangle \otimes U \cong \langle 1\rangle \perp \langle a_1a_2\rangle \perp \langle a_1a_3\rangle \perp \langle a_1a_4\rangle,$$

and

$$\langle a_4\rangle \otimes U \cong \langle 1\rangle \perp \langle a_3a_4\rangle \perp \langle a_1a_4\rangle \perp \langle a_2a_4\rangle$$

are anisotropic. By the claim, we know

$$\begin{cases} (x, -a_1a_2) = 1 & \Rightarrow \quad (x, -a_3a_4) = 1, \\ (x, -a_3a_4) = 1 & \Rightarrow \quad (x, -a_1a_2) = 1. \end{cases}$$

Hence $(x, -a_1a_2) = (x, -a_3a_4)$ for $x \in F^\times$ and then (8) in Theorem 3.3.1 implies $\mathrm{d}\,U = a_1a_2a_3a_4 \in (F^\times)^2$, i.e. $\mathrm{d}\,U = 1$. Then Theorem 3.4.2 yields

$$\begin{aligned} S(U) &= S(U \otimes \langle a_1\rangle) = S(\langle 1\rangle \perp \langle a_1a_2\rangle \perp \langle a_1a_3\rangle \perp \langle a_1a_4\rangle) \\ &= S(\langle a_1\rangle \otimes (\langle a_2\rangle \perp \langle a_3\rangle \perp \langle a_4\rangle)) \\ &= S(\langle a_2\rangle \perp \langle a_3\rangle \perp \langle a_4\rangle) = -(-1, -1) \end{aligned}$$

since $\langle a_2\rangle \perp \langle a_3\rangle \perp \langle a_4\rangle$ is anisotropic. Thus the case of $m = 4$ has been proved.

For $m \geq 5$, we have only to show that a quinary quadratic space is isotropic. Suppose $\dim U = 5$ and U is anisotropic. For $a \in \mathrm{d}\,U$, $\langle a\rangle \otimes U$ is also anisotropic and $\mathrm{d}(\langle a\rangle \otimes U) = a^6 = 1$ in $F^\times/(F^\times)^2$. Hence we may assume $\mathrm{d}\,U = 1$. The discriminant of every quaternary subspace of U is equal to 1 in $F^\times/(F^\times)^2$ by the case of $m = 4$. Hence we have $U \cong \perp_5 \langle 1\rangle$. If $p \neq 2$, then $S(\perp_3 \langle 1\rangle) = 1 = (-1, -1)$ and so $\perp_3 \langle 1\rangle$ is isotropic. This is a contradiction. If $p = 2$, then $-7 \in (\mathbb{Q}_2^\times)^2$ and $1^2 + 1^2 + 1^2 + 2^2 + \sqrt{-7}^2 = 0$ show that $\perp_5 \langle 1\rangle$ is isotropic, which is again a contradiction. Hence, U is isotropic. □

The above proof for $m \leq 4$ shows that the theorem is also valid for $\mathbb{R}$.

Corollary 3.5.1. *For a regular quaternary quadratic space U over $\mathbb{Q}_p$, $Q(U) = \mathbb{Q}_p$ holds.*

Proof. For $a \in \mathbb{Q}_p^\times$, $U \perp \langle -a \rangle$ is a regular quinary quadratic space and hence it is isotropic, which yields $Q(U) \ni a$. □

Corollary 3.5.2. *If $p \neq 2$ and $a_i \in \mathbb{Z}_p^\times$ $(i = 1, 2, 3)$, then the quadratic space $\langle a_1 \rangle \perp \langle a_2 \rangle \perp \langle a_3 \rangle$ over $\mathbb{Q}_p$ is isotropic.*

Proof. The Hasse invariant is equal to $1 = (-1, -1)$ by (6) in Theorem 3.3.1. □

Theorem 3.5.2. *Let U, V be regular quadratic spaces over $\mathbb{Q}_p$. Then $U \cong V$ if and only if $\dim U = \dim V, \mathrm{d}\, U = \mathrm{d}\, V$ and $S(U) = S(V)$.*

Proof. The " only if " part is clear. We use induction on $m := \dim U$ to prove the " if " part. If $m = 1$, then $U \cong V$ follows from $U \cong \langle \mathrm{d}\, U \rangle \cong \langle \mathrm{d}\, V \rangle \cong V$.

Suppose $m = 2$ and put $W = U \perp (\langle -1 \rangle \otimes V)$; then $\mathrm{d}\, W = 1$ and

$$\begin{aligned} S(W) &= S(U)S(\langle -1 \rangle \otimes V)(\mathrm{d}\, U, \mathrm{d}\, V) \\ &= S(U)S(V)(-1, \mathrm{d}\, V)(-1, -1)(\mathrm{d}\, U, \mathrm{d}\, V) = (-1, -1), \end{aligned}$$

using $S(U) = S(V), \mathrm{d}\, U = \mathrm{d}\, V$. Hence W is isotropic by the previous theorem, and there exist $u \in U, v \in V$ $(u \neq 0, v \neq 0)$ such that $Q(u) = Q(v) \neq 0$. Thus

$$U \cong \langle Q(u) \rangle \perp \langle Q(u)\, \mathrm{d}\, U \rangle \cong \langle Q(v) \rangle \perp \langle Q(v)\, \mathrm{d}\, V \rangle \cong V$$

holds.

Suppose $m \geq 3$ and put $W = U \perp (\langle -1 \rangle \otimes V)$; then $\dim W \geq 5$ implies that W is isotropic and so we can take non-zero vectors $u \in U, v \in V$ such that $Q(u) = Q(v) \neq 0$. Write

$$U = \langle Q(u) \rangle \perp U_0, \; V = \langle Q(v) \rangle \perp V_0.$$

Then $\dim U_0 = \dim V_0$ and $\mathrm{d}\, U_0 = \mathrm{d}\, V_0$ are obvious, and

$$S(U) = (Q(u), Q(u))S(U_0)(Q(u), \mathrm{d}\, U_0)$$

and a similar formula for V implies $S(U_0) = S(V_0)$. By the induction hypothesis, we have $U_0 \cong V_0$ and then $U \cong V$. □

Corollary 3.5.3. *Let U be a regular quaternary anisotropic quadratic space over $\mathbb{Q}_p$. Then $U \cong \langle 1\rangle \perp \langle -\delta\rangle \perp \langle p\rangle \perp \langle -\delta p\rangle$ where $\delta \in \mathbb{Z}_p^\times$ such that $(p,\delta) = -1, (\delta, \mathbb{Z}_p^\times) = 1$. In particular,* $\mathrm{d}\,U = 1, S(U) = -(-1,-1)$.

Proof. By Theorem 3.5.1, U is anisotropic if and only if $\mathrm{d}\,U = 1$ and $S(U) = -(-1,-1)$, which are satisfied for the given form in the assertion. And from the theorem it follows that they are unique up to isometry if they exist. Hence we have only to prove the existence of δ in the assertion. If $p = 2$, then $\delta = 5$ satisfies $(2,\delta) = -1, (\delta, \mathbb{Z}_p^\times) = 1$ by (7) of Theorem 3.3.1. If $p > 2$, then we have only to take an integer δ such that $(\frac{\delta}{p}) = -1$ by (6) of the same theorem. Thus the existence of δ has been proved. □

Corollary 3.5.4. *Let U, V be regular quadratic spaces over $\mathbb{Q}_p$ with $\dim U \le \dim V$. Then U is represented by V if and only if $\dim V - \dim U \ge 3$ or*

$$\begin{cases} V \cong U & \text{if } \dim V = \dim U, \\ V \cong U \perp \langle \mathrm{d}\,U \cdot \mathrm{d}\,V\rangle & \text{if } \dim V = \dim U + 1, \\ V \cong U \perp H & \text{if } \dim V = \dim U + 2 \text{ and } \mathrm{d}\,V = -\,\mathrm{d}\,U, \end{cases}$$

where H is the hyperbolic plane.

Proof. If $\dim V - \dim U = 0$ or 1, then the assertion is clear.

Suppose $\dim V - \dim U = 2$. If $\mathrm{d}\,V\,\mathrm{d}\,U = -1$, then the assertion is clear by Proposition 1.2.3. Suppose $\mathrm{d}\,V\,\mathrm{d}\,U \ne -1$; then we take $a, b \in \mathbb{Q}_p^\times$ such that $ab\,\mathrm{d}\,U = \mathrm{d}\,V$ and put $W = U \perp \langle a\rangle \perp \langle b\rangle$. Clearly U is represented by W. Moreover $\dim V = \dim W$ and $\mathrm{d}\,V = \mathrm{d}\,W$. Therefore

$$\begin{aligned} S(W) &= S(U)(\mathrm{d}\,U, ab)(a, ab)(b,b) \\ &= S(U)(\mathrm{d}\,U, \mathrm{d}\,U\,\mathrm{d}\,V)(a, \mathrm{d}\,U\,\mathrm{d}\,V)(a\,\mathrm{d}\,U\,\mathrm{d}\,V, -1) \\ &= S(U)(\mathrm{d}\,U, \mathrm{d}\,U\,\mathrm{d}\,V)(\mathrm{d}\,U\,\mathrm{d}\,V, -1)(a, -\,\mathrm{d}\,U\,\mathrm{d}\,V). \end{aligned}$$

Since $-\,\mathrm{d}\,U\,\mathrm{d}\,V \ne 1$ in $\mathbb{Q}_p^\times/(\mathbb{Q}_p^\times)^2$, we can choose a such that $S(W) = S(V)$. Then $W \cong V$ and so U is represented by V.

Suppose $\dim V - \dim U \ge 3$. We have only to prove the case of $\dim V - \dim U = 3$. Take $a \in \mathbb{Q}_p^\times$ such that $a\,\mathrm{d}\,U\,\mathrm{d}\,V \ne -1$; then

$$\dim V - \dim(U \perp \langle a\rangle) = 2 \text{ and } \mathrm{d}\,V\,\mathrm{d}(U \perp \langle a\rangle) \ne -1$$

are obvious and so by the above argument, $U \perp \langle a\rangle$ (and hence U) is represented by V. □

4

Quadratic forms over $\mathbb{Q}$

4.1 Quadratic forms over $\mathbb{Q}$

In this chapter, we prove the so-called Minkowski-Hasse theorem and the existence of a quadratic space over $\mathbb{Q}$ with prescribed local behaviours. Both are generalized over algebraic number fields. In number theory, we often say that Hasse's principle holds when some assertion over $\mathbb{Q}$ is true if and only if it is true over all local fields. It comes from the Minkowski-Hasse theorem.

We denote by Ω the set consisting of all prime numbers and the formal symbol ∞. We call an element of Ω a place (of $\mathbb{Q}$); a prime number is called a finite place and ∞ is called the infinite place. For a quadratic space U over $\mathbb{Q}$ and place $v \in \Omega$, we write U_v for the scalar extension $\mathbb{Q}_v U$ of the quadratic space U, where we put $\mathbb{Q}_\infty = \mathbb{R}$. We do not use the letters p or q for the infinite place.

Theorem 4.1.1. *Let U be a regular quadratic space over $\mathbb{Q}$. Then U is isotropic if and only if U_v is isotropic for every place $v \in \Omega$.*

Proof. The " only if "-part is clear. We prove the converse by induction on $\dim U$. If $m := \dim U = 1$, then there is nothing to do, since a regular quadratic space with dim $= 1$ is anisotropic.

Suppose $m = 2$ and put $U = \langle a \rangle \perp \langle b \rangle$ $(a, b \in \mathbb{Q}^\times)$. Since U_v is isotropic, $-ab \in (\mathbb{Q}_v^\times)^2$ for every place $v \in \Omega$ by Proposition 1.2.3. Hence we have

$-ab > 0$, taking $v = \infty$ and $\mathrm{ord}_p(-ab)$ is even for v a prime p. Thus $-ab$ is a square in $\mathbb{Q}^\times$, that is U is isotropic.

Suppose $m = 3$; then U is isotropic if and only if $\langle c\rangle \otimes U$ is isotropic for $c \in \mathbb{Q}^\times$. Hence we may suppose

$$U = \langle 1\rangle \perp \langle -a\rangle \perp \langle -b\rangle$$

for $a, b \in \mathbb{Q}^\times$. Moreover multiplying a, b by a square, we may assume that a and b are integers. We prove the case of $m = 3$ using induction on $|a| + |b|$. We note that $\langle 1\rangle \perp \langle -a\rangle \perp \langle -b\rangle$ is isotropic if and only if $x^2 - ay^2 = b$ has a solution.

If $|a| + |b| = 2$, then $a, b = \pm 1$, and since U_∞ is isotropic, a or b must be positive. Hence a or $b = 1$ and then U is isotropic.

Suppose $|a| + |b| > 2$ and $|a| \le |b|$; then we have $|b| \ge 2$. Moreover, dividing by a square if necessary, we may assume a and b are square-free integers. If $a = 1$, then U is obviously isotropic. Hence we assume $a \ne 1$. First we show that $c^2 \equiv a \bmod b$ is soluble with $c \in \mathbb{Z}$. Since b is a product of different primes up to sign, we have only to show $c^2 \equiv a \bmod p$ is soluble for a prime divisor p of b. Since U_p is isotropic, $x^2 - ay^2 - bz^2 = 0$ has a non-trivial solution over $\mathbb{Q}_p$ and we may assume $x, y, z \in \mathbb{Z}_p$ and one of them is in $\mathbb{Z}_p^\times$. Now $b \not\equiv 0 \bmod p^2$ implies $y \in \mathbb{Z}_p^\times$ and then $a \equiv (xy^{-1})^2 \bmod p$ follows, where $xy^{-1} \in \mathbb{Z}_p$. Thus $w^2 \equiv a \bmod p$ is soluble and so is $w^2 \equiv a \bmod b$. Hence there exists an integer c such that $c^2 = a + bd$ for some $d \in \mathbb{Z}$ and we may assume $|c| \le b/2$. Then

$$|d| = |c^2 - a|/|b| \le |c^2|/|b| + |a|/|b| \le |b|/4 + 1 < |b|$$

by virtue of $|b| \ge 2$. Next we show that $x^2 - ay^2 = d$ is soluble over a field F if and only if $X^2 - aY^2 = b$ is soluble over F. This follows from

$$bd(x^2 - ay^2) = (c^2 - a)(x^2 - ay^2) = (cx + ay)^2 - a(x + cy)^2$$

since $\begin{vmatrix} c & a \\ 1 & c \end{vmatrix} = c^2 - a \ne 0$. By the assumption, $\langle 1\rangle \perp \langle -a\rangle \perp \langle -b\rangle$ is isotropic over $\mathbb{Q}_v$ which means $x^2 - ay^2 = b$ is soluble over $\mathbb{Q}_v$, and hence $x^2 - ay^2 = d$ is soluble over $\mathbb{Q}_v$ for every place v. Applying the induction hypothesis to $\langle 1\rangle \perp \langle -a\rangle \perp \langle -d\rangle$ since $|d| < |b|$, we find $x^2 - ay^2 = d$ is soluble over $\mathbb{Q}$ and hence $x^2 - ay^2 = b$ is soluble over $\mathbb{Q}$, which means that U is isotropic over $\mathbb{Q}$.

Suppose $m \ge 4$ and put

$$U = V \perp W,\ V = \langle a_1\rangle \perp \langle a_2\rangle,\ W = \langle a_3\rangle \perp \cdots \perp \langle a_m\rangle,$$

where we may assume all a_i are integers, and put $d = 2a_1 \cdots a_m$. Since U_v is isotropic for every place v, there exists $b_v \in \mathbb{Q}^\times$ such that

$$b_v \in Q(V_v) \cap -Q(W_v) \tag{1}$$

for every place v. Considering $U^{(-1)} := \langle -1 \rangle \otimes U$ if necessary, we may assume $b_\infty > 0$. Write

$$b_p = c_p p^{e(p)}$$

($c_p \in \mathbb{Z}_p^\times, e(p) \in \mathbb{Z}, e(p) \geq 0$) and put

$$c = \prod_{p|d} p^{e(p)}.$$

Now we need Dirichlet's prime number theorem:

Theorem. *Every arithmetic progression $a + kb$, $k = 1, 2, \cdots$ with a and b relatively prime contains infinitely many primes.*

Proof. For the proof see [BS].

Since $c_p(cp^{-e(p)})^{-1}$ is in $\mathbb{Z}_p^\times$ for $p|d$, there is a sufficiently large prime q such that $q \equiv c_p(cp^{-e(p)})^{-1} \bmod p^3$ for every prime divisor p of d by using the above theorem. Put

$$b = qc.$$

Then $b/b_p = qc_p{}^{-1}cp^{-e(p)} \equiv 1 \bmod p^3$ for $p|d$ implies $b/b_p \in (\mathbb{Z}_p^\times)^2$ for $p|d$, and hence (1) yields

$$b \in Q(V_v) \cap -Q(W_v) \tag{2}$$

for $v|d$, and $b_\infty > 0$ yields that (2) is valid for $v = \infty$. If a prime p does not divide qd, then all a_i and b are in $\mathbb{Z}_p^\times$ and then $V \perp \langle -b \rangle$ and $W \perp \langle b \rangle$ are isotropic over $\mathbb{Q}_p$ by virtue of Corollary 3.5.2 and hence (2) holds for such a prime p. Thus (2) holds for all places $v \neq q$, which means that quadratic spaces $V' := V \perp \langle -b \rangle, W' := W \perp \langle b \rangle$ over $\mathbb{Q}$ are isotropic over $\mathbb{Q}_v$ for $v \neq q$. Let us show that V'_q and W'_q are also isotropic. V'_v is a ternary isotropic quadratic space if $v \neq q$. Using (9) in Theorem 3.3.1 and 3.5.1, we have

$$S(V'_q) = \prod_{v \neq q} S(V'_v) = \prod_{v \neq q} (-1, -1)_{\mathbb{Q}_v} = (-1, -1)_{\mathbb{Q}_q}$$

which implies that V'_q is isotropic. If $m \geq 5$, then $\dim W \geq 3$ and $a_i \in \mathbb{Z}_q^\times$ imply that W_q and hence W'_q is isotropic by Corollary 3.5.2. Suppose that $m = 4$; then $\dim W' = 3$ and W'_q is isotropic for the same reason that V'_q is. Thus we have proved that V'_v and W'_v are isotropic for all places. Since $\dim V', \dim W' < \dim U$, we can apply the induction hypothesis to conclude that V' and W' are isotropic which means $b \in Q(V)$, $-b \in Q(W)$. This means that $U = V \perp W$ is isotropic. □

Corollary 4.1.1. *Let U be a regular quadratic space over $\mathbb{Q}$ and $a \in \mathbb{Q}^\times$. Then $a \in Q(U)$ if and only if $a \in Q(U_v)$ for all places v.*

Proof. We have only to apply the theorem to $U \perp \langle -a \rangle$. □

Corollary 4.1.2. *Let U, V be regular quadratic spaces over $\mathbb{Q}$. Then U is represented by V if and only if U_v is represented by V_v for all places v.*

Proof. We only have to show the " if " part. We use induction on $\dim U$. If $\dim U$ is 1, then it is just the previous corollary. Write $U = \langle a \rangle \perp U'$ for $a \in \mathbb{Q}^\times$; then $a \in Q(U_v) \subset Q(V_v)$ holds for all places v and then $a \in Q(V)$ holds. Hence we can write $V = \langle a \rangle \perp V'$. By virtue of Corollary 1.2.3, $U_v \hookrightarrow V_v$ implies $U'_v \hookrightarrow V'_v$ for all places v. The induction hypothesis tells us $U' \hookrightarrow V'$ and so $U \hookrightarrow V$. □

Corollary 4.1.3. *For regular quadratic spaces U and V over $\mathbb{Q}$, $U \cong V$ if and only if $U_v \cong V_v$ for all places.*

Proof. This is a trivial corollary of the previous one. □

Corollary 4.1.4. *Let U and V be regular quadratic spaces over $\mathbb{Q}$ with* $\dim V - \dim U \geq 3$. *Then $U \hookrightarrow V$ if $U_\infty \hookrightarrow V_\infty$.*

Proof. From Corollary 3.5.4 it follows that $U_p \hookrightarrow V_p$ for every prime p, and then Corollary 4.4.2 yields the result. □

Theorem 4.1.2. *Let n be a natural number and $U(v)$ be a regular quadratic space over $\mathbb{Q}_v$ with $\dim U(v) = n$ for all places. Then there exists a regular quadratic space U over $\mathbb{Q}$ such that $U_v \cong U(v)$ for all places v if and only if the following three conditions hold:*

(i) *There exists a non-zero rational number a such that*

$$\mathrm{d}\, U(v) = a \text{ in } \mathbb{Q}_v^\times / (\mathbb{Q}_v^\times)^2$$

for all places v.

(ii) $S(U(v)) = 1$ *for all but finitely many places.*

(iii) $\prod_v S(U(v)) = 1$.

Proof. If there exists a regular quadratic space U over $\mathbb{Q}$ such that $U_v \cong U(v)$ for all places, then (i) is clear and (ii), (iii) follow from (6), (9) in Theorem 3.3.1.

Now suppose (i), (ii) and (iii). If $n = 1$, then we have only to put $U = \langle a \rangle$.

Suppose $n \geq 2$. Define a subset S of Ω by

$$S = \{2, \infty\} \cup \{p \mid a \notin \mathbb{Z}_p^\times\} \cup \{p \mid S(U(p)) = -1\}.$$

S is a finite set. For $v \in S$, we take

$$a_v \in Q(U(v))$$

($a_v \neq 0$) and write

$$a_p = b_p p^{e(p)}$$

($b_p \in \mathbb{Z}_p^\times$, $e(p) \in \mathbb{Z}$) for a prime $p \in S$. Put

$$c = sgn(a_\infty) \prod_{p \in S} p^{e(p)},$$

where $sgn(a_\infty) = a_\infty/|a_\infty|$ and p runs over finite places in S; then $b_p(cp^{-e(p)})^{-1}$ is in $\mathbb{Z}_p^\times$ for $p \in S$ and using Dirichlet's theorem as in the proof of Theorem 4.1.1, we can take a sufficiently large prime number q such that

$$q \equiv b_p(cp^{-e(p)})^{-1} \bmod p^3 \qquad \text{for a prime } p \in S,$$

which implies $qc/a_p \equiv b_p p^{e(p)}/a_p \equiv 1 \bmod p^3$ since $c/a_p \in \mathbb{Z}_p^\times$. Thus we have $qc/a_p \in (\mathbb{Q}_p^\times)^2$ for $p \in S, p \neq \infty$, and $qc/a_\infty > 0$ is clear. So we have shown

(iv) $$qc/a_v \in (\mathbb{Q}_v^\times)^2 \qquad \text{for } v \in S.$$

Suppose $n = 2$ and put $U = \langle qc \rangle \perp \langle qca \rangle$; then $\mathrm{d}U = a = \mathrm{d}U(v)$ in $\mathbb{Q}_v^\times/(\mathbb{Q}_v^\times)^2$ by (1), and for $v \in S$, $U_v = \langle a_v \rangle \perp \langle aa_v \rangle \cong U(v)$ holds by (iv). If $p \notin S$ and $p \neq q$, then $qc, qca \in \mathbb{Z}_p^\times$ implies $S(U_p) = 1 = S(U(p))$ and hence $U_p \cong U(p)$. So,

$$S(U_q) = \prod_{v \neq q} S(U_v) = \prod_{v \neq q} S(U(v)) = S(U(q))$$

implies $U_q \cong U(q)$ which completes the case of $n = 2$.

Suppose $n \geq 3$. We claim first

$$qc \in Q(U(v))$$

that is, $U(v) \perp \langle -qc \rangle$ is isotropic over $\mathbb{Q}_v$ for all places v. If $n \geq 4$, then it follows from Theorem 3.5.1 for finite places and from (iv) for the infinite place.

Suppose $n = 3$. It is clear for $v \in S$ by (iv). Suppose $p \notin S$. If

$$\mathrm{d}(U(p) \perp \langle -qc \rangle) = -aqc \neq 1 \text{ in } \mathbb{Q}_p^\times/(\mathbb{Q}_p^\times)^2,$$

then $U(p) \perp \langle -qc \rangle$ is isotropic by Theorem 3.5.1. Next suppose $\mathrm{d}(U(p) \perp \langle -qc \rangle) = -aqc = 1$ in $\mathbb{Q}_p^\times/(\mathbb{Q}_p^\times)^2$; then

$$\begin{aligned} S(U(p) \perp \langle -qc \rangle) &= S(U(p))(\mathrm{d}\,U(p), -qc)(-qc, -qc) \\ &= (a, -qc)(-qc, -qc) = (-aqc, -qc) \\ &= 1 = (-1, -1) \text{ by } p \neq 2. \end{aligned}$$

Hence $U(p) \perp \langle -qc \rangle$ is isotropic by Theorem 3.5.1 Thus we have verified that $U(v) \perp \langle -qc \rangle$ is isotropic for all places v and the claim is proved.

Therefore, $qc \in Q(U(v))$ and we can write $U(v) = \langle qc \rangle \perp W(v)$ for every place v. Then $\mathrm{d}\,W(v) = aqc$ is clear. From

$$S(U(v)) = (qc, qc)(qc, aqc)S(W(v))$$

it follows that the conditions (ii), (iii) hold for $W(v)$ in place of $U(v)$. Induction then gives the existence of a quadratic space W over $\mathbb{Q}$ such that $W_v \cong W(v)$ for all places v. We have only to put $U = \langle qc \rangle \perp W$. □

5

Quadratic forms over the p-adic integer ring

Throughout this chapter, p is a prime number, and R and F denote the p-adic integer ring $\mathbb{Z}_p$ and the p-adic number field $\mathbb{Q}_p$, respectively. We denote by (a) the principal ideal aR for $a \in F$.

An important property of R is that R is a principal ideal domain, that is R is a commutative ring without zero-divisors and every ideal of R is principal. Then it is known that every (finitely generated) module M over R is isomorphic to $R/I_1 \oplus \cdots \oplus R/I_n$ where the I_k are ideals of R satisfying $I_1 \subset \cdots \subset I_n$ which are uniquely determined by M. Therefore if M is torsion-free, i.e. $ax = 0$ ($a \in R$, $x \in M$) implies a or $x = 0$, then M has a basis over R.

Let V be a vector space over F. A submodule L over R in V is called a *lattice* on V if L is finitely generated over R and contains a basis of V over F. Then there exists a basis $\{v_i\}$ of V over F such that $L = \oplus_i Rv_i$. We can define the *discriminant* $\mathrm{d}\,L$ of L by $\det(B(v_i, v_j)) \times (R^\times)^2$. This does not conflict with the definition in Chapter 1.

By a *regular quadratic lattice* over R, we mean a lattice on some regular quadratic space over F. If K is a torsion-free (finitely generated) module over R, then K is a lattice on FK. In this sense, if there is no confusion, we often refer to a submodule M of possibly smaller rank as a regular (sub)lattice when FM is regular.

5.1 Dual lattices

Let L be a regular quadratic lattice over R, i.e. a lattice on a regular quadratic space V over F. We define the *dual lattice* $L^\sharp$ of L by

$$L^\sharp := \{x \in V \mid B(x, L) \subset R\}.$$

Let $\{v_i\}_{i=1}^n$ ($n := \dim V$) be a basis of L over R and $\{u_i\}$ be the dual basis of $\{v_i\}$, i.e. $(B(v_i, u_j)) = 1_n$. Then it is easy to see

$$L^\sharp = Ru_1 \oplus \cdots \oplus Ru_n$$

and hence

$$(L^\sharp)^\sharp = L.$$

Put

(1) $$(v_1, \cdots, v_n) = (u_1, \cdots, u_n)T \quad \text{for} \quad T \in GL_n(F).$$

Then $1_n = (B(u_i, v_j)) = (B(u_i, \sum_k t_{kj} u_k)) = (B(u_i, u_j))T$ implies

(2) $$T = (B(u_i, u_j))^{-1} = {}^tT$$

and then $(B(v_i, v_j)) = {}^tT(B(u_i, u_j))T = (B(u_i, u_j))^{-1} = T$ by (1) and (2). Thus we have

$$\det T = \det(B(v_i, v_j)) = \mathrm{d}\,L, \;\; \mathrm{d}\,L^\sharp = (\mathrm{d}\,L)^{-1} \;\; \text{and} \;\; L^\sharp \cong \langle (B(v_i, v_j))^{-1} \rangle.$$

If $B(L, L) \subset R$, then $L^\sharp \supset L$ and $[L^\sharp : L] = p^{\mathrm{ord}_p \det T} = p^{\mathrm{ord}_p \mathrm{d}\,L}$. If $L = J \perp K$, then $L^\sharp = J^\sharp \perp K^\sharp$ trivially, where $J^\sharp$ and $K^\sharp$ are dual lattices of J and K on FJ and FK respectively. For lattices L, M on V, it is easy to see

$$L \supset M \Leftrightarrow L^\sharp \subset M^\sharp, \quad (L + M)^\sharp = L^\sharp \cap M^\sharp, \quad L^\sharp + M^\sharp = (L \cap M)^\sharp.$$

For a submodule M of V, $M^\sharp$ is defined in FM as above if FM is regular.

5.2 Maximal and modular lattices

Let L be a regular quadratic lattice over R and $V = FL$. We define the *scale* $\mathrm{s}(L)$ and the *norm* $\mathrm{n}(L)$ by

$$\mathrm{s}(L) = B(L, L), \quad \mathrm{n}(L) = \{\textstyle\sum a_i Q(x_i) \mid a_i \in R,\ x_i \in L\}.$$

Noting $B(x, x) = Q(x)$, $2B(x, y) = Q(x + y) - Q(x) - Q(y)$, we have

$$2\,\mathrm{s}(L) \subset \mathrm{n}(L) \subset \mathrm{s}(L).$$

Hence $\mathrm{n}(L) = \mathrm{s}(L)$ if $p \neq 2$, and $\mathrm{n}(L)$ equals either $\mathrm{s}(L)$ or $2\,\mathrm{s}(L)$ if $p = 2$.

Let $\{v_i\}$ be a basis of L over R; then let us recall that we write $L \cong \langle A \rangle$ for $A = (B(v_i, v_j))$. If $a^{-1}A \in GL_n(R)$ ($n = \mathrm{rank}\, L$) for some $a \in F^\times$, then L is said to be an aR(or (a))-*modular* lattice. If $a \in R^\times$, then it is called *unimodular*. For an (a)-modular lattice L, $\mathrm{s}(L) = (a)$ ($= aR$) is clear and $\mathrm{n}(L)$ equals either (a) or $(2a)$, and $\mathrm{n}(L) = (2a)$ if and only if all diagonals of A are in $(2a)$.

For $L = M \perp N$, L is (a)-modular if and only if M and N are (a)-modular.

Proposition 5.2.1. *Let L be a quadratic lattice over R. Then L is (a)-modular if and only if $L^\sharp = a^{-1}L$. In particular, L is unimodular if and only if $L^\sharp = L$.*

Proof. Let $\{v_i\}$ be a basis of L, and $\{u_i\}$ be the dual basis, i.e. $(B(v_i, u_j)) = 1_n$. Then by the previous section, $\{u_i\}$ is a basis of $L^\sharp$ and $(v_1, \cdots, v_n) = (u_1, \cdots, u_n)(B(v_i, v_j))$ and $(B(u_i, u_j)) = (B(v_i, v_j))^{-1}$.

If L is (a)-modular, then $(B(v_i, v_j)) = aA$ for $A \in GL_n(R)$ and hence

$$(u_1, \cdots, u_n) = (v_1, \cdots, v_n)(B(v_i, v_j))^{-1} = (v_1, \cdots, v_n)a^{-1}A^{-1}$$

which means $L^\sharp = a^{-1}L$.

If $L^\sharp = a^{-1}L$, then

$$(u_1, \cdots, u_n) = (v_1, \cdots, v_n)a^{-1}C$$

for some $C \in GL_n(R)$, which implies

$$1_n = (B(u_i, v_j)) = a^{-1}(B(\sum_k c_{ki}v_k, v_j)) = a^{-1\,t}C(B(v_i, v_j))$$

for $C = (c_{ij})$ and so $(B(v_i, v_j)) = a\,{}^tC^{-1}$. Thus L is (a)-modular. □

Proposition 5.2.2. *Let L be a regular quadratic lattice over R and M be a submodule of L and suppose that M is (a)-modular. Then $L = M \perp N$ holds for a submodule N if and only if $B(M, L) \subset \mathrm{s}(M)$. In particular, $\mathrm{s}(L) = \mathrm{s}(M)$ implies $L = M \perp M^\perp$.*

Proof. The "only if" part is clear. Let us prove the "if" part. For $x \in L$, the mapping $m \mapsto a^{-1}B(m, x)$ is a homomorphism from M to R by virtue of $B(M, L) \subset \mathrm{s}(M) = (a)$. Hence there is an element $y \in M^\sharp = a^{-1}M$ such that $a^{-1}B(m, x) = B(m, y)$ for $m \in M$, which implies $x - ay \in M^\perp \cap L$ because $ay \in M \subset L$. Thus $x \in M \perp (M^\perp \cap L)$ and hence $L \subset M \perp (M^\perp \cap L) \subset L$ holds and we have only to put $N = M^\perp \cap L$. □

For a regular quadratic lattice L over R, L is called (a)-*maximal* if and only if $\mathrm{n}(L) \subset (a)$ and for any lattice M containing L properly, $\mathrm{n}(M) \not\subset (a)$ holds.

Suppose that L is (a)-maximal and $\mathrm{n}(L) \subset (a)$ but $\mathrm{n}(L) \neq (a)$; if $\mathrm{n}(L) \subset (p^2a)$, then $p^{-1}L \supset L$ and $\mathrm{n}(p^{-1}L) \subset (a)$, which contradict the (a)-maximality of L. Hence for an (a)-maximal quadratic lattice L, we have $\mathrm{n}(L) = (a)$ or (pa). The following two theorems are most fundamental.

Theorem 5.2.1. *Let L be a lattice on a regular anisotropic quadratic space V. L is (a)-maximal if and only if $L = \{x \in V \mid Q(x) \in (a)\}$.*

Proof. First we claim that $M := \{x \in V \mid Q(x) \in (a)\}$ is a module over R. Let $x, y \in V$ satisfy $Q(x), Q(y) \in (a), x \neq 0, y \neq 0$. Suppose $2B(x,y) \notin (a)$; Then we define an integer $n \geq 1$ by $(2B(x,y)p^n) = (a)$. It is easy to see

$$Q(x)Q(y)B(x,y)^{-2} \in (a)^2(4p^{2n}a^{-2}) = (4p^{2n}).$$

Hence by virtue of Theorem 3.1.4, $1 - Q(x)Q(y)B(x,y)^{-2}$ is a square of some element $c \in R^\times$ and then $Q(x)Q(y) - B(x,y)^2 = -(cB(x,y))^2 \neq 0$. It means that a binary subspace $F[x,y]$ $(\subset V)$ is a hyperbolic plane and hence it is isotropic, which contradicts the assumption. Thus we have $2B(x,y) \in (a)$ and then $Q(x+y) \in (a)$, meaning $x + y \in M$. Thus M is a module.

Next we show that M is finitely generated. Let $\{v_i\}$ be a basis of V over F such that $v_i \in M$ for all i. Put $K = \oplus R v_i$ and let N be a finitely generated submodule such that $K \subset N \subset M$. Then

$$\mathrm{s}(N) \subset \frac{1}{2}\,\mathrm{n}(N) \subset (\frac{a}{2})$$

implies $(\mathrm{d}\,N) \subset (\frac{a}{2})^m$ $(m := \dim V)$. Since $\mathrm{d}\,K = [N : K]^2\,\mathrm{d}\,N$ implies

$$R \supset ([N : K]^2) = (\mathrm{d}\,K \cdot (\mathrm{d}\,N)^{-1}) \supset (\mathrm{d}\,K)(\frac{a}{2})^{-m},$$

$[N : K]$ is bounded and hence M is finitely generated over R and (a)-maximal.

Let L be an (a)-maximal lattice; then $L \subset M$ is clear, and then the maximality of L and $\mathrm{n}(M) \subset (a)$ imply $L \supset M$ and hence $L = M$. □

Theorem 5.2.2. *Let L be a lattice on a regular quadratic space V over F. If L is (a)-maximal, then there exist hyperbolic planes H_i and an anisotropic subspace V_0 such that*

$$V = \perp_i H_i \perp V_0,\ L = \perp_i (L \cap H_i) \perp (L \cap V_0),\ L \cap H_i \cong \langle \frac{a}{2}\begin{pmatrix} 0 & 1 \\ 1 & 0 \end{pmatrix} \rangle$$

and $L \cap V_0$ is (a)-maximal. In particular, (a)-maximal lattices on V are mutually isometric.

To prove this, we prepare with three lemmas.

Lemma 5.2.1. *Let L be a regular quadratic lattice over R with* $\operatorname{rank} L = m$. *If* $\mathrm{n}(L) \subset (a)$ *and* $((2a^{-1})^m \,\mathrm{d}\, L) = R$ *or* pR, *then L is (a)-maximal.*

Proof. Let M be a lattice on FL such that $M \supset L$ and $\mathrm{n}(M) \subset (a)$. Then we have $\mathrm{d}\, L = [M : L]^2 \,\mathrm{d}\, M$ and

$$(\mathrm{d}\, M) \subset \mathrm{s}(M)^m \subset (2^{-1}\,\mathrm{n}(M))^m \subset (a/2)^m.$$

Thus

$$((2a^{-1})^m \,\mathrm{d}\, L) = ([M : L]^2 (2a^{-1})^m \,\mathrm{d}\, M) \subset [M : L]^2 R$$

and so by the assumption we have $[M : L] = 1$, i.e. $M = L$. Therefore L is (a)-maximal. □

Corollary 5.2.1. *If L is unimodular with* $\mathrm{n}(L) = 2R$, *then L is (2)-maximal.*

Proof. We have only to put $a = 2$ in the lemma. □

Lemma 5.2.2. *For $c \in F^\times$ and a lattice L on the hyperbolic plane H, the following three conditions are equivalent:*

(i) $L \cong \langle \begin{pmatrix} 0 & c \\ c & 0 \end{pmatrix} \rangle$;

(ii) *L is (c)-modular and* $\mathrm{n}(L) \subset (2c)$;

(iii) *L is $(2c)$-maximal.*

Proof. (i) ⇒ (ii) is clear. (ii) ⇒ (iii) follows from the previous lemma, putting $a = 2c$. Let us prove (iii) ⇒ (i). Let $\{e_1, e_2\}$ be a basis of L such that

$$Q(e_1) = 0, \quad B(e_1, e_2) = a, \quad Q(e_2) = b \quad \text{for} \quad a, b \in F.$$

Note that $a \in \mathrm{s}(L) \subset 2^{-1}\,\mathrm{n}(L) \subset (c)$ and $b \in \mathrm{n}(L) \subset (2c)$. We show $(a) = (c)$. If $(a) \neq (c)$, then $a \in (cp)$ holds, and then for $x_1, x_2 \in R$ we have

$$Q(x_1 p^{-1} e_1 + x_2 e_2) = 2x_1 x_2 p^{-1} a + x_2^2 b \in (2c).$$

It means $L \subset R[p^{-1}e_1, e_2]$, $L \neq R[p^{-1}e_1, e_2]$ and $\mathrm{n}(R[p^{-1}e_1, e_2]) \subset (2c)$, which contradicts the maximality of L. Thus we have shown $(a) = (c)$ and can put $c = au$ $(u \in R^\times)$. Then $b(2a)^{-1} \in R$ and so $L = R[ue_1, e_2 - b(2a)^{-1}e_1] \cong \langle \begin{pmatrix} 0 & c \\ c & 0 \end{pmatrix} \rangle$ which completes the proof. □

Note that c in (i) can be replaced by cu for $u \in R^\times$.

Lemma 5.2.3. *Let L be an (a)-maximal lattice on an isotropic regular quadratic space V over F. For an isotropic primitive element x in L, there exists an isotropic element $y \in L$ such that $L = R[x,y] \perp R[x,y]^{\perp}$, $B(x,y) = a/2$ and $R[x,y]$ is (a)-maximal.*

Proof. We show $B(x, L) = (a/2)$. From the definition follows $B(x, L) \subset \mathrm{s}(L) \subset 2^{-1}\,\mathrm{n}(L) \subset (a/2)$. So, we assume $B(x, L) \subset (ap/2)$; then

$$Q(w + p^{-1}x) = Q(w) + 2p^{-1}B(w, x) \in (a)$$

for $w \in L$. Hence $L \subset L + p^{-1}Rx$, $L \neq L + p^{-1}Rx$ and $\mathrm{n}(L + p^{-1}Rx) \subset (a)$, which contradicts the maximality of L. Thus we have shown $B(x, L) = (2^{-1}a)$, and we take $z \in L$ such that $B(x,z) = 2^{-1}a$ and put $y = z - a^{-1}Q(z)x \in L$. Then

$$M := R[x,y] \cong \langle \begin{pmatrix} 0 & 2^{-1}a \\ 2^{-1}a & 0 \end{pmatrix} \rangle$$

and so M is $(2^{-1}a)$-modular and hence $B(M, L) \subset \mathrm{s}(L) \subset (2^{-1}a) = \mathrm{s}(M)$. By Proposition 5.2.2, we have $L = M \perp M^{\perp}$. By the previous lemma, M is (a)-maximal. □

Using Lemma 5.2.3 repeatedly, we have the orthogonal decomposition of L in Theorem 5.2.2. Since L is (a)-maximal, $L \cap V_0$ is also (a)-maximal. The number of hyperbolic planes is $\mathrm{ind}\, V$ and then by Corollary 1.2.3 the isometry class of an anisotropic subspace V_0 is uniquely determined by V. Then Theorem 5.2.1 yields the class of (a)-maximal lattices which are uniquely determined up to isometry. Thus we have completed the proof of Theorem 5.2.2. □

We know the following stronger assertion.

Exercise 1.

Let V be an isotropic regular quadratic space over F. For maximal lattices L, N whose norms are not necessarily equal, show there exists a hyperbolic plane $H \subset V$ such that

$$L = (L \cap H) \perp (L \cap H^{\perp}), \quad N = (N \cap H) \perp (N \cap H^{\perp}).$$

Exercise 2.

For an R-maximal lattice M, $\langle \begin{pmatrix} 0 & 1/2 \\ 1/2 & 0 \end{pmatrix} \rangle \perp M$ is also R-maximal. But in general, the orthogonal sum of R-maximal lattices is not R-maximal.

Exercise 3.

Let V be the 4-dimensional anisotropic quadratic space over F and L be the R-maximal lattice on it. Prove

$$L \cong \begin{cases} \langle 1\rangle \perp \langle -\delta\rangle \perp \langle p\rangle \perp \langle -\delta p\rangle & \text{if } p \neq 2, \\ \langle \begin{pmatrix} 1 & 1/2 \\ 1/2 & 1 \end{pmatrix} \rangle \perp \langle \begin{pmatrix} 2 & 1 \\ 1 & 2 \end{pmatrix} \rangle & \text{if } p = 2, \end{cases}$$

where $\delta \in R^\times$ such that the quadratic residue symbol $(\frac{\delta}{p}) = -1$.

Lemma 5.2.4. *Let V be a regular quadratic space over F. Then for a submodule L over R with $\mathrm{n}(L) \subset (a)$, there exists an (a)-maximal lattice on V which contains L.*

Proof. Let $\{v_i\}_{i=1}^m$ be a basis of V over F such that a subset $\{v_i\}_{i=1}^n$ is a basis of L over R $(n \leq m)$. We put

$$M = \oplus_{i=1}^n R v_i \oplus_{i=n+1}^m R p^t v_i$$

for an integer t. If t is sufficiently large, then $\mathrm{n}(M) \subset (a)$. If a lattice N on V satisfies $N \supset M$ and $\mathrm{n}(N) \subset (a)$, then $\mathrm{d}\, M = [N : M]^2 \,\mathrm{d}\, N$ and $\mathrm{d}\, N \in \mathrm{s}(N)^m \subset (2^{-1}a)^m$. Hence

$$[N : M]^2 R = \mathrm{d}\, M (\mathrm{d}\, N)^{-1} R \ni \mathrm{d}\, M \cdot (2^{-1}a)^{-m}$$

and so $[N : M]$ is bounded. Therefore there exists a maximal lattice K with respect to inclusion such that $K \supset M$ and $\mathrm{n}(K) \subset (a)$. Such a lattice K is an (a)-maximal lattice. □

Theorem 5.2.3. *Let L and M be regular quadratic lattices over R and suppose that* $\mathrm{rank}\, L + 3 \leq \mathrm{rank}\, M$, *$M$ is (a)-maximal and $\mathrm{n}(L) \subset (a)$. Then L is represented by M.*

Proof. Put $V := FL, W := FM$; then by virtue of Corollary 3.5.4, V is represented by W and hence there exists a module K $(\subset W)$ isometric to L. By the previous lemma, there exists an (a)-maximal lattice N on W which contains K. From Theorem 5.2.2, it follows that M and N are isometric, and hence L is represented by M. □

Now we give a few facts about unimodular lattices.

Theorem 5.2.4. *Suppose $p \neq 2$. For a unimodular lattice L, we have*

$$L \cong \langle 1\rangle \perp \cdots \perp \langle 1\rangle \perp \langle \mathrm{d}\, L\rangle.$$

Proof. Since $\mathrm{n}(L) = \mathrm{s}(L) = R$, there is an element $v_1 \in L$ such that $Q(v_1) \in R^\times$, and then Proposition 5.2.2 implies $L = Rv_1 \perp Rv_1^\perp$, where $v_1^\perp$ is unimodular. Repeating this, L has an orthogonal basis, that is $L \cong \langle \epsilon_1\rangle \perp \cdots \perp \langle \epsilon_n\rangle$, $\epsilon_i \in R^\times$. Put $M = \langle 1\rangle \perp \cdots \perp \langle 1\rangle \perp \langle \mathrm{d}\, L\rangle$ with $\mathrm{rank}\, M = \mathrm{rank}\, L$. Hasse invariants of both L and M are 1 by (6) in Theorem 3.3.1, and then by Theorem 3.5.2 we have $FL \cong FM$, and L and M are R-maximal by Lemma 5.2.1. Theorem 5.2.2 yields $L \cong M$. □

Corollary 5.2.2. *Suppose $p \neq 2$ and let K be a unimodular lattice of* $\operatorname{rank} K \geq 3$. *Then $Q(K) = R$ holds.*

Proof. Put

$$L = \langle\begin{pmatrix} 0 & 1 \\ 1 & 0 \end{pmatrix}\rangle \perp \langle 1\rangle \perp \cdots \perp \langle 1\rangle \perp \langle -\mathrm{d}\, K\rangle$$

with $\operatorname{rank} L = \operatorname{rank} K$. By the theorem, $K \cong L$ and hence we have

$$R = Q(\langle\begin{pmatrix} 0 & 1 \\ 1 & 0 \end{pmatrix}\rangle) \subset Q(K) \subset R.$$

□

Lemma 5.2.5. *Let $p = 2$ and L be a unimodular lattice over R. Then L is an orthogonal sum of unimodular lattices of rank* 1 *and binary unimodular lattices with norm* (2).

Proof. Since $\mathrm{s}(L) = R$, we have $\mathrm{n}(L) = R$ or (2). If $\mathrm{n}(L) = R$, then there exists an element $v \in L$ such that $Q(v) \in R^\times$ and Proposition 5.2.2 yields $L = Rv \perp v^\perp$. Suppose $\mathrm{n}(L) = (2)$; since there exist $v_1, v_2 \in L$ such that $B(v_1, v_2) = 1$ by virtue of $\mathrm{s}(L) = R$, $M := R[v_1, v_2]$ is unimodular by $Q(v_i) \in (2)$ $(i = 1, 2)$. The same proposition implies $L = M \perp M^\perp$. Repeating this, we complete the proof. □

Lemma 5.2.6. *Let $p = 2$ and L be a binary unimodular lattice with* $\mathrm{n}(L) = (2)$.

If L is isotropic, then $L \cong \langle\begin{pmatrix} 0 & 1 \\ 1 & 0 \end{pmatrix}\rangle$, $\mathrm{d}\, L = -1$, $S(FL) = -1$ and $Q(L) = (2)$.

If L is anisotropic, then $L \cong \langle\begin{pmatrix} 2 & 1 \\ 1 & 2 \end{pmatrix}\rangle$, $\mathrm{d}\, L = 3$, $S(FL) = 1$ and $Q(L) = \{0\} \cup \{x \in R \mid \operatorname{ord}_2 x \equiv 1 \bmod 2\}$.

Proof. L is (2)-maximal by virtue of Corollary 5.2.1. If L is isotropic, then the assertion follows from Lemma 5.2.2. Suppose that L is anisotropic; then $L \cong \langle\begin{pmatrix} 2a & 1 \\ 1 & 2b \end{pmatrix}\rangle$ for some $a, b \in R$. If $ab \equiv 0 \bmod 2$, then

$$\mathrm{d}\, L = -(1 - 4ab)(R^\times)^2 = -(R^\times)^2$$

and so FL is isotropic, which is a contradiction. Thus $ab \equiv 1 \bmod 2$ and so $\mathrm{d}\, L = 3$. Put $r := 1 - a - b \in R^\times$; then $ar^2 + r + b \equiv 1 \bmod 8$ is easy. Since $Q(L) \ni 2(ar^2 + r + b)$, we have $2 \in Q(L)$. Hence

$$FL \cong \langle 2\rangle \perp \langle 2 \cdot 3\rangle \cong \langle\begin{pmatrix} 2 & 1 \\ 1 & 2 \end{pmatrix}\rangle$$

holds over F, and since L and $\langle\begin{pmatrix}2&1\\1&2\end{pmatrix}\rangle$ are (2)-maximal, Theorem 5.2.2 implies $L\cong\langle\begin{pmatrix}2&1\\1&2\end{pmatrix}\rangle$. It remains to show

$$Q(L)=\{x\in R\mid x=0 \text{ or } \operatorname{ord}_2 x \text{ is odd}\}.$$

For $x\in F^\times$, $x\in Q(FL)\Leftrightarrow FL\perp\langle -x\rangle$ is isotropic $\Leftrightarrow S(FL\perp\langle -x\rangle)=-1$ by Theorem 3.5.1, and $S(FL\perp\langle -x\rangle)=(-3,x)_{\mathbb{Q}_2}=(-1)^n$ for $n=\operatorname{ord}_2 x$. Thus $0\neq x\in Q(FL)\Leftrightarrow\operatorname{ord}_2 x$ is odd. Thus the proof is complete. □

Theorem 5.2.5. *Let $p=2$ and L be a unimodular lattice. If $\mathrm{n}(L)=2R$, then $L\cong H$ or $H\perp\langle\begin{pmatrix}2&1\\1&2\end{pmatrix}\rangle$ where H denotes an orthogonal sum of copies of $\langle\begin{pmatrix}0&1\\1&0\end{pmatrix}\rangle$. If $\mathrm{n}(L)=R$, then L has an orthogonal basis.*

Proof. Suppose $\mathrm{n}(L)=2R$; then L is (2)-maximal by Corollary 5.2.1. By Lemma 5.2.5 L is an orthogonal sum of binary unimodular lattices with norm=(2), which are isometric to $\langle\begin{pmatrix}0&1\\1&0\end{pmatrix}\rangle$ or $\langle\begin{pmatrix}2&1\\1&2\end{pmatrix}\rangle$ by Lemma 5.2.6. To complete the proof in the case of $\mathrm{n}(L)=(2)$, we have only to show

$$\langle\begin{pmatrix}2&1\\1&2\end{pmatrix}\rangle\perp\langle\begin{pmatrix}2&1\\1&2\end{pmatrix}\rangle\cong\langle\begin{pmatrix}0&1\\1&0\end{pmatrix}\rangle\perp\langle\begin{pmatrix}0&1\\1&0\end{pmatrix}\rangle.$$

By Corollary 5.2.1 both are (2)-maximal, and hence we have only to prove that they are isometric over F by virtue of Theorem 5.2.2. It is easily done by comparing discriminants and Hasse invariants.

Next suppose $\mathrm{n}(L)=R$; then by Lemma 5.2.5 L has an orthogonal component of rank 1. Hence we have only to show that $M:=R[v_1,v_2]\perp Rv_3$ with $Q(v_1)=Q(v_2)=2a$ $(=0$ or $2)$, $B(v_1,v_2)=1$, $Q(v_3)\in R^\times$ has an orthogonal basis. It is easy to see

$$M=R[v_1+v_3]\perp R[v_1-2av_2,v_2-Q(v_3)^{-1}v_3],$$

where both orthogonal components are unimodular and the norm of the second one is R. Hence by the proof of Lemma 5.2.5, the second lattice has an orthogonal basis and so M has an orthogonal basis. □

Proposition 5.2.3. *Let $p=2$ and $\epsilon_i\in R^\times$ $(i=1,2,3)$. Then*

$$\langle\epsilon_1\rangle\perp\langle\epsilon_2\rangle\perp\langle\epsilon_3\rangle\cong\langle\begin{pmatrix}2a&1\\1&2a\end{pmatrix}\rangle\perp\langle\epsilon\rangle$$

for $\epsilon \in R^\times$ and $a = 0$ or 1. If $a = 1$, then it is anisotropic.

Proof. Let $L = Rv_1 \perp Rv_2 \perp Rv_3$ and $Q(v_i) = \epsilon_i$; then

$$M := R[v_1 + v_2, v_2 + v_3] \cong \langle \begin{pmatrix} \epsilon_1 + \epsilon_2 & \epsilon_2 \\ \epsilon_2 & \epsilon_2 + \epsilon_3 \end{pmatrix} \rangle \cong \langle \begin{pmatrix} 2a & 1 \\ 1 & 2a \end{pmatrix} \rangle$$

for $a = 0$ or 1 by Lemma 5.2.6. By Proposition 5.2.2, we have $L = M \perp M^\perp$. When $a = 1$, L is anisotropic by Lemma 5.2.6. □

In the above proof, if $\epsilon_1 + \epsilon_2 \equiv 0 \bmod 4$, then M is isotropic by discriminant consideration, and then $a = 0$ holds by Lemma 5.2.6.

Proposition 5.2.4. *If $p = 2$ and L is unimodular with* $\operatorname{rank} L \geq 5$, *then* $L \cong \langle \begin{pmatrix} 0 & 1 \\ 1 & 0 \end{pmatrix} \rangle \perp$ *(some lattice).*

Proof. If $\mathrm{n}(L) = 2R$, then Theorem 5.2.5 implies the assertion. Otherwise, L has an orthogonal basis $\{v_i\}$ $(Q(v_i) = \epsilon_i)$. If $\epsilon_i + \epsilon_j \equiv 0 \bmod 4$ for some i, j, then $M := R[v_i + v_j, v_j + v_k] \cong \langle \begin{pmatrix} 0 & 1 \\ 1 & 0 \end{pmatrix} \rangle$ for $k \neq i, j$ by the note just before the proposition, and M splits L. If $\epsilon_i + \epsilon_j \equiv 2 \bmod 4$, i.e. $\epsilon_i \equiv \epsilon_j \bmod 4$ for all i, j, then

$$M := R[v_1 + v_2 + v_3 + v_4, v_2 + v_3 + v_4 + v_5] \cong \langle \begin{pmatrix} 0 & 1 \\ 1 & 0 \end{pmatrix} \rangle$$

and L is split by M. □

5.3 Jordan decompositions

Let L be a regular quadratic lattice over R. If $\mathrm{s}(L) = \mathrm{n}(L)$, then there exists an element $u \in L$ such that $(Q(u)) = \mathrm{s}(L)$ and by Proposition 5.2.2 we have $L = Ru \perp u^\perp$. If $\mathrm{s}(L) \neq \mathrm{n}(L)$, i.e. $\mathrm{n}(L) = 2\,\mathrm{s}(L)$ and $p = 2$, then there exist $v, w \in L$ such that $Q(v), Q(w) \in 2\,\mathrm{s}(L)$ and $(B(v, w)) = \mathrm{s}(L)$, and then $M := R[v, w]$ is an $\mathrm{s}(L)$-modular lattice and by the same proposition, $L = M \perp M^\perp$ holds. Thus repeating this argument, L is an orthogonal sum of modular lattices and we can write

$$L = L_1 \perp \cdots \perp L_t,$$

where L_i is $\mathrm{s}(L_i)$-modular and $\mathrm{s}(L_1) \supset \cdots \supset \mathrm{s}(L_t)$, where every inclusion is proper, i.e. $\mathrm{s}(L_i) \neq \mathrm{s}(L_{i+1})$. This decomposition is called a *Jordan decomposition*, and each L_i is called a *Jordan component*. In general, a Jordan decomposition is not unique. The following is fundamental.

Theorem 5.3.1. *Let L be a regular quadratic lattice over R, and $L = L_1 \perp \cdots \perp L_t = K_1 \perp \cdots \perp K_u$ be Jordan decompositions. Then we have*
(i) $t = u$,
(ii) $\mathrm{s}(L_i) = \mathrm{s}(K_i)$, and $\operatorname{rank} L_i = \operatorname{rank} K_i$ *for* $1 \le i \le t$
(iii) $\mathrm{n}(L_i) = \mathrm{n}(K_i)$ for $1 \le i \le t$,
(iv) $\mathrm{n}(L_1) = \mathrm{n}(L)$ and $\mathrm{s}(L_1) = \mathrm{s}(L)$.

Proof. For a regular quadratic lattice M over R, we put

$$M(a) := \{x \in M \mid B(x, M) \subset (a)\} = aM^\sharp \cap M \subset M.$$

Then

$$\mathrm{s}(M(a)) \subset (a) \tag{1}$$

is clear. Suppose that M is (s)-modular; then $M(a) = as^{-1}M \cap M$ by Proposition 5.2.1. Hence when $(a) \supset (s)$, $M(a) = M$ is (s)-modular, and when $(a) \subset (s)$, $M(a) = as^{-1}M$ is (a^2s^{-1})-modular. Thus, for an (s)-modular lattice M,
(*): $M(a)$ is (a)-modular if and only if $(s) = (a)$, and $(s) \neq (a)$ implies $\mathrm{s}(M(a)) \subset (pa)$.

Since $L(a) = L_1(a) \perp \cdots \perp L_t(a)$, from (1) and (*) it follows that $\mathrm{s}(L(a)) = (a)$ if and only if $(a) = \mathrm{s}(L_i)$ for some i. Thus the set of $\mathrm{s}(L_i)$ $(1 \le i \le t)$ is characterized by $\mathrm{s}(L(a)) = (a)$. Similarly we have $t = u$ and $\mathrm{s}(L_i) = \mathrm{s}(K_i)$ for $1 \le i \le t$.

Put $\mathrm{s}(L_i) = \mathrm{s}(K_i) = (a)$; then

$$L(a) = L_i \perp (\perp_{j \neq i} L_j(a)) = K_i \perp (\perp_{j \neq i} K_j(a))$$

and $\mathrm{s}(L_j(a)), \mathrm{s}(K_j(a)) \subset (pa) \subset (a) = \mathrm{s}(L_i)$ for $j \neq i$. Thus L_i, K_i give the first component of the Jordan decompositions of $L(a)$ and we may assume $i = 1$ to prove $\operatorname{rank} L_i = \operatorname{rank} K_i$ and the assertion (iii). Moreover, taking scaling by a^{-1}, we may assume $\mathrm{s}(L_1) = \mathrm{s}(K_1) = R$. Then we have $\mathrm{s}(L_i), \mathrm{s}(K_i) \subset (p)$ for $i \ge 2$ and hence $\mathrm{n}(L_i), \mathrm{n}(K_i) \subset (p)$ for $i \ge 2$. Thus we have $\mathrm{n}(L_1) = \mathrm{n}(L) = \mathrm{n}(K_1)$, which is the assertion (iii). L/pL becomes a symmetric bilinear space over $\mathbb{Z}/p\mathbb{Z}$ by $B(x, y) \bmod pR$ and L_1 induces a regular subspace and $\perp_{j \ge 2} L_j$ induces $(L/pL)^\perp$ and

$$\operatorname{rank} L_1 = \operatorname{rank} L - \dim(L/pL)^\perp,$$

and the similar assertion about the decomposition $L = \perp K_i$. Thus we have $\operatorname{rank} L_1 = \operatorname{rank} K_1$. $\mathrm{s}(L_1) = \mathrm{s}(L)$ is clear. Finally

$$\mathrm{n}(L_{i+1}) \subset \mathrm{s}(L_{i+1}) \subset p\,\mathrm{s}(L_i) \subset \mathrm{n}(L_i)$$

implies $\mathrm{n}(L) = \mathrm{n}(L_1)$. □

Theorem 5.3.2. *Suppose $p \neq 2$. For a regular quadratic lattice L over R, let $L = L_1 \perp \cdots \perp L_t = K_1 \perp \cdots \perp K_t$ be Jordan decompositions. Then $L_i \cong K_i$ holds for every i.*

Proof. Put $\mathrm{s}(L_i) = \mathrm{s}(K_i) = (a)$; then as in the proof of the previous theorem $L(a) = L_i \perp \cdots = K_i \perp \cdots$ become Jordan decompositions. Hence we may assume $i = 1$ and $\mathrm{s}(L_1) = \mathrm{s}(K_1) = R$ by scaling. Then L/pL becomes a quadratic space over $F_p =: \mathbb{Z}/p\mathbb{Z}$. While L_1 and K_1 induce regular subspaces, both $\perp_{j\geq 2} L_j$ and $\perp_{j \geq 2} K_j$ become $\mathrm{Rad}(L/pL)$. Thus $L_1/pL_1 \cong K_1/pK_1$ over F_p and implies $\mathrm{d}\, L_1 = \mathrm{d}\, K_1$ in $F_p^\times/(F_p^\times)^2$. Theorem 3.1.4 yields $\mathrm{d}\, L_1 = \mathrm{d}\, K_1$ and then $L_1 \cong K_1$ by Theorem 5.2.4. □

Corollary 5.3.1. *Suppose $p \neq 2$ and let L be a regular quadratic space over R. $L = L_1 \perp L_2 = K_1 \perp K_2$ and $L_1 \cong K_1$ imply $L_2 \cong K_2$.*

Proof. Let $L_i = \perp_j L_{i,j}$ $(i = 1, 2)$ and $K_i = \perp_j K_{i,j}$ $(i = 1, 2)$ be orthogonal decompositions where $L_{i,j}$ and $K_{i,j}$ are (p^j)-modular or 0. Then $L = \perp_j (L_{1,j} \perp L_{2,j}) = \perp_j (K_{1,j} \perp K_{2,j})$. From the theorem it follows that

$$(2) \qquad L_{1,j} \perp L_{2,j} \cong K_{1,j} \perp K_{2,j}, \quad L_{1,j} \cong K_{1,j}.$$

We have only to prove that (2) implies $L_{2,j} \cong K_{2,j}$. Thus, taking scaling of (2), the proof is reduced to the case that they are unimodular. Then the assertion follows from Theorem 5.2.4, comparing discriminants. □

The corollary does not hold for $p = 2$ in general.

Proposition 5.3.1. *For a regular quadratic lattice L over R, $O(L)$ contains a symmetry.*

Proof. Take an element $x \in L$ such that $(Q(x)) = \mathrm{n}(L)$. Then, for $y \in L$, $2B(x,y)Q(x)^{-1} \in 2\,\mathrm{s}(L)\,\mathrm{n}(L)^{-1} \subset R$, and

$$\tau_x(y) = y - 2B(x,y)Q(x)^{-1}x \in L.$$

Thus $\tau_x \in O(L)$. □

Theorem 5.3.3. *Suppose $p \neq 2$. For a regular quadratic lattice L over R, $O(L)$ is generated by symmetries.*

Proof. We use induction on $\mathrm{rank}\, L$. If $\mathrm{rank}\, L = 1$, then $O(L) = \{\pm 1\} = \{\tau_v, \tau_v^2\}$ where v is a basis of L. Suppose $\mathrm{rank}\, L > 1$. Let $L = L_1 \perp \cdots$ be a Jordan decomposition; we may suppose $\mathrm{s}(L_1) = R$ by scaling and then $\mathrm{s}(L) = R$ and L_1 is unimodular. Let $\sigma \in O(L)$, and in Theorem 1.2.2 put $\tilde{R} = R$, $P = (p)$ and $U = H = L$, $M = L_1$, $N = \sigma(L_1)$, and $q(x) = Q(x)$, $b(x,y) = 2B(x,y)$. Then conditions (1), (2), (3) and (5) are satisfied, and therefore there is a $\tau \in O(L)$ such that τ is a product of symmetries in $O(L)$ and $\tau = \sigma$ on L_1. Thus $\tau^{-1}\sigma = \mathrm{id}$ on L_1 and hence $\tau^{-1}\sigma|_{L_1^\perp} \in O(L_1^\perp)$ is a product of symmetries by the induction hypothesis. A symmetry in $O(L_1^\perp)$ induces canonically a symmetry in $O(L)$ which is the identity on L_1. Thus $\tau^{-1}\sigma$ is a product of symmetries in $O(L)$, so σ is a product of symmetries in $O(L)$. □

Henceforth, we study the case of $p = 2$, but we do not necessarily assume $p = 2$.

Theorem 5.3.4. *Let L be a regular quadratic lattice over R and M, N be submodules of L. For $a \in F^\times$, we put*

$$L^a := \{x \in L \mid B(x, L) \subset (a)\},$$
$$L_a := \{x \in L^a \mid Q(x) \subset (2a)\}.$$

Suppose that M and N satisfy

$$(3) \qquad \begin{cases} \mathrm{Hom}_R(M, R) = \{x \mapsto a^{-1}B(x, y) \mid y \in L_a\}, \\ \mathrm{Hom}_R(N, R) = \{x \mapsto a^{-1}B(x, y) \mid y \in L_a\}. \end{cases}$$

If $\sigma : M \cong N$ satisfies

$$(4) \qquad \sigma(x) \equiv x \bmod L^a \text{ for } x \in M,$$

then σ can be extended to an isometry of L which satisfies

$$\sigma(x) \equiv x \bmod L_a \quad \text{for } x \in L.$$

Proof. It is easy to see that L^a and L_a are submodules of L. Putting

$$h := \sigma(x) - x \in L^a \text{ for } x \in M,$$

$Q(\sigma(x)) = Q(h) + 2B(h, x) + Q(x)$ implies $Q(h) = -2B(h, x) \in (2a)$ and hence $h \in L_a$. Thus we may replace L^a in the condition (4) by L_a. Putting $H := L_a$ and taking scaling by $(2a)^{-1}$, we introduce a new quadratic form $q(x) = (2a)^{-1}Q(x)$, and then

$$b(x, y) := q(x + y) - q(x) - q(y) = a^{-1}B(x, y)$$

and $Q(H) \subset (2a)$ yield

$$b(H, L) = a^{-1}B(H, L) \subset R$$

and $q(H) \subset R$. Applying Theorem 1.2.2 to

$$\tilde{R} = R, P = (p), U = L,$$

we complete the proof. □

Lemma 5.3.1. *Let L be a regular quadratic lattice over R and suppose $L = M \perp M_1 = N \perp N_1$ where M and N are unimodular with $\mathrm{n}(M) = \mathrm{n}(N) = (2)$. Then $M_1 \cong N_1$ holds if M and N are isometric.*

Proof. We apply the previous theorem for $a = 1$. To do it, we must verify the conditions (3), (4). Since $L_a \supset M, N$ and $\mathrm{s}(L_a) \subset \mathrm{s}(L^a) \subset (a) = R$, Proposition 5.2.2 implies $L_a = M \perp * = N \perp *$, where $*$ is an appropriate lattice, and then the condition (3) in Theorem 5.3.4 is satisfied. For an isometry $\sigma : M \cong N$, let us verify the condition (4). For $x \in M$ and $y \in L$, we have

$$B(\sigma(x) - x, y) = B(\sigma(x), y) - B(x, y) \in B(N, L) - B(M, L) = R$$

and hence $\sigma(x) - x \in L^a$ for $x \in M$, which is nothing but the condition (4). Thus the theorem implies that σ can be extended to an isometry of L, and we have proved the lemma. □

Theorem 5.3.5. *Let $L = M \perp M_1 = N \perp N_1$ be a regular quadratic lattice over R and suppose that M and N are isometric and for every Jordan component K of M, $\mathrm{n}(K) = 2\,\mathrm{s}(K)$. Then $M_1 \cong N_1$.*

Proof. By scaling, Lemma 5.3.1 holds if M and N are isometric, modular and $\mathrm{n}(M) = 2\,\mathrm{s}(M)$. Applying it repeatedly to each Jordan component of M and N in the theorem, we complete the proof. □

Theorem 5.3.6. *Suppose $p = 2$. Let $L = M \perp M_1 = N \perp N_1$ be a regular quadratic lattice over R. Suppose that $M \cong N \cong \langle \epsilon \rangle$ for $\epsilon \in R^\times$, $\mathrm{s}(M_1) = \mathrm{s}(N_1)$, $\mathrm{n}(M_1) = \mathrm{n}(N_1)$ and either $\mathrm{s}(M_1) \subset (2)$ or $\mathrm{s}(M_1) = \mathrm{n}(M_1) = R$. Then $M_1 \cong N_1$.*

Proof. Put $M := Rv$. First suppose $\mathrm{s}(M_1) \subset (2)$. We apply Theorem 5.3.4 for $a = 2$; then $L^a = 2M \perp M_1 = 2N \perp N_1$ and $L_a \supset 2M, 2N$. Hence (3) is obviously satisfied. Putting $\sigma(v) = bv + w$ ($b \in R$, $w \in M_1$) for an isometry $\sigma : M \cong N$, we have $Q(\sigma(v)) = b^2 Q(v) + Q(w)$ and $(b^2 - 1)Q(v) = -Q(w) \in \mathrm{s}(M_1) \subset (2)$. Since $Q(v) \in R^\times$, we have $b \in R^\times$. Thus, for $y \in L$

$$\begin{aligned} B(\sigma(v) - v, y) = B((b-1)v + w, y) &= (b-1)B(v, y) \\ &+ B(w, y) \in (b-1)B(M, L) + B(M_1, L) \subset (2) \end{aligned}$$

because $\mathrm{s}(M_1) \subset (2)$. This means $\sigma(v) - v \in L^a$, which is the condition (4). By the theorem, σ is extended to an isometry of L and then $M_1 \cong N_1$.

Next suppose $\mathrm{s}(M_1) = \mathrm{n}(M_1) = R$; then $\mathrm{s}(L) = R$. We apply the same theorem for $a = 1$; then clearly $L^a = L$ and hence the condition (4) is also clear for $\sigma : M \cong N$. We have only to verify the condition (3). Now

$\mathrm{n}(M_1) = R$ implies that there exists $x \in M_1$ such that $Q(x) \in R^\times$. Now $v + x \in L_1$ follows from $Q(v+x) = Q(v) + Q(x) \equiv 0 \bmod 2$. By

$$B(v, v+x) = Q(v) \in R^\times,$$

the condition (3) is verified for M and similarly for N. □

Corollary 5.3.2. *Let L be a regular quadratic lattice over R and $L = M_1 \perp M_2 \perp \cdots = N_1 \perp N_2 \perp \cdots$ be Jordan decompositions. If $M_1 \cong N_1$, then $M_1^\perp \cong N_1^\perp$ holds.*

Proof. By scaling, we may assume that M_1 ($\cong N_1$) is unimodular. If $\mathrm{n}(M_1) = \mathrm{n}(N_1) = (2)$, then we have only to apply Lemma 5.3.1. If $\mathrm{n}(M_1) = \mathrm{n}(N_1) = R$, then by Theorem 5.2.5 M_1 and N_1 have an orthogonal basis. Noting $\operatorname{rank} M_1 = \operatorname{rank} N_1$, $\mathrm{n}(M_1^\perp) = \mathrm{n}(N_1^\perp)$ and $\mathrm{s}(M_1^\perp) = \mathrm{s}(N_1^\perp) \subset (2)$ by Theorem 5.3.1, Theorem 5.3.6 yields the corollary inductively on $\operatorname{rank} M_1$. □

Theorem 5.3.7. *Let V be a regular quadratic space over F. Suppose that L and M are unimodular lattices on V with $\mathrm{n}(L) = \mathrm{n}(M)$. Then $L \cong M$ holds.*

Proof. First, we note that $\mathrm{d}\, L = \mathrm{d}\, M$ since they are units and equal to $\mathrm{d}\, V$ in $F/(F^\times)^2$. If $p \neq 2$, then Theorem 5.2.4 completes the proof.

Assume $p = 2$. Now we claim, first that if K is a unimodular lattice on a hyperbolic space, then

$$\langle\epsilon\rangle \perp K \cong \langle\epsilon\rangle \perp (\perp \langle\begin{pmatrix} 0 & 1 \\ 1 & 0 \end{pmatrix}\rangle) \text{ for } \epsilon \in R^\times.$$

Let us prove it by induction on $\operatorname{rank} K$. Suppose $\operatorname{rank} K = 2$. Using either Proposition 5.2.3 if $\mathrm{n}(K) = R$, or Lemma 5.2.6 if $\mathrm{n}(K) = 2R$, we have

$$\langle\epsilon\rangle \perp K \cong \langle\begin{pmatrix} 2 & 1 \\ 1 & 2 \end{pmatrix}\rangle \perp \langle -3\epsilon\rangle \text{ or } \langle\begin{pmatrix} 0 & 1 \\ 1 & 0 \end{pmatrix}\rangle \perp \langle\epsilon\rangle.$$

Now

$$Q(\langle\begin{pmatrix} 2 & 1 \\ 1 & 2 \end{pmatrix}\rangle) = \{x \in R \mid x = 0 \text{ or } \operatorname{ord}_2 x \text{ is odd}\}$$

in Lemma 5.2.6 yields that $\langle\begin{pmatrix} 2 & 1 \\ 1 & 2 \end{pmatrix}\rangle \perp \langle -3\epsilon\rangle$ is anisotropic. Since K is isotropic, it is a contradiction and so $\langle\epsilon\rangle \perp K \cong \langle\begin{pmatrix} 0 & 1 \\ 1 & 0 \end{pmatrix}\rangle \perp \langle\epsilon\rangle$. If $\operatorname{rank} K \geq 4$, then by Proposition 5.2.4, $\langle\epsilon\rangle \perp K \cong (\perp \langle\begin{pmatrix} 0 & 1 \\ 1 & 0 \end{pmatrix}\rangle) \perp H$, where $\operatorname{rank} H = 1$ or 3 and H is unimodular. If $\operatorname{rank} H = 1$, then we have $H \cong \langle \epsilon \rangle$, comparing discriminants. If $\operatorname{rank} H = 3$, then H is unimodular and isotropic, comparing Witt indices. Moreover H has an orthogonal basis by Theorem 5.2.5 and then Proposition 5.2.3 implies $H \cong \langle \begin{pmatrix} 0 & 1 \\ 1 & 0 \end{pmatrix} \rangle \perp \langle \epsilon \rangle$. Thus we have proved the claim.

If $\mathrm{n}(L) = \mathrm{n}(M) = (2)$, then Theorem 5.2.5 completes the proof. Hence we may assume $\mathrm{n}(L) = \mathrm{n}(M) = R$; then L and M have an orthogonal basis. Applying the above to $K = M \perp L^{(-1)}$ and $L \perp L^{(-1)}$, we have

$$L \perp (M \perp L^{(-1)}) \cong L \perp (\perp_{\operatorname{rank} L} \langle \begin{pmatrix} 0 & 1 \\ 1 & 0 \end{pmatrix} \rangle),$$
$$M \perp (L \perp L^{(-1)}) \cong M \perp (\perp_{\operatorname{rank} L} \langle \begin{pmatrix} 0 & 1 \\ 1 & 0 \end{pmatrix} \rangle).$$

Since the left-hand sides are isometric, we have $L \cong M$ by Lemma 5.3.1. □

Proposition 5.3.2. *Let N be a maximal lattice isometric to the orthogonal sum of n copies of $\langle \begin{pmatrix} 0 & 1 \\ 1 & 0 \end{pmatrix} \rangle$ and M be a regular quadratic lattice over R with* $\operatorname{rank} M = n$ *and* $\mathrm{s}(M) \subset 2R$. *Then there is a primitive submodule M_0 in N such that $M_0 \cong M$ and $M_0^{\perp} \cong M^{(-1)}$.*

Proof. By using a Jordan splitting of M, we may assume that either $n = 1$ and $M = \langle a \rangle$ or $n = 2$ and $M = \langle 2^b \begin{pmatrix} 2c & 1 \\ 1 & 2c \end{pmatrix} \rangle$, where the latter occurs only when $p = 2$ and $c = 0$ or 1. Suppose $M \cong \langle a \rangle$ and $N = R[e, f]$ with $Q(e) = Q(f) = 0$, $B(e, f) = 1$. Put $v = e + a/2 \cdot f$; then $Q(v) = a$ and $v^{\perp} = R[e - a/2 \cdot f] \cong \langle -a \rangle$.

Next, suppose $p = 2$ and $M = R[z_1, z_2] = \langle 2^b \begin{pmatrix} 2c & 1 \\ 1 & 2c \end{pmatrix} \rangle$ and $N = R[u_1, \cdots, u_4]$ such that $Q(\sum x_i u_i) = 2(x_1 x_2 + x_3 x_4)$. Putting

$$v_1 := u_1, \quad v_2 := u_1 + 2^b u_2 - 2^b u_3 + u_4,$$
$$w_1 := u_1 + 2^b u_2, \quad w_2 := 2^b u_2 + u_3 + 2^b u_4,$$

then

$$M_1 := R[v_1, v_2] \cong \langle 2^b \begin{pmatrix} 0 & 1 \\ 1 & 0 \end{pmatrix} \rangle,$$
$$M_2 := R[w_1, w_2] \cong \langle 2^b \begin{pmatrix} 2 & 1 \\ 1 & 2 \end{pmatrix} \rangle$$

are clear. M_1, M_2 are primitive and

$$M_1^\perp = R[u_1 - 2^b u_3, 2^b u_3 + u_4] \cong \langle \begin{pmatrix} 0 & -2^b \\ -2^b & 2^{b+1} \end{pmatrix} \rangle \cong M_1^{(-1)},$$
$$M_2^\perp = R[u_1 - 2^b u_2 - 2^b u_4, u_3 - 2^b u_4] \cong \langle \begin{pmatrix} -2^{b+1} & -2^b \\ -2^b & -2^{b+1} \end{pmatrix} \rangle \cong M_2^{(-1)}.$$

Now, we have only to take $M_0 = M_1$ and M_2 for $c = 0$ and 1 respectively. □

Proposition 5.3.3. *Let N be a regular quadratic lattice over R with $\mathrm{s}(N) \subset R$ and let L be a regular submodule of N. Then $\mathrm{d}\,L^\perp$ divides $\mathrm{d}\,N\,\mathrm{d}\,L$, and the number of submodules of N which are isometric to L and are not mutually transformed by isometries of N is finite.*

Proof. For $x \in N$, $z \mapsto B(x,z)$ is a homomorphism from L to R and hence there exists $y \in L^\sharp$ such that $B(x,z) = B(y,z)$ for $z \in L$. Define the mapping ϕ from N to $L^\sharp$ by $\phi(x) = y$. Since $\ker\phi = L^\perp$ and $\phi(x) = x$ for $x \in L$, $\phi^{-1}(L) = L \perp L^\perp$ and implies $[N : L \perp L^\perp] = [\phi(N) : L]$, which divides $[L^\sharp : L]$. Then

$$\mathrm{ord}_p(\mathrm{d}\,L\,\mathrm{d}\,L^\perp) = \mathrm{ord}_p([N : L \perp L^\perp]^2\,\mathrm{d}\,N) \le \mathrm{ord}_p([L^\sharp : L]^2\,\mathrm{d}\,N)$$

implies $0 \le \mathrm{ord}_p\,\mathrm{d}\,L^\perp \le \mathrm{ord}_p(\mathrm{d}\,N\,\mathrm{d}\,L)$ from which it follows, upon considering the various Jordan decompositions of $L^\perp$ with $\mathrm{s}(L^\perp) \subset R$, that there are only a finite number of isometry classes of $L^\perp$. For a given K which is one of the possibilities for $L^\perp$, there are only finitely many lattices isometric to N which contain $L \perp K$. This completes the proof. □

5.4 Extension theorems

In this section, we give some results of Witt type theorems. The following is essentially due to Eichler.

Theorem 5.4.1. *Let L be a regular quadratic lattice over R, and M and N be isometric submodules of L. We put $\tilde{M} := FM \cap L$ and $\tilde{N} := FN \cap L$ and suppose*

$$\tilde{M}/M \cong \tilde{N}/N \cong R/(p^{a_1}) \oplus \cdots \oplus R/(p^{a_n})$$

for $0 \le a_1 \le \cdots \le a_n$ as modules. Take $a \in F^\times$ such that $a\,\mathrm{n}(L^\sharp) \subset (2)$. If $\sigma : M \cong N$ satisfies $\sigma(x) \equiv x \bmod ap^{a_n} L^\sharp$ for every $x \in M$, then σ can be extended to an isometry of L such that

$$\sigma(x) \equiv x \bmod aL^\sharp \quad \text{for } x \in L.$$

Proof. Putting

$$L^a := \{x \in L \mid B(x, L) \subset (a)\} \text{ and } L_a := \{x \in L^a \mid Q(x) \in (2a)\},$$

we show

$$(1) \qquad L^a = L_a = aL^\sharp.$$

Since $B(aL^\sharp, L^\sharp) = a\,\mathrm{s}(L^\sharp) \subset \frac{a}{2}\,\mathrm{n}(L^\sharp) \subset R$, we have

$$(2) \qquad aL^\sharp \subset (L^\sharp)^\sharp = L.$$

Hence $L^a = L \cap aL^\sharp = aL^\sharp$ holds. If $x \in L^a$, then putting $x = ay$, $y \in L^\sharp$, $Q(x) = a^2 Q(y) \in a^2\,\mathrm{n}(L^\sharp) \subset (2a)$ holds, which means $L^a \subset L_a$ and hence $L_a = L^a = aL^\sharp$.

Next we show $\sigma(\tilde{M}) = \tilde{N}$, where we regard σ as an isometry from FM on FN. Let $\{v_i\}$ be a basis of $\tilde{M}$ such that $\{p^{a_i} v_i\}$ is a basis of M; then $\sigma(p^{a_i} v_i) \equiv p^{a_i} v_i \bmod ap^{a_n} L^\sharp$ is valid for every i. Thus we have

$$(3) \qquad \sigma(v_i) \equiv v_i \bmod ap^{a_n - a_i} L^\sharp \equiv v_i \bmod L \text{ by (2)},$$

and then $\sigma(v_i) \in L$ and $\sigma(v_i) \in \sigma(\tilde{M}) \cap L \subset FN \cap L = \tilde{N}$. It means $\sigma(\tilde{M}) \subset \tilde{N}$. Furthermore $[\tilde{M} : M] = [\tilde{N} : N]$ and $\sigma(M) = N$ imply $\sigma(\tilde{M}) = \tilde{N}$. From (3) follows $\sigma(x) \equiv x \bmod aL^\sharp \equiv x \bmod L^a$ for $x \in \tilde{M}$, which is the condition (4) in Theorem 5.3.4 replacing M there by $\tilde{M}$. To complete the proof, we have only to verify the condition (3) there. Since $\tilde{M}$ is primitive in L, we can extend the basis $\{v_i\}$ of $\tilde{M}$ to a basis of L. Hence we can choose the dual basis $\{u_i\}$ ($u_i \in L^\sharp$) of $\{v_i\}$, i.e. $B(v_i, u_j) = \delta_{ij}$; then $a^{-1}B(v_i, au_j) = \delta_{ij}$ holds, and it confirms the condition (3) in Theorem 5.3.4 for $\tilde{M}$ since $\{v_i\}$ is a basis of $\tilde{M}$ and $au_j \in aL^\sharp = L_a$ by (1), and similarly for $\tilde{N}$. Thus it completes the proof. □

Corollary 5.4.1. *Let L be a unimodular lattice over R with $\mathrm{n}(L) = (2)$. If M and N are isometric submodules of L and they are direct summands of L, then they are transformed by an isometry of L.*

Proof. Although this is a direct corollary of Theorem 1.2.2, this follows from the theorem, noting $\tilde{M} = M, \tilde{N} = N, L^\sharp = L, a = 1$. □

Exercise 1.

Let L be a regular quadratic lattice over R, and $M = R[v_1, \cdots, v_m]$ be a submodule of L. If $u_i \in L$ is sufficiently close to v_i for $i = 1, \cdots, m$ and $B(u_i, u_j) = B(v_i, v_j)$ for $1 \le i, j \le m$, show there exists an isometry σ of L such that $\sigma(u_i) = v_i$ for $1 \le i \le m$.

The following is due to Kneser.

Theorem 5.4.2. *Let V and W be quadratic spaces over F and L a lattice on V. Suppose that the following condition $\natural(\sigma, k)$ is satisfied for $\sigma \in \mathrm{Hom}_R(L, W)$, an integer k and a submodule G over R in W:*

$$\natural(\sigma, k) : \begin{cases} \text{(i) } p^{k-1}\,\mathrm{n}(G) \subset (2) \\ \text{(ii) } \mathrm{Hom}_R(L, R) = \{x \mapsto B(\sigma(x), g) \mid g \in G\} + \mathrm{Hom}_R(L, (p)) \\ \text{(iii) } Q(\sigma(x)) \equiv Q(x) \bmod 2p^k R \text{ for every } x \in L. \end{cases}$$

Then there exists a homomorphism $\eta \in \mathrm{Hom}_R(L, W)$ such that $\eta(x) \equiv \sigma(x) \bmod p^k G$ for $x \in L$ and the condition $\natural(\eta, k+1)$ is satisfied. Moreover the number of such homomorphisms which are mutually different modulo $p^{k+1}G$ is $p^{vg-(v+1)v/2}$ where $v = \dim V = \mathrm{rank}\, L$ and $g = \mathrm{rank}\, G$. If V is regular in addition, then there is an isometry σ_0 from L to W such that $\sigma_0(x) \equiv \sigma(x) \bmod p^k G$ for $x \in L$.

Proof. Let $\eta \in \mathrm{Hom}_R(L, W)$ such that $\eta(x) \equiv \sigma(x) \bmod p^k G$ for $x \in L$; then the condition (i) of $\natural(\eta, k+1)$ follows trivially from (i) of $\natural(\sigma, k)$. Next we show that the condition (ii) is satisfied. Write

$$\eta(x) = \sigma(x) + p^k u(x) \quad \text{for} \quad u \in \mathrm{Hom}_R(L, G);$$

then for $x \in L$, $g \in G$ we have

$$B(\eta(x), g) - B(\sigma(x), g) = p^k B(u(x), g) \in p^k\,\mathrm{s}(G) \subset \frac{1}{2} p^k\,\mathrm{n}(G) \subset (p)$$

by (i). Hence the mapping

$$x \mapsto B(\eta(x), g) - B(\sigma(x), g)$$

is in $\mathrm{Hom}_R(L, (p))$. Thus the condition (ii) of $\natural(\eta, k+1)$ is satisfied.

Lastly let us show that the number of homomorphisms η such that they are mutually different modulo $p^{k+1}G$, $\eta(x) \equiv \sigma(x) \bmod p^k G$ and the condition $\natural(\eta, k+1)$ is $p^{vg-v(v+1)/2}$. To do it, we introduce a new bilinear form $a(x, y)$ associated with Q as follows: Let $\{x_i\}_{i=1}^v$ ($v = \mathrm{rank}\, L = \dim V$) be a basis of L and define a bilinear form $a(x, y)$ ($x, y \in L$) by

$$\begin{aligned} a\left(\sum m_i x_i, \sum n_i x_i\right) := & \frac{1}{2} p^{-k} \sum_{i=1}^{v} (Q(\sigma(x_i)) - Q(x_i)) m_i n_i \\ & + p^{-k} \sum_{1 \le i < j \le v} (B(\sigma(x_i), \sigma(x_j)) - B(x_i, x_j)) m_i n_j \end{aligned}$$

where $m_i, n_j \in R$. By virtue of the condition (iii), $a(x,y) \in R$ for $x, y \in L$ is easy to see, and $Q(\sigma(x)) - Q(x) = 2p^k a(x,x)$ is clear.

Put $\eta = \sigma + p^k u$ as above. For $x \in L$,

(iv) $$Q(\eta(x)) \equiv Q(x) \bmod 2p^{k+1}R$$

is equivalent to

$$Q(\sigma(x)) - Q(x) + 2p^k B(\sigma(x), u(x)) + p^{2k} Q(u(x)) \equiv 0 \bmod 2p^{k+1}R.$$

Here we note $p^{2k}Q(u(x)) \in p^{2k}\,\mathrm{n}(G) \subset 2p^{k+1}R$ by (i). Then the condition (iv) is equivalent to $a(x,x) \equiv -B(\sigma(x), u(x)) \bmod pR$ for $x \in L$. Hence the number required is the number of $u \in \mathrm{Hom}_R(L,G) \bmod pG$ satisfying $B(\sigma(x), u(x)) \equiv -a(x,x) \bmod pR$ for $x \in L$.

We associate $u \in \mathrm{Hom}_R(L, G/pG)$ with the mapping $x \mapsto B(\sigma(x), u(x)) \bmod p$, which is a quadratic form on L/pL over R/pR since $B(\sigma(x), G) \subset R$ by the condition (ii). Thus the number which we want is the number of the inverse images of $-a(x,x)$ by this mapping. We will show that every quadratic form on L/pL is given like this and the number of the inverse images of each quadratic form is constant. The latter is easy; because u and v give the same quadratic form if and only if $B(\sigma(x), (u-v)(x)) \equiv 0 \bmod p$ for $x \in L$. Therefore, once the surjectivity is shown, the number of the inverse images which we want is

$$\frac{\sharp\,\mathrm{Hom}_R(L, G/pG)}{\sharp\{\text{quadratic forms on } L/pL\}} = p^{vg}(p^{v(v+1)/2})^{-1}.$$

Now we must show the surjectivity. Let q be a quadratic form on L/pL and define the bilinear form b by $b(x,y) = q(x+y) - q(x) - q(y)$. Define the bilinear form A on L/pL by

$$A(\sum m_i x_i, \sum n_i x_i) = \sum q(x_i) m_i n_i + \sum_{i<j} b(x_i, x_j) m_i n_j \quad \text{for } m_i, n_j \in R$$

where $\{x_i\}$ is a basis of L as above; then $A(x,x) = q(x)$ is valid for $x \in L$. For $x_i \in L$, the linear mapping $x \mapsto A(x, x_i)$ from L to $R/(p)$ is induced by $x \mapsto B(\sigma(x), g) \bmod p$ for some $g \in G$ by virtue of (ii) in $\natural(\sigma, k)$. We denote the element g by $u(x_i)$. Extending u linearly, we can define a homomorphism $u \in \mathrm{Hom}(L/pL, G/pG)$ such that $A(x,y) \equiv B(\sigma(x), u(y)) \bmod p$ for $x, y \in L$ and so $q(x) \equiv B(\sigma(x), u(x)) \bmod p$. Thus the surjectivity has been shown.

It remains to show the existence of the isometry σ_0 in the assertion. Suppose that V is regular. By what we have shown, there exists a homomorphism $\eta_h \in \mathrm{Hom}(L, W)$ such that $\eta_{h+1}(x) \equiv \eta_h(x) \bmod p^{k+h}G$ ($\eta_0 = \sigma$) and $Q(\eta_h(x)) \equiv Q(x) \bmod 2p^{k+h}$ for $x \in L$. Since the module G is finitely generated as assumed throughout this book, η_k has a limit linear mapping σ_0 and it satisfies $Q(\sigma_0(x)) = Q(x)$ and $\sigma_0(x) \equiv \sigma(x) \bmod p^k G$. The regularity of V implies that σ_0 is an isometry. □

Remark 1.

If V is regular and M is a submodule over R in W such that $\sigma(L),\ p^kG \subset M$ in the theorem, then $\eta(L) \subset M$ and $\sigma_0(L) \hookrightarrow M$.

Remark 2.

If M is a submodule of W such that

$$\mathrm{Hom}_R(L, R) = \{x \mapsto B(\sigma(x), m) \mid m \in M\}$$

and $p^{k-1}\,\mathrm{n}(M) \subset (2)$ in the theorem, then we can put $G = M$. In particular, if M is unimodular with $\mathrm{n}(M) = (2)$ and $\sigma(L)$ is a direct summand of M as a module, then conditions (i), (ii) of $\natural(\sigma, k)$ are satisfied for $G = M$ and $k = 1$.

Corollary 5.4.2. *Let L and M be quadratic lattices over R and suppose that L is regular. Fix an integer h such that $p^hQ(L^\sharp) \subset (2)$. If $\sigma \in \mathrm{Hom}_F(FL, FM)$ satisfies*

$$\sigma(L) \subset M \quad \textit{and} \quad Q(\sigma(x)) \equiv Q(x) \bmod 2p^{h+1}R \quad \textit{for} \quad x \in L,$$

then there exists an isometry $\eta : L \hookrightarrow M$ such that

$$\eta(L) = \sigma(L), \quad \eta(L^\sharp) = \sigma(L^\sharp) \quad \textit{and} \quad \eta(x) \equiv \sigma(x) \bmod p^{h+1}\sigma(L^\sharp)$$

for $x \in L$.

Proof. First we show that $p^hQ(L^\sharp) \subset (2)$ implies

$$(4) \qquad p^hL^\sharp \subset L.$$

We have, for $x, y \in L^\sharp$, $2B(p^hx, y) \in 2p^h\,\mathrm{s}(L^\sharp) \subset p^h\,\mathrm{n}(L^\sharp) \subset (2)$ by our choice of h. Hence $p^hL^\sharp \subset (L^\sharp)^\sharp = L$ holds.

Next we verify the condition $\natural(\sigma, h+1)$ for $G := \sigma(L^\sharp)$.

Let $x \in L^\sharp$; then $p^hx \in L$ by (4) and hence $Q(\sigma(p^hx)) \equiv Q(p^hx) \bmod 2p^{h+1}R$, i.e. $p^hQ(\sigma(x)) \equiv p^hQ(x) \bmod 2pR$. By our choice of h, we have $p^hQ(x) \subset (2)$ and hence $p^hQ(\sigma(x)) \in (2)$. This is simply the condition (i) in $\natural(\sigma, h+1)$.

Let us examine the condition (ii). Let $\phi \in \mathrm{Hom}_R(L, R)$; then there exists $y \in L^\sharp$ such that $\phi(x) = B(x, y)$ for $x \in L$, since FL is regular. Condition (4) implies $p^hy \in L$ and then for $x \in L$

$$\begin{aligned} p^h\phi(x) = B(x, p^hy) &\equiv B(\sigma(x), \sigma(p^hy)) \bmod p^{h+1}R \\ &\equiv p^hB(\sigma(x), \sigma(y)) \bmod p^{h+1}R, \end{aligned}$$

which means $\phi(x) \equiv B(\sigma(x), \sigma(y)) \bmod pR$ and the condition (ii) is satisfied.

The condition (iii) is simply the definition of σ. Hence, by the theorem there exists an isometry η from L to FM such that $\eta(x) \equiv \sigma(x) \bmod p^{h+1} \sigma(L^\sharp)$ for $x \in L$. It implies

$$(5) \qquad \eta(x) \equiv \sigma(x) \bmod p\sigma(L)$$

by (4). Hence $\eta(L) \subset \sigma(L) \subset M$ holds. Let us show $\sigma(L) \subset \eta(L)$, which yields $\eta(L) = \sigma(L)$. Take $x_0 \in L$ and define $x_1 \in L$ by $\sigma(x_0) = \eta(x_0) + p\sigma(x_1)$ by virtue of (5), and inductively define $x_{n+1} \in L$ by $\sigma(x_n) = \eta(x_n) + p\sigma(x_{n+1})$; then

$$\sigma(x_0) = \sum_{t=0}^{\infty} p^t \eta(x_t) \in \eta(L),$$

i.e. $\sigma(L) \subset \eta(L)$. It remains to show $\eta(L^\sharp) = \sigma(L^\sharp)$. For $y \in L^\sharp$, (4) implies $p^h y \in L$ and then as was shown

$$\eta(p^h y) \equiv \sigma(p^h y) \bmod p^{h+1} \sigma(L^\sharp)$$

and hence $\eta(y) \equiv \sigma(y) \bmod p\sigma(L^\sharp)$. This yields $\eta(L^\sharp) = \sigma(L^\sharp)$, like $\eta(L) = \sigma(L)$. □

Corollary 5.4.3. *Let $L = R[v_1, \cdots, v_n]$ and $M = R[w_1, \cdots, w_n]$ be quadratic lattices over R with $\operatorname{rank} L = \operatorname{rank} M = n$ and suppose that FL is regular. If $B(v_i, v_j) \equiv B(w_i, w_j) \bmod p^{h+1}R$ and $Q(v_i) \equiv Q(w_i) \bmod 2p^{h+1}R$ for an integer h such that $p^h Q(L^\sharp) \subset (2)$, then there exists an isometry $\eta : L \cong M$ satisfying $\eta(v_i) \equiv w_i \bmod p^{h+1} M^\sharp$.*

Proof. Define a homomorphism $\sigma \in \mathrm{Hom}_F(FL, FM)$ by $\sigma(v_i) = w_i$. Apply the previous corollary and let η be the isometry there, then, $\sigma(L) = M$ implies $\eta(L) = \sigma(L) = M$. Since $\eta : FL \cong FM$ and $\eta(L) = M$, $\sigma(L^\sharp) = \eta(L^\sharp) = \eta(L)^\sharp = M^\sharp$ holds and then $\eta(x) \equiv \sigma(x) \bmod p^{h+1} M^\sharp$ for $x \in L$ from the previous corollary. □

Corollary 5.4.4. *Let $A = {}^tA \in GL_n(F)$ and take an integer h such that $p^h A^{-1} \in M_n(R)$ and all diagonals of $p^h A^{-1}$ are in (2). If $B = {}^tB \in M_n(F)$ satisfies $B \equiv A \bmod p^{h+1} M_n(R)$ and all diagonals of $B - A$ are in $(2p^{h+1})$, then there exists a matrix $U \in GL_n(R)$ such that $B = A[U]$ and $U \equiv 1_n \bmod p^{h+1} A^{-1} M_n(R)$.*

Proof. Let $L = R[v_1, \cdots, v_n]$ and $M = [w_1, \cdots, w_n]$ be quadratic lattices over R defined by $(B(v_i, v_j)) = A$ and $(B(w_i, w_j)) = B$; then L is regular.

Since $L^\sharp \cong \langle A^{-1}\rangle$, we have $p^h Q(L^\sharp) \subset (2)$. By the previous corollary, there exists an isometry $\eta : L \cong M$ such that $\eta(v_i) \equiv w_i \bmod p^{h+1}M^\sharp$. Put $(\eta(v_1), \cdots, \eta(v_n)) = (w_1, \cdots, w_n)X$ for $X \in GL_n(R)$; then $A = B[X]$ and $(w_1, \cdots, w_n)(X - 1_n) \equiv 0 \bmod p^{h+1}M^\sharp$. Since $(w_1, \cdots, w_n)B^{-1}$ is a basis of $M^\sharp$ by 5.1, we have $X - 1_n \in p^{h+1}B^{-1}M_n(R)$. Putting $U = X^{-1}$ and multiplying X^{-1} from the left, we have $1_n - U \in p^{h+1}X^{-1}B^{-1}M_n(R)$ and then $A = B[X]$ implies $1_n - U \in p^{h+1}A^{-1t}XM_n(R) \subset p^{h+1}A^{-1}M_n(R)$. □

Corollary 5.4.5. *Let N be a regular quadratic lattice over R, and $L = R[v_1, \cdots, v_n]$ be a regular submodule of N. If $w_i \in N$ is sufficiently close to v_i for $i = 1, \cdots, n$, then there exists an isometry $\sigma \in O(N)$ such that $\sigma(L) = R[w_1, \cdots, w_n]$.*

Proof. If w_i is sufficiently close to v_i, then the existence of an integer h in the assumption of Corollary 5.4.3 is clear. Therefore there is an isometry $\eta : L \cong R[w_1, \cdots, w_n]$ such that $\eta(v_i)$ is sufficiently close to w_i, and hence $\eta(v_i)$ and v_i are sufficiently close. Then we can apply Theorem 5.4.1 and complete the proof. □

5.5 The spinor norm

Recall from section 1.6 that θ and τ_v denote the spinor norm and the symmetry, respectively, and the image of θ may be identified with the inverse image by $F^\times \to F^\times/(F^\times)^2$ if there is no confusion.

Proposition 5.5.1. *Let V be a regular quadratic space over F with $\dim V \geq 3$. Then $\theta(O^+(V)) = F^\times$.*

Proof. Since $Q(V) = F$ for $\dim V \geq 4$ by virtue of Corollary 3.5.1, for $a \in F^\times$ there exist x and $y \in V$ such that $Q(x) = a$, $Q(y) = 1$. Then $\theta(\tau_x\tau_y) = a$ is clear.

Suppose $\dim V = 3$. As $\sharp(F^\times/(F^\times)^2) \geq 4$ follows from Corollary 3.1.4, for any given $y \in F^\times$, there exists $z \in F^\times$ such that $z \neq y\,\mathrm{d}V, \mathrm{d}V$. Then both $x := z$ and yz satisfy $x\,\mathrm{d}V \neq 1 \bmod (F^\times)^2$ and so then $V \perp \langle x\rangle$ is isotropic by Theorem 3.5.1, which means $-x \in Q(V)$. Thus we have $-yz, -z \in Q(V)$, that is there exist v and $w \in V$ such that $Q(v) = -yz$ and $Q(w) = -z$, which implies $\theta(\tau_v\tau_w) = y$. □

Proposition 5.5.2. *If V is a regular anisotropic binary quadratic space over F, then $\theta(O^+(V)) = \mathrm{N}_{K/F}(K^\times)$ where $K = F(\sqrt{-\mathrm{d}V})$.*

Proof. Writing $V = \langle a\rangle \perp \langle ad\rangle$ for $a, d \in F^\times$, we have $-d \notin (F^\times)^2$, since V is anisotropic. Hence we have $Q(V) = a\,\mathrm{N}_{K/F}(K^\times)$ where $K = F(\sqrt{-d})$ and it completes the proof. □

Proposition 5.5.3. *Let L be a maximal lattice on a regular quadratic space V over F with* $\dim V \geq 3$. *Then $\theta(O^+(L)) \supset R^\times$ holds.*

Proof. If V is anisotropic, then Theorem 5.2.1 implies $O^+(L) = O^+(V)$ and so Proposition 5.5.1 gives $\theta(O^+(L)) \supset R^\times$. If V is isotropic, then $L \cong \langle \frac{1}{2}\begin{pmatrix} 0 & a \\ a & 0 \end{pmatrix} \rangle \perp$ (some lattice) holds by Theorem 5.2.2. Put

$$M = R[u,v] \cong \langle \frac{1}{2}\begin{pmatrix} 0 & a \\ a & 0 \end{pmatrix} \rangle;$$

then $\theta(O^+(L)) \supset \theta(O^+(M)) \supset R^\times$ since $\tau_{u+bv} \in O(M)$ and $\theta(\tau_{u+bv}) = ab$ for $b \in R^\times$. □

Proposition 5.5.4. *Let L be a modular lattice over R. Then we have $\theta(O^+(L)) \supset R^\times$ if either*
(i) $\mathrm{n}(L) = 2\,\mathrm{s}(L)$ *and* $\operatorname{rank} L \geq 2$ *or*
(ii) $p = 2$ *and* $\operatorname{rank} L \geq 3$ *holds.*

Proof. By scaling, we may suppose that L is unimodular. First suppose $p \neq 2$ and $\operatorname{rank} L \geq 2$. Let $a \in R^\times$; then by Theorem 5.2.4,

$$L \cong \langle a \rangle \perp \langle a\,\mathrm{d}L \rangle \perp \langle 1 \rangle \perp \cdots.$$

If $v \in L$ gives a basis of $\langle a \rangle$, then $\tau_v \in O(L)$ and $\theta(\tau_v) = a$. Thus we have $\theta(O^+(L)) \supset R^\times$.

Suppose $p = 2$. Let $M \cong \langle \begin{pmatrix} 2c & 1 \\ 1 & 2c \end{pmatrix} \rangle$ for $c = 0$ or 1; then Lemma 5.2.6 implies $Q(M) \supset 2R^\times$. For $a \in R^\times$, we can take $v \in M$ such that $Q(v) = 2a$, and then $\tau_v \in O(M)$ is easy to see. $\theta(\tau_v) = 2a$ yields $\theta(O^+(M)) \supset R^\times$. L is split by M for $c = 0$ or 1 either in the case of (i) or (ii) with $\mathrm{n}(L) = 2\,\mathrm{s}(L)$ by Theorem 5.2.5, and in the case of (ii) with $\mathrm{n}(L) = \mathrm{s}(L)$ by Proposition 5.2.3. Thus we have $\theta(O^+(L)) \supset \theta(O^+(M)) \supset R^\times$. □

Proposition 5.5.5. *If L is unimodular with* $\operatorname{rank} L \geq 2$ *and $p \neq 2$, then $\theta(O^+(L)) = R^\times(F^\times)^2$.*

Proof. By Theorem 5.3.3, $O(L)$ is generated by symmetries. Let $\tau_v \in O(L)$ and we may assume that v is a primitive element in L since $\tau_{av} = \tau_v$ for $a \in F^\times$. Now $\tau_v(x) = x - 2B(x,v)Q(v)^{-1}v \in L$ for $x \in L$ implies $B(L,v)Q(v)^{-1} \in R$, and then Proposition 5.2.2 implies $L = Rv \perp v^\perp$. Thus $Q(v) \in R^\times$ and hence $\theta(O^+(L)) \subset R^\times(F^\times)^2$. The converse inclusion follows from the previous proposition. □

5.6 Local densities

In this section, we introduce the notion of local density, which is an important invariant of representations of quadratic forms over R. By Siegel's theorem, which will be proved in the next chapter, it turns out to be important for representations of quadratic forms over $\mathbb{Z}$.

First, we show the following fundamental lemma.

Lemma 5.6.1. *Let S and T be regular symmetric matrices over R of degree s and t respectively. Put*

$$E_t(R) := \{B = {}^tB \in M_t(R) \mid \text{ all diagonals of } B \in (2)\},$$

and take an integer h such that $p^h T^{-1} \in E_t(R)$. For $G \in M_{s,t}(R)$ and integers $r, e \geq 0$, we put

$$A_{p^r}(T, S; G, p^e) := \{X \in M_{s,t}(R) \bmod p^r \mid S[X] \equiv T \bmod p^r E_t(R), X \equiv G \bmod p^e\}.$$

If $r \geq h + \max(e, 1)$, then

$$(p^{r+1})^{t(t+1)/2-st} \sharp A_{p^{r+1}}(T, S; G, p^e) = (p^r)^{t(t+1)/2-st} \sharp A_{p^r}(T, S; G, p^e)$$

holds.

Proof. Since

$$S[x + p^r y] = S[x] + p^r({}^txSy + {}^tySx) + p^{2r} S[y]$$

for $x, y \in M_{s,t}(R)$, the set $A_{p^r}(T, S; G, p^e)$ is well-defined if $r \geq \max(e, 1)$. Since T is integral, the integer h is non-negative and so the sets $A_{p^r}, A_{p^{r+1}}$ are well-defined for $r \geq h + \max(e, 1)$. Let $\{T_i\}_{i=1}^a$ be a complete set of representatives of symmetric matrices $T' \bmod p^{r+1} E_t(R)$ such that

$$T' \equiv T \bmod p^r E_t(R).$$

The number a of representations is equal to $[E_t(R) : pE_t(R)] = p^{t(t+1)/2}$. We define the mapping ϕ from the disjoint union of $A_{p^{r+1}}(T_i, S; G, p^e)$ to $A_{p^r}(T, S; G, p^e)$ by

$$\phi(X) = X \bmod p^r.$$

This is well-defined. For $X \in A_{p^r}(T, S; G, p^e)$ and $Y \in M_{s,t}(R)$, we put $X' = X + p^r Y$; then

$$S[X'] = S[X] + p^r({}^tXSY + {}^tYSX) + p^{2r} S[Y] \equiv S[X] \equiv T \bmod p^r E_t(R)$$

and hence $S[X'] \equiv T_j \bmod p^{r+1}E_t(R)$ for some j because of the definition of the set $\{T_i\}$. Thus ϕ is surjective and the number of inverse images is p^{st} and so

$$p^{st}\sharp A_{p^r}(T,S;G,p^e) = \sum \sharp A_{p^{r+1}}(T_i,S;G,p^e).$$

Now let us show

$$\sharp A_{p^{r+1}}(T_i,S;G,p^e) = \sharp A_{p^{r+1}}(T,S;G,p^e).$$

There exists $U_i \in GL_t(R)$ such that $T_i = T[U_i]$ and $U_i \equiv 1 \bmod p^r T^{-1}M_t(R)$, applying Corollary 5.4.4 to $A := T$, $B := T_i$ and $h := r-1$. Since $r \geq h+e$, $p^rT^{-1} \equiv 0 \bmod p^e$ holds and hence $U_i \equiv 1 \bmod p^e$. Thus we have $\sharp A_{p^{r+1}}(T_i,S;G,p^e) = \sharp A_{p^{r+1}}(T,S;G,p^e)$. Hence we have

$$p^{st}\sharp A_{p^r}(T,S;G,p^e)=\sum_i \sharp A_{p^{r+1}}(T_i,S;G,p^e)=p^{t(t+1)/2}\sharp A_{p^{r+1}}(T,S;G,p^e),$$

which implies the assertion. □

Lemma 5.6.2. *Let S,T,G and $A_{p^r}(T,S;G,p^e)$ be those in Lemma 5.6.1. For a non-negative integer a, we have*

$$\sharp A_{p^{r+a}}(p^aT,p^aS;G,p^e) = p^{ast}\sharp A_{p^r}(T,S;G,p^e) \quad \text{for } r \geq h+\max(e,1).$$

Proof. It is easy to see

$$\begin{aligned}
&\sharp A_{p^{r+a}}(p^aT,p^aS;G,p^e)\\
&\quad=\sharp\{X \in M_{s,t}(R) \bmod p^{r+a}M_{s,t}(R) \mid p^aS[X] \equiv p^aT \bmod p^{r+a}E_t(R),\\
&\qquad\qquad X \equiv G \bmod p^e\}\\
&\quad=\sharp\{X \in M_{s,t}(R) \bmod p^{r+a}M_{s,t}(R) \mid S[X] \equiv T \bmod p^rE_t(R),\\
&\qquad\qquad X \equiv G \bmod p^e\}\\
&\quad=p^{ast}\sharp A_{p^r}(T,S;G,p^e).
\end{aligned}$$

□

Lemma 5.6.3. *Let S and T be regular symmetric matrices over F of degree s and t respectively, and $G \in M_{s,t}(R)$. Put*

$$\begin{aligned}
&B_{p^r}(T,S;G,p^e)\\
&\quad:= \{X \in M_{s,t}(R) \bmod p^rS^{-1}M_{s,t}(R) \mid S[X] \equiv T \bmod p^rE_t(R),\\
&\qquad\qquad X \equiv G \bmod p^e\}.
\end{aligned}$$

Let h' be an integer such that $p^{h'}S^{-1} \in E_s(R)$. Then $B_{p^r}(T,S;G,p^e)$ is well-defined for $r \geq h'+e$. For an integer a and a sufficiently large r, we have

$$\sharp B_{p^{r+a}}(p^aT,p^aS;G,p^e) = \sharp B_{p^r}(T,S;G,p^e).$$

Moreover if S and T are integral and $r \geq h'+\max(e,1)$, then we have

$$\sharp A_{p^r}(T,S;G,p^e) = p^{t\,\mathrm{ord}_p(\det S)}\sharp B_{p^r}(T,S;G,p^e).$$

Proof. For $X, Y \in M_{s,t}(R)$, it is easy to see

$$\begin{aligned} S[X+p^rS^{-1}Y] &= S[X]+p^r({}^tXY+{}^tYX)+p^{2r}S^{-1}[Y] \\ &\equiv S[X] \bmod p^rE_t(R) \text{ for } r \geq h', \end{aligned}$$

and if $r \geq h'+e$, then

$$p^rS^{-1}M_{s,t}(R) \subset p^e(p^{h'}S^{-1})M_{s,t}(R) \subset p^eM_{s,t}(R).$$

Hence for $r \geq h'+e$, $B_{p^r}(T,S;G,p^e)$ is well-defined. For an integer a and a sufficiently large integer r, we have

$$\begin{aligned} &\sharp B_{p^{r+a}}(p^aT,p^aS;G,p^e) \\ &\quad =\sharp\{X \in M_{s,t}(R) \bmod p^rS^{-1}M_{s,t}(R) \mid p^aS[X] \equiv p^aT \bmod p^{r+a}E_t(R), \\ &\qquad\qquad X \equiv G \bmod p^e\} \\ &\quad =\sharp B_{p^r}(T,S;G,p^e). \end{aligned}$$

Suppose that S and T are integral and $r \geq h'+\max(e,1)$; then

$$\sharp A_{p^r}(T,S;G,p^e) = \sharp B_{p^r}(T,S;G,p^e)[S^{-1}M_{s,t}(R) : M_{s,t}(R)]$$

is easy to derive. It gives the required equation. □

Lemma 5.6.4. *Let S and T be regular symmetric matrices over F of degree s and t respectively, and $G \in M_{s,t}(R)$. Then for a sufficiently large r, $(p^r)^{t(t+1)/2-st}\sharp B_{p^r}(T,S;G,p^e)$ is independent of r.*

Proof. Let a be an integer such that p^aT and p^aS are integral; then for a sufficiently large r, the previous lemma yields

$$\begin{aligned} \sharp B_{p^r}(T,S;G,p^e) &= \sharp B_{p^{r+a}}(p^aT,p^aS;G,p^e) \\ &= p^{-t\,\mathrm{ord}_p(\det(p^aS))}\sharp A_{p^{r+a}}(p^aT,p^aS;G,p^e). \end{aligned}$$

Lemma 5.6.1 implies the assertion. □

Before the definition of our local density, we give one more lemma, which connects ours with Siegel's original.

Lemma 5.6.5. *Let S and T be regular symmetric matrices over R of degree s and t respectively and $G \in M_{s,t}(R)$. Put*

$$A'_{p^r}(T, S; G, p^e) := \{X \in M_{s,t}(R) \bmod p^r M_{s,t}(R) \mid S[X] \equiv T \bmod p^r,\\ X \equiv G \bmod p^e\}.$$

Then we have

$$\sharp A'_{p^r}(T, S; G, p^e) = 2^{t\delta_{2p}} \sharp A_{p^r}(T, S; G, p^e)$$

for a sufficiently large r where δ is Kronecker's delta function.

Proof. When $p \neq 2$, $E_t(R)$ is simply $\{A = {}^tA \in M_t(R)\}$. Hence $A'_{p^r} = A_{p^r}$ holds. So we assume $p = 2$. Let $\{T_i\}$ be a complete set of representatives of integral symmetric matrices $T' \bmod 2^{r+1}$ such that $T' \equiv T \bmod 2^r E_t(R)$. The number of them is

$$[2^r E_t(R) : \{x \in 2^{r+1} M_t(R) \mid x = {}^tx\}] = 2^{t(t-1)/2}.$$

Define a mapping

$$\phi : \sqcup_i A'_{2^{r+1}}(T_i, S; G, 2^e) \to A_{2^r}(T, S; G, 2^e)$$

by $\phi(X) \equiv X \bmod 2^r$. For $X \in A_{2^r}(T, S; G, 2^e)$ and $Y \in M_{s,t}(R)$, we have

$$S[X + 2^r Y] = S[X] + 2^r({}^tXSY + {}^tYSX) + 2^{2r} S[Y]\\ \equiv S[X] \bmod 2^r E_t(R) \equiv T \bmod 2^r E_t(R).$$

Thus $X + 2^r Y$ is in $A'_{2^{r+1}}(T_i, S; G, 2^e)$ for some i and so ϕ is surjective and the number of inverse images is 2^{ts}. Using Corollary 5.4.4 as in the proof of Lemma 5.6.1, we have

$$\sharp A'_{2^{r+1}}(T_i, S; G, 2^e) = \sharp A'_{2^{r+1}}(T, S; G, 2^e)$$

for a sufficiently large integer r. Thus

$$2^{t(t-1)/2} \sharp A'_{2^{r+1}}(T, S; G, 2^e) = \sum_i \sharp A'_{2^{r+1}}(T_i, S; G, 2^e)\\ = 2^{st} \sharp A_{2^r}(T, S; G, 2^e)\\ = 2^{st + t(t+1)/2 - st} \sharp A_{2^{r+1}}(T, S; G, 2^e)$$

holds by Lemma 5.6.1. This completes the proof. □

Let us define the local density. Let M and N be regular quadratic lattices over R with $\operatorname{rank} M = m$ and $\operatorname{rank} N = n$ and $\{m_i\}$, $\{n_i\}$ be bases of M, N respectively. Let g be a homomorphism from M to N; then

$$(1)\quad \begin{cases} M = R[m_1, \cdots, m_m] \quad \text{and} \quad N = R[n_1, \cdots, n_n] \\ (g(m_1), \cdots, g(m_m)) = (n_1, \cdots, n_n)G \ \text{ for some } G \in M_{n,m}(R). \end{cases}$$

Putting

$$(2)\qquad T = (B(m_i, m_j)) \quad \text{and} \quad S = (B(n_i, n_j)),$$

the dual basis of N is given by $(n_1, \cdots, n_n)S^{-1}$. Then the set

$$\begin{aligned} &\tilde{B}_{p^r}(M, N; g, p^e) \\ &\quad := \{\sigma : M \to N/p^r N^\sharp \mid Q(\sigma(x)) \equiv Q(x) \bmod 2p^r, \\ &\qquad\qquad \sigma(x) \equiv g(x) \bmod p^e N \text{ for } x \in M\} \end{aligned}$$

is canonically identified with $B_{p^r}(T, S; G, p^e)$ through matrix representation. Hence by Lemma 5.6.4, the following definition of the *local density* is well-defined.

$$\begin{aligned} \beta_p(M, N; g, p^e) :=& p^{m \operatorname{ord}_p \mathrm{d} N} \lim_{r \to \infty} (p^r)^{m(m+1)/2 - mn} \sharp \tilde{B}_{p^r}(M, N; g, p^e) \\ =& p^{m \operatorname{ord}_p \mathrm{d} N} \lim_{r \to \infty} (p^r)^{m(m+1)/2 - mn} \sharp B_{p^r}(T, S; G, p^e) \\ =& \lim_{r \to \infty} (p^r)^{m(m+1)/2 - mn} \sharp A_{p^r}(T, S; G, p^e), \end{aligned}$$

where in the last row S and T are supposed to be integral and we used Lemma 5.6.3. When $e = 0$, the additional condition $\sigma(x) \equiv g(x) \bmod p^e N$ is trivially satisfied and we put

$$\begin{aligned} A_{p^r}(T, S) &:= A_{p^r}(T, S; G, p^0), \\ A'_{p^r}(T, S) &:= A'_{p^r}(T, S; G, p^0), \\ \beta_p(M, N) &:= \beta_p(M, N; g, p^0). \end{aligned}$$

Through matrix representation,

$$A_{p^r}(M, N) := \{\sigma : M \to N/p^r N \mid Q(\sigma(x)) \equiv Q(x) \bmod 2p^r \text{ for } x \in M\}$$

is identified with $A_{p^r}(T, S)$ if $S(M)$, $S(N) \subset R$. Siegel's original definition of local density is

$$\begin{aligned} \alpha_p(S, T) &:= 2^{-\delta_{m,n}} \lim_{r \to \infty} (p^r)^{m(m+1)/2 - mn} A'_{p^r}(T, S) \\ &:= 2^{m\delta_{2,p} - \delta_{m,n}} \beta_p(M, N) \\ &:= \alpha_p(M, N). \end{aligned}$$

Note that the positions of S and T are reversed. When we adopt the matrix notation $S[X] = T$, it is natural that S is on the left, but when we use the lattice notation $M \to N$, it is natural that S is on the right.

Proposition 5.6.1. *Let M, N, g, T, S, G be those as above. Then*

(i)

$$\begin{aligned}
&\beta_p(M, N; g, p^e) \\
&= (p^r)^{m(m+1)/2-mn} \sharp A_{p^r}(T, S; G, p^e) \\
&= (p^r)^{m(m+1)/2-mn} \sharp\{\sigma : M \to N/p^r N \mid Q(\sigma(x)) \equiv Q(x) \bmod 2p^r, \\
&\qquad \sigma(x) \equiv g(x) \bmod p^e N \text{ for } x \in M\}
\end{aligned}$$

if S and T are integral and $r \geq h + \max(e, 1)$ for an integer h satisfying $p^h Q(M^\sharp) \subset (2)$,

(ii) $$\beta_p(M, N; g, p^e) = 2^{-m\delta_{2,p}} (p^r)^{m(m+1)2-mn} \sharp A'_{p^r}(T, S; G, p^e)$$

if S and T are integral and r is sufficiently large,

(iii) $$\beta_p(M^{(p^a)}, N^{(p^a)}; g, p^e) = p^{am(m+1)/2} \beta_p(M, N; g, p^e)$$

if S and T are integral and $a \geq 0$. where $M^{(p^a)} \cong \langle p^a T \rangle$ and $N^{(p^a)} \cong \langle p^a S \rangle$ are scalings of M and N by p^a.

Proof. The assertions (i), (ii) and (iii) follows from Lemma 5.6.1, 5.6.5 and 5.6.2 respectively. □

Exercise.

Show $\beta_p(M, N; g, p^e) \neq 0$ if and only if there exists an isometry σ from M to N such that $\sigma(x) \equiv g(x) \bmod p^e N$ for $x \in M$. In particular, $\operatorname{rank} M > \operatorname{rank} N$ implies $\beta_p(M, N; g, p^e) = 0$.

It is not easy to evaluate local densities. But there are several cases in which we know the explicit formula. For example, if M is unimodular with $\mathrm{n}(M) \subset (2)$, and N is a regular quadratic lattice with $\mathrm{n}(N) \subset (2)$, then Proposition 5.6.1 implies

$$\begin{aligned}
\beta_p(M, N) &= p^{m(m+1)/2-mn} \sharp A_p(T, S) \\
&= p^{m(m+1)/2-mn} \sharp\{ \text{ isometries from } \tilde{M} \text{ to } \tilde{N}\},
\end{aligned}$$

where $\tilde{M}$ and $\tilde{N}$ are regular quadratic spaces over $R/(p)$ defined by $(M/pM, \frac{1}{2}Q)$ and $(N/pN, \frac{1}{2}Q)$ respectively. If, moreover N is unimodular with $\mathrm{n}(N) \subset (2)$, then the number of isometries is given in Theorem 1.3.2.

For regular quadratic lattices M, N over R and a homomorphism g from M to N, we still keep (1) and (2). We define an integer h' such that $p^{h'} S^{-1} \in E_n(R)$, i.e.

$$p^{h'} Q(N^\sharp) \subset (2),$$

which implies $p^{h'}N^\sharp \subset N$ as in the proof of Corollary 5.4.2. Applying Theorem 5.4.2 to $L := M$, $G := N^\sharp$, $k \geq h' + \max(e,1)$,

$$(p^{-k})^{mn-m(m+1)/2} \sharp\{\sigma : M \to N/p^k N^\sharp \mid *(k)\}$$

is independent of k, where the condition $*(k)$ is:

$$*(k) : Q(\sigma(x)) \equiv Q(x) \bmod 2p^k,\ \sigma(x) \equiv g(x) \bmod p^e N \text{ for } x \in M$$

and σ induces an injective mapping from M/pM to N/pN.
Hence we can define the *primitive local density* $d_p(M, N; g, p^e)$ by

$$\lim_{k\to\infty} (p^k)^{m(m+1)/2-mn} \sharp\{\sigma : M \to N/p^k N \mid *(k)\}$$
$$= (p^{r_0})^{m(m+1)/2-mn} p^{m \operatorname{ord}_p \mathrm{d} N} \sharp\{\sigma : M \to N/p^{r_0} N^\sharp \mid *(r_0)\}$$

for $r_0 := h' + \max(e, 1)$.
When $e = 0$, we denote $d_p(M, N; g, p^e)$ by $d_p(M, N)$.

If N is unimodular with $\mathrm{n}(N) \subset (2)$, then $d_p(M, N)$ is the product of $p^{m(m+1)/2-mn}$ with the number of isometries from M/pM to N/pN, where quadratic forms are defined by $\frac{1}{2}Q$.

If $\mathrm{s}(M), \mathrm{s}(N) \subset R$, then we have, for integers $e \geq 0$, $r \geq h' + \max(e, 1)$,

$$\sharp\{X \in A_{p^r}(T, S; G, p^e) \mid X : \text{ primitive }\}$$
$$= \sharp\{\sigma : M \to N/p^r N \mid \quad *(r) \quad \}$$
$$= [N^\sharp : N]^m \sharp\{\sigma : M \to N/p^r N^\sharp \mid *(r)\}.$$

If the three sets are well-defined, then the equalities are clear. The first is well-defined if $r \geq \max(e, 1)$ as at the beginning of the proof of Lemma 5.6.1. The second is the translation of the first to lattices. The third one is well-defined if $p^r N^\sharp \subset pN$, $p^r N^\sharp \subset p^e N$ and $r \geq h'$, which, in order, correspond to the injectivity, $\sigma(x) \equiv g(x) \bmod p^e N$ and the dependence of $Q(\sigma(x))$ only on $\sigma \bmod p^r N^\sharp$, as at the beginning of the proof of Lemma 5.6.3. Then the three sets are well-defined and the equalities hold if $r \geq h' + \max(e, 1)$. Thus we have

$$d_p(M, N; g, p^e) =$$
$$\lim_{r\to\infty} (p^r)^{m(m+1)/2-mn} \sharp\{X \in A_{p^r}(T, S; G, p^e) \mid X : \text{ primitive}\}.$$

We note that r_0 in the above is independent of M and hence there is a positive number c_N dependent only on N such that $d_p(M, N; g, p^e) > c_N$ if it is not zero, and that $d_p(M, N; g, p^e) = 0$ unless there exists an isometry $\sigma : M \to N$ such that $\sigma(M)$ is primitive in N and $\sigma(x) \equiv g(x) \bmod p^e N$ for $x \in M$.

Now we give two reduction formulas. The following is due to Siegel.

Theorem 5.6.1. *Let M and N be regular quadratic lattices over R, then we have*

$$\beta_p(M,N) = \sum_{FM \supset M' \supset M} [M' : M]^{m+1-n} d_p(M', N),$$

where $m = \operatorname{rank} M \leq n = \operatorname{rank} N$. The summation on M' is finite.

Proof. We still keep (1) and (2). The property similar to (iii) in Proposition 5.6.1 holds for primitive local density since

$$\begin{aligned}
&\sharp\{\sigma : M^{(p^a)} \to N^{(p^a)}/p^r N^{(p^a)} \mid *(r)\} \\
&= \sharp\{\sigma : M \to N/p^r N \mid *(r-a)\} \\
&= p^{amn} \sharp\{\sigma : M \to N/p^{r-a} N \mid *(r-a)\}
\end{aligned}$$

and hence we may suppose that S and T are integral by scaling. Suppose

$$r > \operatorname{ord}_p \det T \quad \text{and} \quad S[X] \equiv T \bmod p^r E_m(R) \quad \text{for} \quad X \in M_{n,m}(R).$$

Then we will show that $\operatorname{rank} X = m$ and writing $X = X_1 K$ by virtue of elementary divisor theory, where $X_1 \in M_{n,m}(R)$ is primitive and $K \in M_m(R)$, $GL_m(R)K$ is uniquely determined by $X \bmod p^r$ and

$$(3) \qquad 2 \operatorname{ord}_p \det K \leq \operatorname{ord}_p \det T < r.$$

Now $S[X] \equiv T \bmod p^r E_m(R)$ implies $\det S[X] \equiv \det T \bmod p^r$, which implies $\det S[X] \not\equiv 0 \bmod p^r$ by $r > \operatorname{ord}_p \det T$, and hence $\operatorname{rank} X = m$. Then

$$\det S[X] = \det S[X_1](\det K)^2 \equiv \det T \bmod p^r$$

implies (3) and

$$p^r K^{-1} \equiv 0 \bmod p.$$

Moreover (3) and $S[X_1] \equiv T[K^{-1}] \bmod p^r E_m(R)[K^{-1}]$ imply that $T[K^{-1}]$ is integral. Now suppose

$$X' \equiv X \bmod p^r$$

and $X' = X_1' K'$ is a similar decomposition. Then

$$\begin{aligned}
X_1' K' K^{-1} &= X' K^{-1} = X K^{-1} + (X' - X) K^{-1} \\
&= X_1 + (X' - X) K^{-1} \equiv X_1 \bmod p
\end{aligned}$$

implies that $X_1' K' K^{-1}$ is integral and hence $K' K^{-1}$ is integral since X_1' is primitive. Similarly $K(K')^{-1} = (K' K^{-1})^{-1}$ is integral and thus

$K'K^{-1} \in GL_m(R)$ follows. Thus $GL_m(R)K$ has been uniquely determined by $X \bmod p^r$.

Next, let $\{J_i\}_{i=1}^h$ be a complete set of representatives of

$$\{p^r J[K^{-1}] \mid J \in E_m(R)\} \bmod p^r E_m(R)$$

for a regular integral matrix K satisfying (3). Then $J_i \in E_m(R)$ is clear. Put

$$M(T,S;K) := \{X \in M_{n,m}(R) \bmod p^r M_{n,m}(R)K \mid S[X] \equiv T \bmod p^r E_m(R), \quad XK^{-1} : \text{ integral and primitive}\},$$

and

$$M_i(T,S;K) := \{X \in M_{n,m}(R) \bmod p^r M_{n,m}(R) \mid S[X] \equiv T[K^{-1}] + J_i \bmod p^r E_m(R), \ X : \text{ integral and primitive}\},$$

and define a mapping ϕ from $M(T,S;K)$ to $\sqcup_i M_i(T,S;K)$ by $\phi(X) = XK^{-1}$. We will show that ϕ is bijective. Both sets are clearly well-defined and ϕ is well-defined. If $X \in M_i(T,S;K)$, then

$$S[XK] \equiv T + J_i[K] \bmod p^r E_m(R) \equiv T \bmod p^r E_m(R).$$

Thus $XK \in M(T,S;K)$ and $\phi(XK) = X$, which yields that ϕ is surjective. The injectivity is obvious. Thus we have

$$\sharp M(T,S;K) = \sum_i \sharp M_i(T,S;K).$$

Next, we show that $\sharp M_i(T,S;K)$ is independent of i. Put

$$f = \operatorname{ord}_p \det K.$$

Then writing $J_i = p^r J[K^{-1}]$ for $J \in E_m(R)$,

$$J_i = p^{r-2f} J[p^f K^{-1}] \equiv 0 \bmod p^{r-2f} E_m(R).$$

Applying Corollary 5.4.4 to $A := T[K^{-1}]$ and $B := T[K^{-1}] + J_i$ for a large integer r there is a matrix $U \in GL_m(R)$ such that

$$T[K^{-1}] + J_i = T[K^{-1}][U].$$

Hence we have

$$\sharp M_i(T,S;K) = \sharp\{X \in A_{p^r}(T[K^{-1}],S) \mid X: \text{ primitive }\}.$$

Since

$$\sharp M(T,S;K) = p^{nf}\sharp\{X \in A_{p^r}(T,S) \mid XK^{-1} \text{ integral and primitive}\},$$

we have

$$\sharp\{X \in A_{p^r}(T,S) \mid XK^{-1} \text{ integral and primitive}\} = p^{-nf}\sharp M(T,S;K)$$
$$= p^{-nf}\sum_i \sharp M_i(T,S;K) = p^{-nf}h\sharp\{X \in A_{p^r}(T[K^{-1}],S)|X \text{ primitive}\}.$$

Let us evaluate the number h. Put $K = UK_1V$ $(U,V \in GL_m(R))$ and $K_1 = \mathrm{diag}(p^{f_1},\cdots,p^{f_m})$; then

$$\begin{aligned} h &= \sharp(\{p^r J[K^{-1}] \mid J \in E_m(R)) \bmod p^r E_m(R)\} \\ &= [E_m(R)[K_1^{-1}] : E_m(R)] = p^{(m+1)f} \end{aligned}$$

since $f = \sum f_i$. Thus we have obtained

$$\begin{aligned} &\sharp\{X \in A_{p^r}(T,S) \mid XK^{-1} \text{ integral and primitive}\} \\ &\quad = p^{(m-n+1)\,\mathrm{ord}_p \det K}\sharp\{X \in A_{p^r}(T[K^{-1}],S) \mid X \text{ primitive}\}. \end{aligned}$$

Therefore we have, for a sufficiently large integer r

$$\begin{aligned} &\sharp A_{p^r}(T,S) \\ &\quad = \sum_K p^{(m-n+1)\,\mathrm{ord}_p \det K}\sharp\{X \in A_{p^r}(T[K^{-1}],S) \mid X \text{ primitive}\}, \end{aligned}$$

where K ranges over regular matrices in $GL_m(R)\backslash M_m(R)$ such that $T[K^{-1}]$ is integral. Considering the correspondence between K and the lattice M' defined by $(m_1,\cdots,m_m)K^{-1}$, we have $[M' : M] = p^{\mathrm{ord}_p \det K}$. Thus we have obtained the equation in the assertion. If $d_p(M',N) \neq 0$, then there is an isometry $\sigma : M' \hookrightarrow N$. Hence $\mathrm{d}\,M' \subset (\mathrm{s}(N))^m$ holds and $\mathrm{d}\,M = [M' : M]^2\,\mathrm{d}\,M'$ implies

$$0 \le \mathrm{ord}_p[M' : M]^2 = \mathrm{ord}_p(\mathrm{d}\,M/\mathrm{d}\,M') \le \mathrm{ord}_p(\mathrm{d}\,M(\mathrm{s}(N))^{-m}).$$

Thus the summation on M' is finite. □

Now let us give another reduction formula.

Let M, N be regular quadratic lattices over R with $\operatorname{rank} M = m$ and $\operatorname{rank} N = n$ respectively and $S(M)$, $S(M) \subset R$; let K be a submodule of N isometric to M. For a basis $\{m_i\}$ of M, we put

$$C_{p^r}(M, N; K) := \{\sigma : M \to N/p^r N \mid Q(\sigma(x)) \equiv Q(x) \bmod 2p^r \text{ and } (**)\ \}$$

where the condition $(**)$ is

(**): there exists an isometry $\eta \in O(N)$ such that $\eta(K) = R[\sigma'(m_1), \cdots, \sigma'(m_m)]$, where $\sigma'(m_i)$ is an element of N which represents $\sigma(m_i) \bmod p^r N$.

Let us verify that the condition $(**)$ is independent of the choice of $\sigma'(m_i)$ for a sufficiently large r, which means that $C_{p^r}(M, N; K)$ is well-defined for a sufficiently large integer r. To do it, we have only to show the following:

If $v_i \in N$ $(i = 1, \cdots, m)$ and $\eta : K \cong K' := R[v_1, \cdots, v_m]$, then for any given $x_i \in N$, there is an isometry

$$\eta' : K \cong K'' := R[v_1 + p^r x_1, \cdots, v_m + p^r x_m]$$

with $\eta' \in O(N)$ when r is sufficiently large. This is simply Corollary 5.4.5. Thus $C_{p^r}(M, N; K)$ is well-defined.

Let us give $C_{p^r}(M, N; K)$ in terms of matrices. Let $\{k_i\}$ and $\{n_i\}$ be bases of K and N respectively and put

$$\begin{aligned}
(k_1, \cdots, k_m) &= (n_1, \cdots, n_n)Y \quad (\, Y \in M_{n,m}(R)\ ),\\
(\sigma'(m_1), \cdots, \sigma'(m_m)) &= (n_1, \cdots, n_n)X \quad (\, X \in M_{n,m}(R)\ ),\\
(\eta(k_1), \cdots, \eta(k_m)) &= (\sigma'(m_1), \cdots, \sigma'(m_m))G\\
&= (n_1, \cdots, n_n)XG \quad \text{for} \quad G \in GL_m(R)\\
(\eta(n_1), \cdots, \eta(n_n)) &= (n_1, \cdots, n_n)A \quad (\, A \in GL_n(R)\ )
\end{aligned}$$

and finally

$$S := (B(n_i, n_j)) \quad \text{and} \quad T := (B(m_i, m_j)).$$

Then $Q(\sigma(x)) \equiv Q(x) \bmod 2p^r$ is equivalent to $S[X] \equiv T \bmod p^r E_m(R)$, $\eta \in O(N)$ means $S[A] = S$, and lastly $\eta(K) = R[\sigma'(m_1), \cdots, \sigma'(m_m)]$, i.e.

$$(\eta(k_1), \cdots, \eta(k_m)) = (\sigma'(m_1), \cdots, \sigma'(m_m))G$$

implies

$$\begin{aligned}
(n_1, \cdots, n_n)AY &= (\eta(n_1), \cdots, \eta(n_n))Y\\
&= (\eta(k_1), \cdots, \eta(k_m)) = (n_1, \cdots, n_n)XG
\end{aligned}$$

and hence $AY = XG$. Thus $C_{p^r}(M, N; K)$ is, in terms of matrices,

$$C_{p^r}(T, S; Y) = \{X \in M_{n,m}(R) \bmod p^r M_{n,m}(R) \mid S[X] \equiv T \bmod p^r E_m(R),\\ S[A] = S, AY = XG \text{ for some } A \in GL_n(R) \text{ and } G \in GL_m(R)\}.$$

Note that this set is well-defined when r is sufficiently large by the above lattice-theoretic consideration, that is the condition on X is independent of the choice of $X \bmod p^r M_{n,m}(R)$.

Lemma 5.6.6. *Keeping the above,*

$$\beta_p(M, N; K) := (p^r)^{m(m+1)/2-mn} \sharp C_{p^r}(M, N; K)$$

is independent of r if r is sufficiently large.

Proof. As above, let $\{T_i\}$ be a complete set of representatives of symmetric matrices $T' \bmod p^{r+1} E_m(R)$ such that $T' \equiv T \bmod p^r E_m(R)$. The number of representatives is $p^{m(m+1)/2}$. We define a mapping ϕ from $\sqcup_i C_{p^{r+1}}(T_i, S; Y)$ to $C_{p^r}(T, S; Y)$ by $\phi(X) = X \bmod p^r$. Obviously ϕ is well-defined and surjective. Let $X \in C_{p^r}(T, S; Y)$ and $Y \in M_{n,m}(R)$; regarding $X \in M_{n,m}(R)$, $X + p^r Y$ gives the same element of $C_{p^r}(T, S; Y)$ and $X + p^r Y$ is in $C_{p^{r+1}}(T_i, S; Y)$ for some i. Thus the number of the inverse images is p^{mn}. By virtue of Corollary 5.4.4, there exists a unimodular matrix $U_i \in GL_m(R)$ such that $T_i = T[U_i]$. Hence

$$\sharp C_{p^{r+1}}(T_i, S; Y) = \sharp C_{p^{r+1}}(T, S; Y),$$

and so

$$\sharp C_{p^{r+1}}(T, S; Y) = p^{nm-m(m+1)/2} \sharp C_{p^r}(T, S; Y).$$

□

Lemma 5.6.7. *Suppose $M = M_1 \perp M_2$, $m_i := \operatorname{rank} M_i > 0$. Then we have*

$$\sharp A_{p^r}(M, N) = \\ \sum_{i=1}^{s} \sharp C_{p^r}(M_1, N; K_i) \\ \times \sharp\{\sigma_2 \in A_{p^r}(M_2, N) \mid B(K_i, \sigma_2(M_2)) \equiv 0 \bmod p^r\}$$

for a sufficiently large r, where the set $\{K_i\}$ is a complete set of submodules of N isometric to M_1 which are not transformed into each other by *isometries of* N. *If* $M_2 = 0$, *then the number of* σ_2 *in the above equation should be 1.*

Proof. First, we define a mapping ϕ from $A_{p^r}(M, N)$ to $O(N)$ as follows: For $\sigma_1 \in A_{p^r}(M_1, N)$ we consider that the image of σ_1 is in N, taking any fixed representatives. Then by virtue of Corollary 5.4.2, M_1 is isometric to $\sigma_1(M_1)$ when r is sufficiently large. Hence there exists $\alpha \in O(N)$ such that $\alpha(\sigma_1(M_1)) = K_i$ for some i. We fix any such α for σ_1 and put $\phi(\sigma_1) = \alpha \ (\in O(N))$.

Let $\sigma \in A_{p^r}(M, N)$ and put

$$\sigma_1 = \sigma|_{M_1} \in A_{p^r}(M_1, N) \quad \text{and} \quad \sigma_2 = \phi(\sigma_1)(\sigma|_{M_2}) \in A_{p^r}(M_2, N).$$

Now $B(M_1, M_2) = 0$ implies $B(\sigma(M_1), \sigma(M_2)) \equiv 0 \bmod p^r$ and hence $B(\phi(\sigma_1)\sigma(M_1), \phi(\sigma_1)\sigma(M_2)) \equiv 0 \bmod p^r$. Noting $\phi(\sigma|_{M_1})\sigma(M_1) = K_i$ for some i, we have $B(K_i, \sigma_2(M_2)) \equiv 0 \bmod p^r$. Now we define a mapping

$$\begin{aligned} \eta : A_{p^r}(M, N) \to \sqcup_{i=1}^{s} & C_{p^r}(M_1, N; K_i) \\ & \times \{\sigma_2 \in A_{p^r}(M_2, N) \,|B(K_i, \sigma_2(M_2)) \equiv 0 \bmod p^r\} \end{aligned}$$

by $\eta(\sigma) = (\sigma_1, \sigma_2)$ as above. The injectivity of η is obvious. For a given (σ_1, σ_2), we define $\sigma \in A_{p^r}(M, N)$ by $\sigma|_{M_1} = \sigma_1$ and $\sigma|_{M_2} = \phi(\sigma_1)^{-1}\sigma_2$. Then $\eta(\sigma) = (\sigma_1, \sigma_2)$ is clear. □

Lemma 5.6.8.

$$\begin{aligned} &\sharp\{\sigma_2 \in A_{p^r}(M_2, N) \mid B(K_i, \sigma_2(M_2)) \equiv 0 \bmod p^r\} \\ &\quad = ([K_i^\sharp : K_i]/[N : K_i \perp K_i^\perp])^{m_2} \sharp A_{p^r}(M_2, K_i^\perp), \end{aligned}$$

where r *is sufficiently large.*

Proof. First we show

$$\{x \in N \mid B(K_i, x) \equiv 0 \bmod p^r\} = p^r K_i^\sharp \perp K_i^\perp.$$

The left-hand side obviously contains the right-hand side. Let $x \in N$ satisfy $B(K_i, x) \equiv 0 \bmod p^r$. Write $x = x_1 + x_2$ ($x_1 \in FK_i$ and $x_2 \in FK_i^\perp$). Then $B(K_i, x_1) = B(K_i, x) \equiv 0 \bmod p^r$ implies $x_1 \in p^r K_i^\sharp \subset K_i \subset N$ and hence $x_2 = x - x_1 \in N \cap FK_i^\perp = K_i^\perp$. Thus the above equality has been proved and we have

$$\begin{aligned} &\sharp\{\sigma_2 \in A_{p^r}(M_2, N) \mid B(K_i, \sigma_2(M_2)) \equiv 0 \bmod p^r\} \\ &\quad = \sharp\{\sigma_2 \in A_{p^r}(M_2, N) \mid \sigma_2(M_2) \subset p^r K_i^\sharp \perp K_i^\perp\} \\ &\quad = \sharp\{\sigma_2 : M_2 \to p^r K_i^\sharp \perp K_i^\perp \bmod p^r N \mid Q(\sigma_2(x)) \equiv Q(x) \bmod 2p^r \\ &\qquad\quad \text{for } x \in M_2\}. \end{aligned}$$

Here we fix any positive integer a such that $p^a K_i^\sharp \subset K_i$ and then the above is equal to

$$[N : p^a K_i^\sharp \perp K_i^\perp]^{-m_2}\{\sigma_2 : M_2 \to p^r K_i^\sharp \perp K_i^\perp \bmod p^r(p^a K_i^\sharp \perp K_i^\perp) \mid Q(\sigma_2(x)) \equiv Q(x) \bmod 2p^r \text{ for } x \in M_2\}.$$

Also, we write

$$\sigma_2(x) = \gamma_1(x) + \gamma_2(x),$$
$$\gamma_1(x) \in p^r K_i^\sharp \bmod p^{r+a} K_i^\sharp,$$
$$\gamma_2(x) \in K_i^\perp \bmod p^r K_i^\perp.$$

Since $Q(p^r K_i^\sharp) = p^{2r} Q(K_i^\sharp)$, we have for a sufficiently large r,

$$Q(\gamma_1(x)) \in Q(p^r K_i^\sharp) \subset (2p^r)$$

and hence the condition $Q(\sigma_2(x)) \equiv Q(x) \bmod 2p^r$ is equivalent to $Q(\gamma_2(x)) \equiv Q(x) \bmod 2p^r$. Thus the number required is equal to

$$\begin{aligned}&[N : p^a K_i^\sharp \perp K_i^\perp]^{-m_2}[p^r K_i^\sharp : p^{r+a} K_i^\sharp]^{m_2}\\ &\quad \times \sharp\{\gamma_2 : M_2 \to K_i^\perp \bmod p^r K_i^\perp \mid Q(\gamma_2(x)) \equiv Q(x) \bmod 2p^r \text{ for } x \in M_2\}\\ &\quad = [N : p^a K_i^\sharp \perp K_i^\perp]^{-m_2} p^{am_1m_2} \sharp A_{p^r}(M_2, K_i^\perp),\end{aligned}$$

since $\operatorname{rank} K_i = \operatorname{rank} M_1 = m_1$. On the other hand, we have

$$\begin{aligned}&p^{am_1}[N : p^a K_i^\sharp \perp K_i^\perp]^{-1}\\ &\quad = p^{am_1}[N : K_i \perp K_i^\perp]^{-1}[K_i \perp K_i^\perp : p^a K_i^\sharp \perp K_i^\perp]^{-1}\\ &\quad = p^{am_1}[N : K_i \perp K_i^\perp]^{-1}[K_i : p^a K_i^\sharp]^{-1}\\ &\quad = [K_i^\sharp : p^a K_i^\sharp][N : K_i \perp K_i^\perp]^{-1}[K_i : p^a K_i^\sharp]^{-1}\\ &\quad = [K_i^\sharp : K_i][N : K_i \perp K_i^\perp]^{-1},\end{aligned}$$

which completes the proof. □

Summarizing, we have the second reduction

Theorem 5.6.2. *Let $M = M_1 \perp M_2$ and N be regular quadratic lattices over R and $m_i = \operatorname{rank} M_i > 0$. Let $\{K_i\}$ be a complete set of submodules of N isometric to M_1 which are not transformed into each other by isometries of N. Moreover we assume $S(M)$, $S(N) \subset R$. Then we have*

$$\beta_p(M, N) = \sum_i ([M_1^\sharp : M_1]/[N : K_i \perp K_i^\perp])^{m_2} \beta_p(M_1, N; K_i) \beta_p(M_2, K_i^\perp).$$

If $m_2 = 0$, then we put $\beta_p(M_2, K_i^\perp) = 1$.

Proof. Since $K_i \cong M_1$, we have $[K_i^\sharp : K_i] = [M_1^\sharp : M_1]$. For

$$\begin{aligned}m = \operatorname{rank} M = m_1 + m_2,\ & m(m+1)/2 - mn\\ &= (m_1(m_1+1)/2 - m_1 n) + (m_2(m_2+1)/2 - m_2(n - m_1))\end{aligned}$$

holds and then the above lemmas complete the proof. □

Corollary 5.6.1. *In Theorem 5.6.2, assume that all submodules of N isometric to M_1 are transformed into each other by $O(N)$. Then we have*

$$\beta_p(M,N) = ([M_1^\sharp : M_1]/[N : K \perp K^\perp])^{m_2}\beta_p(M_1,N)\beta_p(M_2,K^\perp),$$

where K is a submodule of N isometric to M_1.

Proof. We note that Corollary 5.4.2 implies $M \cong R[\sigma'(m_1),\cdots,\sigma'(m_m)]$ in the definition of $C_{p^r}(M,N;K)$ when r is sufficiently large, and so if the submodules of N isometric to M are mutually transformed by $O(N)$, then $\beta_p(M,N;K) = \beta_p(M,N)$. The corollary is now obvious. □

Using this corollary, we can evaluate the local density $\beta_p(N,N)$ (see [Ko], [W3]).

Theorem 5.6.3. *Let $N = \perp_{j\in\mathbb{Z}} M_j$ be a regular quadratic lattice over R, where M_j is either (p^j)-modular or $\{0\}$, and $M_j \neq \{0\}$ occurs only for finitely many integers j, and write $M_j = N_j^{(p^j)}$ for the scaling of a unimodular lattice N_j by p^j. Put*

$$n_j := \operatorname{rank} N_j = \operatorname{rank} M_j, \quad w := \sum_j jn_j\{\sum_{k>j} n_k + (n_j+1)/2\},$$

$$P(m) := \prod_{i=1}^{m}(1-p^{-2i}) \ (=1 \ \textit{if} \ m=0).$$

With, for a regular quadratic space U over $\mathbb{Z}/p\mathbb{Z}$,

$$\chi(U) := \begin{cases} 0 & \text{if } \dim U \text{ is odd,} \\ 1 & \text{if } U \text{ is a hyperbolic space,} \\ -1 & \text{otherwise,} \end{cases}$$

and for a unimodular lattice M over R with $\mathrm{n}(M) = 2\,\mathrm{s}(M)$, we define a regular quadratic space $\bar{M} := (M/pM, \frac{1}{2}Q)$ over $\mathbb{Z}/p\mathbb{Z}$, and put

$$\chi(M) := \chi(\bar{M}).$$

(I) The case of $p \neq 2$.

$$\beta_p(N,N) = 2^s p^w PE,$$

where

$$s := \sharp\{j \mid M_j \neq \{0\}\},$$
$$P := \prod_j P([n_j/2]),$$
$$E := \prod_{j:M_j\neq 0}(1+\chi(N_j)p^{-n_j/2})^{-1},$$

and $[x]$ denotes the largest integer which does not exceed x.

(II) The case of $p=2$.

$$\beta_2(N,N) = 2^{w-q}PE.$$

We define q, P, E as follows: For a unimodular lattice M over $\mathbb{Z}_2$, we say that M is even if $\mathrm{n}(M) = (2)$, odd otherwise; for the sake of convenience, $\{0\}$ is also called even. With

$$q_j := \begin{cases} 0 & \text{if } N_j \text{ is even,} \\ n_j & \text{if } N_j \text{ is odd and } N_{j+1} \text{ is even,} \\ n_j+1 & \text{if } N_j \text{ and } N_{j+1} \text{ are odd,} \end{cases}$$

we put

$$q := \sum_j q_j.$$

We write for a unimodular lattice M, $M = M(e) \perp M(o)$, where $M(e)$ is even and $M(o)$ is either odd or $\{0\}$ with $\operatorname{rank} M(o) \leq 2$. Then we put

$$P := \prod_j P(\frac{1}{2}\operatorname{rank} N_j(e)).$$

Defining E_j by

$$\begin{cases} \frac{1}{2}(1+\chi(N_j(e))2^{-\operatorname{rank} N_j(e)/2}) & \text{if both } N_{j-1} \text{ and } N_{j+1} \text{ are even} \\ & \text{and unless } N_j(o) \cong \langle\epsilon_1\rangle \perp \langle\epsilon_2\rangle \text{ with} \\ & \epsilon_1 \equiv \epsilon_2 \bmod 4, \\ \frac{1}{2} & \text{otherwise,} \end{cases}$$

we put

$$E := \prod_j E_j^{-1}.$$

Before giving the proof, we note that q_j depends on N_j and N_{j+1}, and that

$$q_j = 0 \quad \text{if} \quad N_j = \{0\}.$$

Moreover E_j depends on N_{j-1}, N_j and N_{j+1} and

$$E_j = 1 \quad \text{if} \quad N_{j-1} = N_j = N_{j+1} = \{0\}$$

but

$$E_j = \begin{cases} 1 & \text{if} \quad \text{either} \quad (N_{j-1} = N_j = 0 \text{ and } N_{j+1} \text{ is even) or} \\ & \qquad\qquad\quad (N_{j-1} \text{ is even and } N_j = N_{j+1} = \{0\}), \\ \frac{1}{2} & \text{if} \quad \text{either} \quad (N_{j-1} = N_j = \{0\} \text{ and } N_{j+1} \text{ is odd) or} \\ & \qquad\qquad\quad (N_{j-1} \text{ is odd and } N_j = N_{j+1} = \{0\}). \end{cases}$$

Let us give a sketch of the proof.

We use induction on rank N. First we note that our formula satisfies the condition (iii) in Proposition 5.6.1. The values except w are invariant by scaling. Thus we may assume $N_0 \neq \{0\}$ and $N_j = 0$ if $j < 0$ by scaling. Let us denote the formula for the local density given in the Theorem by β'; we will verify $\beta = \beta'$.

Suppose $p \neq 2$; when rank $N = 1$, the Theorem is easy and since every submodule isometric to N_0 is transformed to N_0 by an isometry of N, by virtue of Proposition 5.2.2 and Theorem 5.3.2, applying Corollary 5.6.1 to $M_0 \perp M_0^\perp$, we have

$$\beta_p(N, N) = \beta_p(M_0, N)\beta_p(M_0^\perp, M_0^\perp).$$

Proposition 5.6.1 implies

$$\beta_p(M_0, N) = p^{n_0(n_0+1)/2 - n_0 n} r(\tilde{M}_0, \tilde{N})$$

where $n = \operatorname{rank} N$, and $\tilde{M}_0, \tilde{N}$ are quadratic spaces $(M_0/pM_0, \frac{1}{2}Q)$, $(N/pN, \frac{1}{2}Q)$ over $\mathbb{Z}/p\mathbb{Z}$ respectively and the function r is the number of all isometries from $\tilde{M}_0$ to $\tilde{N}$. Using Proposition 1.3.3, we have

$$\beta_p(M_0, N) = p^{n_0(1-n_0)/2} r(\tilde{M}_0, \tilde{M}_0).$$

The formula for $r(\tilde{M}_0, \tilde{M}_0)$ in Theorem 1.3.2 shows

$$\beta_p(M_0, N) = \beta'_p(N, N)/\beta'_p(M_0^\perp, M_0^\perp) = 2(1 - \chi(M_0)p^{-n_0/2})P([(n_0 - 1)/2])$$

and hence

$$\beta_p(N, N)/\beta_p(M_0^\perp, M_0^\perp) = \beta_p(M_0, N) = \beta'_p(N, N)/\beta'_p(M_0^\perp, M_0^\perp)$$

Hence the induction hypothesis yields $\beta_p(N, N) = \beta'_p(N, N)$.

Suppose $p = 2$. It is easy to verify $\beta = \beta'$ if N is even unimodular or $N = \langle\epsilon\rangle$ for $\epsilon \in \mathbb{Z}_2^\times$, using Proposition 5.6.1. The following is the fundamental reduction formula needed to prove the theorem:

(3)
$$\beta_2(N,N) = \begin{cases} \beta_2(M_0(e),N)\beta_2(M_0(e)^\perp, M_0(e)^\perp) & \text{if } M_0(e) \neq \{0\}, \\ \beta_2(\langle\epsilon_1\rangle,N)\beta_2(\langle\epsilon_2\rangle \perp M_0^\perp, \langle\epsilon_2\rangle \perp M_0^\perp) & \text{if } M_0 = \langle\epsilon_1\rangle \perp \langle\epsilon_2\rangle, \\ \beta_2(\langle\epsilon\rangle,N)\beta_2(M_0^\perp, M_0^\perp) & \text{if } M_0 = \langle\epsilon\rangle, \end{cases}$$

using Theorem 5.3.5 for the first and Theorem 5.3.6 for the others to apply Corollary 5.6.1. We will verify the following three formulas:

(4)
$$\begin{cases} \beta_2(M_0(e),N) & = \beta_2'(N,N)/\beta_2'(M_0(e)^\perp, M_0(e)^\perp) \text{ if } M_0(e) \neq 0, \\ \beta_2(\langle\epsilon_1\rangle,N) & = \beta_2'(N,N)/\beta_2'(\langle\epsilon_2\rangle \perp M_0^\perp, \langle\epsilon_2\rangle \perp M_0^\perp) \\ & \qquad \text{if} \quad M_0 = \langle\epsilon_1\rangle \perp \langle\epsilon_2\rangle, \\ \beta_2(\langle\epsilon\rangle,N) & = \beta_2'(N,N)/\beta_2'(M_0^\perp, M_0^\perp) \text{ if } M_0 = \langle\epsilon\rangle, \end{cases}$$

Once they have been proved, our theorem is proved by induction on rank N as follows:

If $M_0(e) \neq 0$, then use the first equations of (3) and (4). So suppose $M_0(e) = 0$. If $n_0 = 1$, then use the third equations of (3) and (4). If $n_0 = 2$, then use the second equations of (3) and (4).

(i) The case of $M_0(e) \neq \{0\}$. We must verify

(5)
$$\begin{aligned} \beta_2(M_0(e),N) &= \beta_2'(N,N)/\beta_2'(M_0(e)^\perp, M_0(e)^\perp) \\ &= 2^{q_0'-q_0} P(\operatorname{rank} M_0(e)/2) E_{-1}' E_0' E_1' (E_{-1}E_0E_1)^{-1}, \end{aligned}$$

where q_0', E_i' mean the invariants q_0, E_i for $M_0(e)^\perp$. Putting rank $M_0(e) = k$ and rank $N = n$, we have by Proposition 5.6.1

$$\begin{aligned} &\beta_2(M_0(e),N) \\ &= 2^{k(k+1)/2-kn} \sharp\{\sigma : M_0(e) \to N/2N \mid Q(\sigma(x)) \equiv Q(x) \bmod 4 \text{ for } x \in M_0(e)\} \end{aligned}$$

Put

$$\begin{aligned} M :=& \{x \in N \mid Q(x) \equiv 0 \bmod 2\} \\ =& \begin{cases} N & \text{if } M_0 \text{ is even}, \\ M_0(e) \perp 2M_0(o) \perp M_0^\perp & \text{if } \operatorname{rank} M_0(o) = 1, \\ M_0(e) \perp (\mathbb{Z}_2[v_1+v_2] + 2M_0(o)) \perp M_0^\perp & \text{if } M_0(o) = \mathbb{Z}_2[v_1, v_2] \\ & \quad \text{with } B(v_1,v_2) = 0, \end{cases} \end{aligned}$$

Since $x \in M_0(e)$ implies $Q(x) \equiv 0 \bmod 2$, we have

$$\beta_2(M_0(e), N) = 2^{k(k+1)/2-kn}\sharp\{\sigma : M_0(e) \to M/2N | Q(\sigma(x)) \equiv Q(x) \bmod 4 \text{ for } x \in M_0(e)\},$$

and

$$M/2N = \begin{cases} N/2N & \text{if } M_0 \text{ is even,} \\ (M_0(e)/2M_0(e)) \perp (M_0^\perp/2M_0^\perp) & \text{if rank } M_0(o) = 1, \\ (M_0(e)/2M_0(e)) \perp \langle \epsilon_1 + \epsilon_2 \rangle \perp (M_0^\perp/2M_0^\perp) & \text{if } M_0(o) \cong \langle \epsilon_1 \rangle \perp \langle \epsilon_2 \rangle. \end{cases}$$

Thus

$$\beta_2(M_0(e), N) = 2^{k(k+1)/2-kn} \begin{cases} r(\overline{M_0(e)}, \overline{M_0(e)} \perp \overline{M_0^\perp}) & \text{if rank } M_0(o) \le 1, \\ r(\overline{M_0(e)}, \overline{M_0(e)} \perp \langle \frac{1}{2}(\epsilon_1 + \epsilon_2) \rangle \perp \overline{M_0^\perp}) & \text{if rank } M_0(o) = 2, \end{cases}$$

where for a quadratic lattice K over $\mathbb{Z}_2$ with $\mathrm{n}(K) \subset 2\mathbb{Z}_2$, $\bar{K}$ denotes the quadratic space $(K/2K, \frac{1}{2}Q)$ over $\mathbb{Z}/2\mathbb{Z}$ and $r(U, V)$ is the number of isometries from U to V.

Put

$$U := \overline{M_0(e)} \perp \overline{M_0^\perp} \text{ and } V := U \perp \langle \frac{1}{2}(\epsilon_1 + \epsilon_2) \rangle;$$

then $U^\perp = \overline{M_0^\perp}$, $V^\perp = \overline{M_0^\perp} \perp \langle \frac{1}{2}(\epsilon_1 + \epsilon_2) \rangle$ follow at once from

$$\frac{1}{2}(Q(x+y) - Q(x) - Q(y)) = B(x, y),$$

and $\mathrm{Rad}\, U = U^\perp$ if and only if N_1 is even. Lastly $\mathrm{Rad}\, V = V^\perp$ if and only if $\epsilon_1 \not\equiv \epsilon_2 \bmod 4$ and N_1 is even. Now we can verify (5) using Theorem 1.3.2 and Proposition 1.3.3.

(ii) The case of $M_0 = \langle \epsilon \rangle$ ($\epsilon \in \mathbb{Z}_2^\times$).

In this case we must verify

$$\begin{aligned} \beta_2(M_0, N) &= \beta_2'(N, N)/\beta_2'(M_0^\perp, M_0^\perp) \\ &= 2^{q_0' - q_0} E_{-1}' E_0' E_1' (E_{-1} E_0 E_1)^{-1}, \end{aligned} \tag{6}$$

where the dashed invariants are those for $M_0^\perp$. Putting $m := \mathrm{rank}\, M_0^\perp = n - 1$, we have by Proposition 5.6.1

$$\begin{aligned} &\beta_2(M_0, N) \\ &= (2^2)^{1-(1+m)}\sharp\{\sigma : M_0 \to N/4N \mid Q(\sigma(x)) \equiv Q(x) \bmod 8 \text{ for } x \in M_0\} \\ &= 2^{-2m}\sharp\{a \bmod 4, x \in M_0^\perp/4M_0^\perp \mid a^2\epsilon + Q(x) \equiv \epsilon \bmod 8\} \end{aligned}$$

(here we note $Q(x) \equiv 0 \bmod 2$ implies $a \equiv 1 \bmod 2$ and so $a^2 \equiv 1 \bmod 8$)

$$
\begin{aligned}
&= 2^{1-2m}\sharp\{x \in M_0^\perp/4M_0^\perp \mid Q(x) \equiv 0 \bmod 8\} \\
&= 2^{1-m}\sharp\{x \in M_0^\perp/2M_0^\perp \mid Q(x) \equiv 0 \bmod 8\} \text{ (by } \mathrm{s}(M_0^\perp) \subset 2\mathbb{Z}_2) \\
&= 2^{1-n_1-n_2}\sharp\{x \in M_1/2M_1 \perp M_2/2M_2 \mid Q(x) \equiv 0 \bmod 8\} \\
&= 2^{1-n_1-n_2}\sharp\{x \in N_1/2N_1 \perp N_2^{(2)}/2N_2^{(2)} \mid Q(x) \equiv 0 \bmod 4\}.
\end{aligned}
$$

We note from the above that $\beta_2(M_0, N) = 2$ if $M_1 = M_2 = 0$, and so (6) is true in that case.

(ii.a) Suppose that N_2 and N_1 are odd. Putting $N_1 = \perp \langle u_i \rangle, N_2 = \perp \langle v_i \rangle$, we have

$$
\begin{aligned}
&\beta_2(M_0, N) \\
&= 2^{1-n_1-n_2}\sharp\{a_i, b_j \bmod 2 \mid \sum_{i=1}^{n_1} u_i a_i^2 + 2\sum_{i=1}^{n_2} v_i b_i^2 \equiv 0 \bmod 4\} \\
&= 2^{1-n_1-n_2} \sum_{\substack{a_i \bmod 2 \\ \sum u_i a_i^2 \equiv 0 \bmod 2}} \sharp\{b_j \bmod 2 \mid v_1 b_1^2 \equiv -\sum_{i \geq 2} v_i b_i^2 - \frac{1}{2}\sum u_i a_i^2 \bmod 2\} \\
&= 2^{1-n_1-n_2} 2^{n_2-1} \sharp\{a_i \bmod 2 \mid \sum u_i a_i^2 \equiv 0 \bmod 2\} \\
&= 2^{-n_1} 2^{n_1-1} = 2^{-1},
\end{aligned}
$$

which yields (6).

(ii.b) Suppose that N_2 is odd but N_1 is even; then we have, putting $N_2 = \perp \langle v_i \rangle$,

$$
\begin{aligned}
&\beta_2(M_0, N) \\
&= 2^{1-n_1-n_2}\sharp\{x \in N_1/2N_1, b_i \bmod 2 \mid Q(x) + 2\sum v_i b_i^2 \equiv 0 \bmod 4\} \\
&= 2^{1-n_1-n_2}\sharp\{x \in N_1/2N_1, b_i \bmod 2 \mid v_1 b_1^2 \equiv -2^{-1}Q(x) + \sum_{i \geq 2} v_i b_i^2 \bmod 2\} \\
&= 1,
\end{aligned}
$$

which yields (6).

Suppose that N_2 is even hereafter in the case of $M_0 = \langle \epsilon \rangle$; then we have

$$\beta_2(M_0, N) = 2^{1-n_1}\sharp\{x \in N_1/2N_1 \mid Q(x) \equiv 0 \bmod 4\}.$$

(ii.c) If N_1 is even, then

$$\begin{aligned}
&\beta_2(M_0, N)\\
&= 2^{1-n_1}\sharp\{x \in N_1/2N_1 \mid \frac{1}{2}Q(x) \equiv 0 \bmod 2\}\\
&= 2^{1-n_1}(2^{n_1-1} - \chi(N_1)2^{n_1/2-1} + \chi(N_1)2^{n_1/2}) \quad \text{by Lemma 1.3.1}\\
&= 1 + \chi(N_1)2^{-n_1/2},
\end{aligned}$$

which yields (6).

(ii.d) Suppose N_1 is odd and $N_1(o) = \langle u \rangle$, then we have

$$\begin{aligned}
&\beta_2(M_0, N)\\
&= 2^{1-n_1}\sharp\{x \in N_1(e)/2N_1(e), a \bmod 2 \mid Q(x) + a^2u \equiv 0 \bmod 4\}\\
&= 2^{1-n_1}\sharp\{x \in N_1(e)/2N_1(e) \mid Q(x) \equiv 0 \bmod 4\}\\
&= 2^{1-n_1}(2^{n_1-2} - \chi(N_1(e))2^{(n_1-1)/2-1} + \chi(N_1(e))2^{(n_1-1)/2})\\
&\quad \text{by Lemma 1.3.1}\\
&= 2^{-1}(1 + \chi(N_1(e))2^{-(n_1-1)/2}),
\end{aligned}$$

which yields (6).

(ii.e) If $N_1(o) = \langle u_1 \rangle \perp \langle u_2 \rangle$, then we have

$$\begin{aligned}
&\beta_2(M_0, N)\\
&= 2^{1-n_1}\sharp\{a_1, a_2 \bmod 2, x \in N_1(e)/2N_1(e) \mid u_1a_1^2 + u_2a_2^2 + Q(x) \equiv 0 \bmod 4\}\\
&= 2^{1-n_1}\sharp\{x \in N_1(e)/2N_1(e) \mid \frac{1}{2}Q(x) \equiv 0 \bmod 2\}\\
&\quad (\text{when } a_1 \equiv a_2 \equiv 0 \bmod 2)\\
&+ 2^{1-n_1}\sharp\{x \in N_1(e)/2N_1(e) \mid \frac{1}{2}Q(x) \equiv \frac{1}{2}(u_1 + u_2) \bmod 2\}\\
&\quad (\text{when } a_1 \equiv a_2 \equiv 1 \bmod 2)\\
&= 2^{1-n_1}\begin{cases} 2(2^{n_1-3} - \chi(N_1(e))2^{(n_1-2)/2-1} + \chi(N_1(e))\, 2^{(n_1-2)/2}) \\ \qquad \text{if } u_1 \not\equiv u_2 \bmod 4, \\ 2^{n_1-2} \qquad \text{if } u_1 \equiv u_2 \bmod 4, \end{cases}\\
&= \begin{cases} 2^{-1}(1 + \chi(N_1(e))2^{-(n_1-2)/2}) & \text{if } u_1 \not\equiv u_2 \bmod 4, \\ 2^{-1} & \text{if } u_1 \equiv u_2 \bmod 4, \end{cases}
\end{aligned}$$

which yields (6) and completes the case of rank $M_0 = 1$.

(iii) The case of $M_0 = \langle \epsilon_1 \rangle \perp \langle \epsilon_2 \rangle$ $(\epsilon_i \in \mathbb{Z}_2^\times)$.

As noted, we have only to verify the second equality in (4). Putting $m := \operatorname{rank} M_0^\perp = n - 2$, Proposition 5.6.1 yields

$$\begin{aligned}
&\beta_2(\langle\epsilon_1\rangle, N)\\
&=(2^2)^{1-(m+2)}\sharp\{\sigma : \langle\epsilon_1\rangle \to N/4N \mid Q(\sigma(x)) \equiv Q(x) \bmod 8\}\\
&=2^{-2-2m}\sharp\{a, b \bmod 4, x \in M_0^\perp/4M_0^\perp \mid a^2\epsilon_1 + b^2\epsilon_2 + Q(x) \equiv \epsilon_1 \bmod 8\}\\
&=2^{-1-2m}\sharp\{b \bmod 4, x \in M_0^\perp/4M_0^\perp \mid b^2\epsilon_2 + Q(x) \equiv 0 \bmod 8\}\\
&\qquad (\text{when } a \equiv 1 \bmod 2)\\
&\quad + 2^{-1-2m}\sharp\{a \bmod 4, x \in M_0^\perp/4M_0^\perp \mid a^2\epsilon_1 + Q(x) \equiv \epsilon_1 - \epsilon_2 \bmod 8\}\\
&\qquad (\text{when } a \equiv 0 \bmod 2)\\
&=2^{-1-2m}(\sharp\{x \in M_0^\perp/4M_0^\perp \mid Q(x) \equiv 0 \bmod 4\}\\
&\qquad\qquad + \sharp\{x \in M_0^\perp/4M_0^\perp \mid Q(x) \equiv \epsilon_1 - \epsilon_2 \bmod 4\})\\
&=2^{-1-m}(\sharp\{x \in M_0^\perp/2M_0^\perp \mid \frac{1}{2}Q(x) \equiv 0 \bmod 2\}\\
&\qquad\qquad + \sharp\{x \in M_0^\perp/2M_0^\perp \mid \frac{1}{2}Q(x) \equiv \frac{1}{2}(\epsilon_1 - \epsilon_2) \bmod 2\})\\
&=\begin{cases} 2^{-1} & \text{if} \quad \epsilon_1 \not\equiv \epsilon_2 \bmod 4,\\ 2^{-m}\sharp\{x \in M_0^\perp/2M_0^\perp \mid \frac{1}{2}Q(x) \equiv 0 \bmod 2\} & \text{if} \quad \epsilon_1 \equiv \epsilon_2 \bmod 4 \end{cases}\\
&=\begin{cases} 2^{-1} & \text{if} \quad \epsilon_1 \not\equiv \epsilon_2 \bmod 4,\\ 1 & \text{if} \quad \epsilon_1 \equiv \epsilon_2 \bmod 4 \text{ and } N_1 \text{ is even},\\ 2^{-1} & \text{if} \quad \epsilon_1 \equiv \epsilon_2 \bmod 4 \text{ and } N_1 \text{ is odd}, \end{cases}
\end{aligned}$$

which yields the second equality of (4) and completes the proof of the theorem. □

Let us study quantitative properties of local densities. They may suggest something global and indeed play a substantial role in global representation problems.

Theorem 5.6.4. *Let M, M', N and N' be regular quadratic lattices over R and $m = \operatorname{rank} M = \operatorname{rank} M' \le n = \operatorname{rank} N = \operatorname{rank} N'$. Then we have:*

(a) $\beta_p(p^r M, p^r N) = p^{r(m+1)m}\beta_p(M, N)$.

(b) *If $N \subset N'$, then*

$$\beta_p(M, N) \le [N' : N]^m \beta_p(M, N').$$

(c) *If $N \subset N'$, $p^r N' \subset N$, then*

$$\beta_p(M, N') \le p^{-rm(m+1)}[N : p^r N']^m \beta_p(p^r M, N).$$

(d) *If $M \subset M'$, then*

$$\beta_p(M', N) \le [M' : M]^{n-m-1} \beta_p(M, N).$$

(e) *If $N' \subset N$ and $\eta(M) \subset N'$ for every isometry $\eta : M \hookrightarrow N$, then*

$$\beta_p(M, N) = [N : N']^{-m} \beta_p(M, N').$$

In particular, if N is maximal and anisotropic, and $M \hookrightarrow N$, then

$$\beta_p(pM, N) = p^{m(m+1-n)} \beta_p(M, N).$$

Proof. (a) follows from (iii) in Proposition 5.6.1. By virtue of (a), we may assume that the scales of M, M', N, N' are in R in the rest of the proof.
(b): Let $i : N/p^t N \to N'/p^t N'$ be the canonical inclusion mapping. For

$$\phi \in A_{p^t}(M, N) := \{\sigma : M \to N/p^t N \mid Q(\sigma(x)) \equiv Q(x) \bmod 2p^t\},$$

$i \circ \phi$ is clearly contained in $A_{p^t}(M, N')$.

$$\sharp\{i \circ \phi \mid \phi \in A_{p^t}(M, N)\} \le \sharp A_{p^t}(M, N')$$

is also clear. Noting that a linear mapping $\psi : M \to N/p^t N$ satisfies $i \circ \psi = 0$ if and only if $\psi(M) \subset p^t N' \cap N$, we have

$$\begin{aligned}\sharp\{\psi : M \to N/p^t N \mid i \circ \psi = 0\} &= \sharp\{\psi : M \to (p^t N' \cap N)/p^t N\} \\ &= [p^t N' \cap N : p^t N]^m \le [N' : N]^m.\end{aligned}$$

Grouping elements ϕ of $A_{p^t}(M, N)$ by the inverse image of $\phi \to i \circ \phi$, we have $\sharp A_{p^t}(M, N) \le [N' : N]^m \sharp A_{p^t}(M, N')$.
(c) is proved by using (a) and (b) as follows:

$$\begin{aligned}\beta_p(M, N') &= p^{-rm(m+1)} \beta_p(p^r M, p^r N') \\ &\le p^{-rm(m+1)} [N : p^r N']^m \beta_p(p^r M, N).\end{aligned}$$

(d): By Theorem 5.6.1, we have

$$\begin{aligned}&\beta_p(M,N) \\ &= \sum_{FM \supset M_0 \supset M} [M_0 : M]^{m+1-n} d_p(M_0, N) \\ &\ge \sum_{FM \supset M_0 \supset M'} ([M_0 : M'][M' : M])^{m+1-n} d_p(M_0, N) \\ &= [M' : M]^{m+1-n} \beta_p(M', N).\end{aligned}$$

(e): Regarding $\sigma \in A_{p^t}(M, N)$ as a linear mapping from M to N and then applying Corollary 5.4.2, there exists an isometry $\eta : M \hookrightarrow N$ such that $\eta(M) = \sigma(M)$ if t is sufficiently large. By the assumption, $\sigma(M) = \eta(M) \subset N'$ holds and then we have

$$\begin{aligned}
&\sharp A_{p^t}(M, N) \\
&\quad = \sharp\{\sigma : M \to N'/p^t N \mid Q(\sigma(x)) \equiv Q(x) \bmod 2p^t\} \\
&\quad = [N : N']^{-m} \sharp\{\sigma : M \to N'/p^t N' \mid Q(\sigma(x)) \equiv Q(x) \bmod 2p^t\} \\
&\quad = [N : N']^{-m} \sharp A_{p^t}(M, N').
\end{aligned}$$

Suppose that $M \hookrightarrow N$ and N is maximal anisotropic; then $\mathrm{n}(M) \subset \mathrm{n}(N)$. For an isometry $\eta : pM \hookrightarrow N$, $\mathrm{n}(\eta(pM)) = \mathrm{n}(pM) \subset \mathrm{n}(pN)$ holds and then Theorem 5.2.1 yields $\eta(pM) \subset pN$. Applying the above, we have

$$\beta_p(pM, N) = [N : pN]^{-m} \beta_p(pM, pN) = p^{-mn+m(m+1)} \beta_p(M, N).$$

□

Theorem 5.6.5. *Let M and N be regular quadratic lattices over R and $m := \operatorname{rank} M \leq n := \operatorname{rank} N$. Suppose M is represented by N, then the following are true:*

(a) *If a submodule M_0 of N is isometric to M, then*

$$\beta_p(M, N) > c[FM_0 \cap N : M_0]^{m+1-n},$$

where c is a positive number dependent only on N.

(b) *If $\operatorname{ind} FN \geq m$ and $\beta_p(M, N) \neq 0$, then*

$$\beta_p(M, N) \geq c_1,$$

where c_1 is a positive number dependent only on N.

(c) *If $n \geq 2m + 1$, then*

$$\beta_p(M, N) < c_2 \min_{M_0} [M_0 : M]^{2m-n}$$

where c_2 is a positive number dependent only on N, and M_0 ranges over lattices on FM such that $M_0 \supset M$ and $d_p(M_0, N) \neq 0$.

Proof. (a): Put $M' := FM_0 \cap N$ and applying (d) in the previous theorem, we have

$$\begin{aligned}
\beta_p(M, N) = \beta_p(M_0, N) &\geq [M' : M_0]^{m+1-n} \beta_p(M', N) \\
&\geq [M' : M_0]^{m+1-n} d_p(M', N).
\end{aligned}$$

Since M' is primitive in N, $d_p(M', N) \neq 0$ and hence by the remark before Theorem 5.6.1, there is a positive number c dependent only on N such that $d_p(M', N) > c$.

(b): Since $\beta_p(M, N) \neq 0$, we may assume $M \subset N$, and we use induction on rank N. We fix any maximal lattice N_0 in N.

Suppose $\mathrm{n}(M) \subset \mathrm{n}(N_0)$; then by virtue of Theorem 5.2.2 and Proposition 5.3.2 there is a submodule M_0 isometric to M which is primitive in N_0. Then

$$\begin{aligned}[FM_0 \cap N : M_0] &= [FM_0 \cap N : FM_0 \cap N_0][FM_0 \cap N_0 : M_0] \\ &= [FM_0 \cap N : FM_0 \cap N_0] \leq [N : N_0]\end{aligned}$$

and (a) implies

$$\beta_p(M, N) > c[N : N_0]^{m+1-n}$$

since $\operatorname{ind} FN \geq m$ implies $n \geq 2m \geq m+1$, i.e. $m+1-n \leq 0$.

Suppose $\mathrm{n}(M) \supset \mathrm{n}(N_0)$; then let $M = M_1 \perp \cdots \perp M_t$ be a Jordan splitting with $\mathrm{s}(M_1) \supset \cdots \supset \mathrm{s}(M_t)$, where all $\mathrm{s}(M_i)$ are different. We note that M_1 is modular and $M_1 \subset M \subset N$ implies $\mathrm{n}(N) \supset \mathrm{n}(M_1) = \mathrm{n}(M) \supset \mathrm{n}(N_0)$. Since the number of the isometry classes of modular lattices with $\mathrm{n}(N) \supset \text{norm} \supset \mathrm{n}(N_0), \text{rank} \leq m$ is finite, by Proposition 5.3.3 a modular submodule M' of N with $\mathrm{n}(M') \supset \mathrm{n}(N_0)$ is transformed by an isometry of N to one of a finite set S of submodules of N. S depends only on N. By Theorem 5.6.2 we have

$$\begin{aligned}&\beta_p(M, N) \\ &= \sum_i ([M_1^\sharp : M_1]/[N : K_i \perp K_i^\perp])^{\operatorname{rank} M - \operatorname{rank} M_1} \beta_p(M_1, N : K_i) \beta_p(M_1^\perp, K_i^\perp),\end{aligned}$$

where K_i are a complete set of submodules of N isometric to M_1 which are not transformed into each other by isometries of N and hence we may assume $K_i \in S$. Thus we have $\beta_p(M, N) \geq c \sum_i \beta_p(M_1^\perp, K_i^\perp)$ for a positive number c dependent only on N. We show $\operatorname{ind} K_i^\perp \geq \operatorname{rank} M_1^\perp$. Put $FN = U$, $FM = V$, $FM_1 = W$, $V = W \perp X$ and $U = W \perp Y$; then we want to see $\operatorname{ind} Y \geq \dim X$. We decompose W, Y as $W = W_1 \perp W_2 \perp H_k$, $Y = Y_1 \perp Y_2 \perp H_h$, where $k = \operatorname{ind} W$, $h = \operatorname{ind} Y$, $W_1 \cong Y_1^{(-1)}$, H_i is a hyperbolic space of $\dim = 2i$ and $W_2 \perp Y_2$ is anisotropic. Then the assumption $\operatorname{ind} U - \dim V \geq 0$ implies

$$0 \leq k + h + \dim W_1 - \dim W - \dim X \leq h - \dim X, \text{ i.e. } \operatorname{ind} Y \geq \dim X.$$

Then the induction assumption implies $\beta_p(M_1^\perp, K_i^\perp) > c(K_i^\perp)$ for some positive number $c(K_i^\perp)$ dependent only on $K_i^\perp$, which completes the proof.

(c): We may suppose $\beta_p(M,N) \neq 0$. Then there is a lattice M' such that $M' \supset M$ and $d_p(M',N) \neq 0$ by Theorem 5.6.1. We can put $r = \min \operatorname{ord}_p[M' : M]$ where M' ranges over lattices on FM such that $M' \supset M$ and $d_p(M',N) \neq 0$. By Theorem 5.6.1, we have

$$\begin{aligned}\beta_p(M,N) &= \sum_{FM \supset M' \supset M} [M' : M]^{m+1-n} d_p(M',N) \\ &\leq \max_{FM \supset M' \supset M} d_p(M',N) \sum_{s \geq r} p^{s(m+1-n)} A(m,s),\end{aligned}$$

where $A(m,s)$ is the number of modules with rank m over R which contains a given module with index p^s. Let us evaluate the number $A(m,s)$. Put $L = R[u_1, \cdots, u_m] \subset L' = R[v_1, \cdots, v_m]$ and $[L' : L] = p^s$. By $(v_1, \cdots, v_m) = (u_1, \cdots, u_n)a$ $(a \in GL_m(F))$, L' corresponds to $aGL_m(R)$ bijectively and $[L' : L] = p^s$ means $a^{-1} \in M_m(R)$ and $\det a^{-1} \in p^s R^\times$. Thus we have

$$A(m,s) = \sharp(GL_m(R)\backslash\{b \in M_m(R) \mid \det b \in p^s R^\times\})$$

and taking upper triangular matrices as representatives of the left cosets, we can get $A(m,s) = \sum_{\substack{s_i \geq 0 \\ s_1 + \cdots + s_m = s}} \prod p^{s_i(i-1)}$. Then

$$\begin{aligned}A(m,s)p^{-(m-1)s} &< \sum_{s_i \geq 0} \prod_{i=1}^{m} p^{s_i(i-m)} \\ &< \sum_{\substack{s_i \geq 0 \\ i<m}} \prod_{i<m} p^{-s_i} = (1-p^{-1})^{m-1}\end{aligned}$$

is easy. Thus we have

$$\begin{aligned}\beta_p(M,N) &< \max_{FM \supset M' \supset M} d_p(M',N) \cdot (1-p^{-1})^{m-1} \sum_{s \geq r} p^{(2m-n)s} \\ &= \max_{FM \supset M' \supset M} d_p(M',N) \cdot (1-p^{-1})^{m-1}(1-p^{2m-n})^{-1} p^{(2m-n)r},\end{aligned}$$

which completes the proof. □

Note that $n \geq 2m+3$ implies the condition $\operatorname{ind} FN \geq m$ in (b). In this case we can show a similar estimate for $\beta_p(M,N;g,p^e)$ instead of $\beta_p(M,N)$ although we need more information on the existence of a certain kind of submodule. This estimate is necessary for a global representation problem. And note also that (a) and (c) mean, roughly speaking, that if $n = 2m+1$ or $2m+2$, then $\beta_p(M,N)$ is bounded away from 0 if and only if there is a submodule M_0 of N isometric to M such that $[FM_0 \cap N : M_0]$ is bounded. However this is false when $n \leq 2m$: When $m+1 \leq n \leq 2m$, we have $\beta_p(p^t M, N) \to \infty$ as $t \to \infty$ unless $n = 2m = 2 \operatorname{ind} FN + 4$ by the exercise below. But $\operatorname{ind} FN \geq m$ follows if $p^t M$ is primitively represented by N for a large integer t.

Exercise.

Let $M \subset N$ be regular quadratic lattices with $\operatorname{rank} N = n$, $\operatorname{rank} M = m \geq \operatorname{ind} N = r$ and $n \geq m+1$. Show

$$\beta_p(p^t M, N) > c(M,N)p^{t(m-r)(m+r+1-n)}$$

for $t \geq 0$ and a positive number $c(M,N)$ dependent on M and N. If $n = 2m = 2r+4$, which is the only case for $(m-r)(m+r+1-n) < 0$ when $n \leq 2m$, show $c(M,N)p^{-2t} < \beta_p(p^t M, N) < cp^{(\epsilon-2)t}$ for any $\epsilon > 0$ and c dependent on ϵ, M, N.
(Hint : see [AK], [Ki16])

Now let us study the infinite product of local densities $\beta_p(\mathbb{Z}_p L, \mathbb{Z}_p M)$ for quadratic lattices L and M over $\mathbb{Z}$. As a preparation we give two lemmas.

Lemma 5.6.9. *Let F_q be a finite field with q elements. Let U be a totally isotropic quadratic space, i.e. $q(U) = 0$, and V be a regular quadratic space over F_q and suppose that U is represented by V. Then the number $r(U,V)$ of the isometries from U to V is*

$$q^{u(u+1)/2-uv} r(U,V)$$
$$= \begin{cases} \prod_{i=0}^{u-1}(1-q^{-(v-2u+2i+1)}) & \text{if } v \equiv 1 \bmod 2, \\ \prod_{i=0}^{u-1}(1-\chi(V)q^{-(v-2u+2i+2)/2})(1+\chi(V)q^{-(v-2u+2i)/2}) & \text{if } v \equiv 0 \bmod 2, \end{cases}$$

where χ is the function defined in 1.3 and $u = \dim U$, $v = \dim V$.

Proof. We use induction on u. When $u = 1$, $r(U,V)$ is the number of isotropic vectors in V, which is equal to $q^{v-1} - \chi(V)q^{v/2-1} + \chi(V)q^{v/2} - 1$ by Lemma 1.3.1. This completes the case of $u = 1$. Suppose $u > 1$, and write $U = F_q x \perp U_1$. Since isotropic vectors in V are mutually transformed by $O(V)$, for any fixed isotropic vector y in V

$$\begin{aligned} r(U,V) &= r(F_q x, V)\sharp\{\sigma : U \hookrightarrow V \mid \sigma(x) = y\} \\ &= r(F_q x, V)\sharp\{\sigma : U_1 \hookrightarrow y^\perp \mid \text{the orthogonal} \\ &\qquad \text{sum of } x \mapsto y \text{ and } \sigma \text{ is an isometry}\}. \end{aligned}$$

Take a vector z in V such that $b(y,z) = 1$; then

$$F_q[y,z] = F_q[y, z - q(z)y]$$

is a hyperbolic plane. Let us evaluate the number of isometries

$$\sigma : U_1 \hookrightarrow y^\perp$$

such that for $a \in F_q$, $w \in U_1$, the mapping $\tilde{\sigma} : ax + w \to ay + \sigma(w)$ is an isometry from $Fx \perp U_1$ to V. Write

$$V = F_q[y, z] \perp V_0;$$

then $y^\perp = F_q y \perp V_0$ is clear. For an isometry σ from U_1 to $y^\perp$, we write $\sigma(w) = f(w)y + \eta(w)$, $f(w) \in F_q$, $\eta(w) \in V_0$. Clearly $q(w) = q(\sigma(w)) = q(\eta(w))$. Suppose that $\tilde{\sigma}$ is an isometry and $\eta(w) = 0$. Then $\tilde{\sigma}(-f(w)x + w) = 0$ implies $w = 0$ and η is injective and so an isometry from U_1 to V_0. Conversely for any $f \in \mathrm{Hom}(U_1, F_q)$ and $\eta : U_1 \hookrightarrow V_0$, the mapping $w \mapsto f(w)y + \eta(w)$ is an isometry from U_1 to $y^\perp$ such that $\tilde{\sigma}$ is an isometry from $Fx \perp U_1$ to V. Hence we have $r(U, V) = q^{u-1} r(F_q x, V) r(U_1, V_0)$. Applying the induction hypothesis, we complete the proof. □

Lemma 5.6.10. *Let p be an odd prime, and U, V be quadratic spaces over a finite field F_p with p elements. Suppose that $U \hookrightarrow V$ and V is regular. Write $U = U_0 \perp U^\perp$ and put $t = \dim U_0$, $u = \dim U$, $v = \dim V$. Suppose $u < v$ and $v > 2$. Then we have*

$$p^{u(u+1)/2 - uv} r(U, V) = \prod_{i=1}^{a} (1 \pm p^{-r_i}) \times \begin{cases} 2 & \text{if } v - 2u + t = 0, \\ 1 + \chi(W)p^{-1} & \text{if } v - 2u + t = 2, \\ 1 & \text{otherwise,} \end{cases}$$

where $2 \le r_i \le b$ and a, b are smaller than a number dependent only on v, W is defined by $V \cong U_0 \perp W$, and χ is as in the previous lemma.

Proof. If $u = t$, then U is regular and Theorem 1.3.2 yields our assertion. Hence we assume

$$u > t \tag{7}$$

Since V is regular, by Corollary 1.2.3 we have

$$r(U, V) = r(U_0, V) r(U^\perp, W)$$

and then by Theorem 1.3.2 and the previous lemma, we have

$$\begin{aligned} r(U,V) &= p^{tv - t(t+1)/2} (1 - \chi(V)p^{-v/2})(1 + \chi(U_0 \perp V^{(-1)})p^{(t-v)/2}) \\ &\times \prod_{\substack{v-t+1 \le e \le v-1 \\ e:\text{even}}} (1 - p^{-e}) \times p^{(u-t)(v-t) - (u-t)(u-t+1)/2} \\ &\times \begin{cases} \prod_{i=0}^{u-t-1} (1 - p^{-(v-t-2(u-t)+2i+1)}) & \text{if } v \not\equiv t \bmod 2, \\ \prod_{i=0}^{u-t-1} \{(1 - \chi(W)p^{-(v-t-2(u-t)+2i+2)/2}) \\ \quad \times (1 + \chi(W)p^{-(v-t-2(u-t)+2i)/2})\} & \text{if } v \equiv t \bmod 2. \end{cases} \end{aligned}$$

Since $q(U^\perp) = 0$,

$$(8) \qquad U^\perp \hookrightarrow W,$$

which follows from the assumption $U_0 \perp U^\perp = U \hookrightarrow V \cong U_0 \perp W$. We have $\operatorname{ind} W \geq \dim U^\perp$, yielding

$$(9) \qquad v - t = \dim W \geq 2 \dim U^\perp = 2(u-t).$$

Denoting by P the product of factors $1 \pm p^{-a}$ for $r(U, V)$ such that a is greater than 1, we have

(10)

$$\begin{aligned} &p^{u(u+1)/2-uv} r(U,V) \\ &\quad = P(1 + \chi(U_0 \perp V^{(-1)}) p^{(t-v)/2}) \\ &\quad \times \begin{cases} 1 - p^{-(v-t-2(u-t)+1)} & \text{if } v \not\equiv t \bmod 2, \\ (1 - \chi(W) p^{-(v-t-2(u-t)+2)/2}) \times & \\ (1 + \chi(W) p^{-(v-t-2(u-t))/2}) & \text{if } v \equiv t \bmod 2, \end{cases} \\ &\quad \times \begin{cases} 1 + \chi(W) p^{-(v-t-2(u-t)+2)/2} & \text{if } v - t = 2(u-t) \geq 4, \\ 1 & \text{otherwise,} \end{cases} \end{aligned}$$

where the last factor comes from the case of $i = 1$ and $v \equiv t \bmod 2$, which implies $u - t \geq 1 + i = 2$. Since $U_0 \perp V^{(-1)} \cong U_0 \perp U_0^{(-1)} \perp W^{(-1)}$ and $U_0 \perp U_0^{(-1)}$ is a hyperbolic space,

$$\chi(U_0 \perp V^{(-1)}) = \chi(W^{(-1)}) = \chi(W).$$

Suppose $v - t = 2(u-t)$ i.e. $v - 2u + t = 0$; then (8) and (9) imply that W is hyperbolic and then

$$\chi(U_0 \perp V^{(-1)}) = \chi(W) = 1.$$

By the assumption $v - t = 2(u-t)$ implies $(t-v)/2 = u - v < 0$. If $u - v = -1$, then we have $t - v = -2$, $u - t = 1$ and then (10) implies

$$p^{u(u+1)/2-uv} r(U,V) = P(1 + p^{-1})(1 - p^{-1}) \cdot 2 = 2P(1 - p^{-2}).$$

If $u - v \leq -2$, then $(t-v)/2 = u - v \leq -2$ and $u - t = v - u \geq 2$ yield

$$\begin{aligned} p^{u(u+1)/2-uv} r(U,V) &= P(1 + p^{(t-v)/2})(1 - p^{-1}) \cdot 2(1 + p^{-1}) \\ &= 2P(1 + p^{(t-v)/2})(1 - p^{-2}). \end{aligned}$$

Thus the assertion of the lemma is true when $v - t = 2(u - t)$.

Suppose $v-t = 2(u-t)+1$, i.e. $v-2u+t = 1$; then $\dim(U_0 \perp V^{(-1)}) = t+v$ is odd and hence $\chi(U_0 \perp V^{(-1)}) = 0$ by definition. Thus we have

$$p^{u(u+1)/2-uv}r(U,V) = P(1-p^{-(v-t-2(u-t)+1)}) = P(1-p^{-2}),$$

completing the proof of this case.

Suppose $v - t = 2(u-t)+2$, i.e. $v - 2u + t = 2$; then we have

$$(t-v)/2 = u+1-v$$

and

$$p^{u(u+1)/2-uv}r(U,V) = P(1+\chi(W)p^{u+1-v})(1-\chi(W)p^{-2})\times(1+\chi(W)p^{-1}).$$

Now $u < v$ implies $u+1-v \le 0$. If $u+1-v \ge -1$, then $u \ge v-2 = 2u-t$ which contradicts (7). Thus $u+1-v \le -2$ and then we complete the proof of the case $v - 2u + t = 2$.

Suppose $v - t - 2(u-t) \ge 3$, i.e. $v - 2u + t \ge 3$; then $u > t$ implies $(t-v)/2 < (t-v+2u-2t)/2 \le -1.5$, and other powers of p in (10) are less than -1. Thus we have completed the proof of this final case and hence also the proof of the lemma. □

Let U, V be regular quadratic spaces over $\mathbb{Q}$ and M, N be lattices on U and V respectively, that is M, N are finitely generated modules over $\mathbb{Z}$ and they contain a basis of U, V respectively. We denote quadratic lattices $\mathbb{Z}_pM$, $\mathbb{Z}_pN$ over $\mathbb{Z}_p$ by M_p, N_p. We study the infinite product of local densities. It is known that the behaviour of such products over all primes suggests something global in some cases through the Siegel formula (see the next chapter), although it is not fully satisfactory.

Let M, N be as above and $\operatorname{rank} M = m \le \operatorname{rank} N = n$, and define two infinite sets of primes

$$P(N) := \{p \mid p \ne 2,\ N_p \text{ is unimodular}\},$$
$$P(M,N) := \{p \mid p \ne 2,\ N_p \text{ and } M_p \text{ are unimodular}\},$$

For a prime $p \in P(M,N)$, we have by Proposition 5.6.1

$$\beta_p(M_p,N_p) = p^{m(m+1)/2-mn}\sharp\{\sigma : M_p \to N_p/pN_p \mid Q(\sigma(x)) \equiv Q(x) \bmod 2p\}.$$

Denoting a quadratic space $(M_p/pM_p, \frac{1}{2}Q)$ over $\mathbb{Z}/p\mathbb{Z}$ by $\bar{M}_p$ and defining similarly $\bar{N}_p$, we have

$$\beta_p(M_p,N_p) = p^{m(m+1)/2-mn}\sharp\{\sigma : \bar{M}_p \hookrightarrow \bar{N}_p\},$$

and then Theorem 1.3.2 yields

$$\beta_p(M_p, N_p) = (1-\chi_p(\bar{N}_p)p^{-n/2})(1+\chi_p(\bar{M}_p \perp \bar{N}_p^{(-1)})p^{(m-n)/2}) \prod_{\substack{n-m+1\le e\le n-1\\ e:\ \text{even}}} (1-p^{-e}),$$

where χ_p denotes χ in Theorem 1.3.2.

As before, we put

$$\alpha_p(M_p, N_p) = 2^{m\delta_{2,p}-\delta_{m,n}}\beta_p(M_p, N_p).$$

Proposition 5.6.2. *Keeping the above notation, we have the following: Put*

$$\alpha'(M,N) := \prod_{p\in P(M,N)} \alpha_p(M_p, N_p).$$

(i) *If $n = m+2$ and $-\mathrm{d}\,M\,\mathrm{d}\,N$ is a square in $\mathbb{Q}$, then $\alpha'(M,N)$ diverges to infinity.*

(ii) *If $n > m+2$, $n = m+1 > 2$ or if $n = m \neq 2$ with $M_p \cong N_p$ for every $p \in P(M,N)$, then there exist positive numbers c_1, c_2 dependent only on n such that*

$$c_1 < \alpha'(M,N) < c_2.$$

(iii) *If $n = m+2$ and $-\mathrm{d}\,M\,\mathrm{d}\,N$ is not a square in $\mathbb{Q}$, then there exist positive numbers c_3, c_4 dependent only on n such that*

$$c_3 \prod (1-\xi_p(-\mathrm{d}\,M\,\mathrm{d}\,N)p^{-1})^{-1} < \alpha'(M,N) < c_4 \prod (1-\xi_p(-\mathrm{d}\,M\,\mathrm{d}\,N)p^{-1})^{-1},$$

where ξ_p is the quadratic residue symbol at p and p ranges over $P(M,N)$.

(iv) *If either $n = m = 2$ with $M_p \cong N_p$ for every $p \in P(M,N)$ or $n = m+1 = 2$ and if $-\mathrm{d}\,N \notin (\mathbb{Q}^\times)^2$, then there exist positive numbers c_5, c_6 such that*

$$c_5 \prod (1-\xi_p(-\mathrm{d}\,N)p^{-1}) < \alpha'(M,N) < c_6 \prod (1-\xi_p(-\mathrm{d}\,N)p^{-1}),$$

where p ranges over $P(M,N)$.

Proof. To prove the proposition, we note that for a subset S of prime numbers and $a > 1$, we have

$$\prod_p (1-p^{-a}) \le \prod_{p\in S} (1-p^{-a}) < 1,$$

$$1 < \prod_{p\in S}(1+p^{-a}) = \prod_{p\in S}(1-p^{-2a}) \cdot \prod_{p\in S}(1-p^{-a})^{-1} < \prod_p (1-p^{-a})^{-1}.$$

Here we note that all infinite products are convergent since $\sum p^{-a}$ is convergent for $a > 1$. Thus $\prod_{p\in S}(1+p^{-a})$ and $\prod_{p\in S}(1-p^{-a})$ are bounded from above and below by positive numbers independent of S but dependent on a. Let q be a natural number greater than 1 and χ be a homomorphism from $(\mathbb{Z}/q\mathbb{Z})^\times$ to $\mathbb{C}^\times$, and define

$$L(s,\chi) := \prod_p (1-\chi(p)p^{-s})^{-1},$$

where we put $\chi(p) = 0$ if p divides q. If the real part of s is greater than 1, then $L(s,\chi)$ is absolutely convergent and $L(s,\chi)$ has a meromorphic continuation on $\mathbb{C}$. If $\chi \neq 1$, then the infinite product $\prod_p(1-\chi(p)p^{-1})^{-1}$ is convergent and it is equal to $L(1,\chi)$ (see §109 in [Lan]). Thus to prove the proposition we can neglect factors $1 \pm p^{-a}$ $(a > 1)$ in the formula of $\beta_p(M_p, N_p)$ and we have only to take care over the factor

$$E_p := \epsilon(1-\chi_p(\bar{N})p^{-n/2})(1+\chi_p(\bar{M} \perp \bar{N}^{(-1)})p^{(m-n)/2})$$

where $\epsilon = 1$ if $n > m$, and $= \frac{1}{2}$ if $n = m$. If moreover $n/2$ or $(n-m)/2 > 1$, then the corresponding factor in E_p can be replaced by 1. We note that $\chi_p(\bar{N}) = 0$ if rank N is odd, and is equal to $\xi_p((-1)^{n/2}\,\mathrm{d}\,N)$ otherwise. The case in which we can neglect the factor $1-\chi_p(\bar{N})p^{-n/2}$ is that either n is odd or $n > 2$. The case in which the factor $\epsilon(1+\chi_p(\bar{M} \perp \bar{N}^{(-1)})p^{(m-n)/2})$ can be neglected is that $m+n$ is odd, $n-m > 2$ or $m = n$ with $M_p \cong N_p$. Thus the case (ii) is over.

Let us consider the case (iii). Then

$$(-1)^{(m+n)/2}\,\mathrm{d}\,M\,\mathrm{d}\,N^{(-1)} = -\,\mathrm{d}\,M\,\mathrm{d}\,N$$

and

$$\begin{aligned} E_p &= (1-\chi_p(\bar{N})p^{-n/2})(1+\xi_p(-\,\mathrm{d}\,M\,\mathrm{d}\,N)p^{-1}) \\ &= (1-\chi_p(\bar{N})p^{-n/2})(1-p^2)(1-\xi_p(-\,\mathrm{d}\,M\,\mathrm{d}\,N)p^{-1})^{-1} \end{aligned}$$

implies this case.

In the case of (iv), $E_p = 1-\xi_p(-\,\mathrm{d}\,N)p^{-1}$ and this case is over.

In the case of (i) $E_p = (1-\chi_p(\bar{N})p^{-n/2})(1+p^{-1})$ and this completes the proof of the case (i) and so the proof of the proposition. □

In (iv), $-\,\mathrm{d}\,N \in (\mathbb{Q}^\times)^2$ if and only if the infinite product diverges to 0.

Corollary 5.6.2. *Let M, N be lattices on regular quadratic spaces over $\mathbb{Q}$ and $n = \operatorname{rank} N, m = \operatorname{rank} M$. If $n \geq 2m+3$ and $M_p \hookrightarrow N_p$ for every prime p, then there exist positive numbers c_1, c_2 dependent only on N such that*

$$c_1 < \prod_p \alpha_p(M_p, N_p) < c_2.$$

Proof. Using (ii) in the last proposition, we have only to consider the products of $\alpha_p(M_p, N_p)$ for $p \notin P(M,N)$. Since $n \geq 2m+3$ implies $\operatorname{ind} \mathbb{Q}_p N \geq m$, (b) in Theorem 5.6.5 implies

$$\prod_{p \notin P(N)} \beta_p(M_p, N_p) > c_1$$

for some positive number c_1 dependent only on N. Suppose $p \in P(N)$ but $p \notin P(M,N)$; then $p \neq 2$ and N_p is unimodular. By Lemma 5.2.1, N_p is $\mathbb{Z}_p$-maximal and hence there is a primitive submodule M' isometric to M_p in N_p because of Theorem 5.2.2, $\operatorname{ind} N_p \geq m$ and Proposition 5.3.2. Theorem 5.6.1 implies

$$\beta_p(M_p, N_p) \geq d_p(M', N_p) = p^{m(m+1)/2 - mn} r(M_p/pM_p, N_p/pN_p),$$

where r is the number of isometries from $(M_p/pM_p, \frac{1}{2}Q)$ to $(N_p/pN_p, \frac{1}{2}Q)$. Then Lemma 5.6.10 together with the consideration of the infinite product of $1 \pm p^{-a}$ $(a > 1)$ given at the beginning of the proof of the last proposition would yield

$$\prod_{\substack{p \in P(N) \\ p \notin P(M,N)}} \beta_p(M_p, N_p) > c_2$$

for some positive constant c_2 dependent only on n. Thus the estimate from below has been proved.

Let us estimate from above. Using the estimate of the proof of (c) in Theorem 5.6.5, we have

$$\beta_p(M_p, N_p) \leq \max_{\mathbb{Q}_p M \supset M' \supset M_p} d_p(M', N_p) \sum_{s \geq 0} p^{s(m+1-n)} A(m,s),$$

where $A(m,s) := \sum_{\substack{s_i \geq 0 \\ s_1 + \cdots + s_m = s}} \prod p^{s_i(i-1)}$. Hence

$$\sum_{s \geq 0} p^{s(m+1-n)} A(m,s) = \prod_{i=1}^{m} \sum_{s_i \geq 0} p^{s_i(m-n+i)} = \prod_{i=1}^{m} (1 - p^{m-n+i})^{-1}$$

and $m - n + i \le m - (2m+3) + m \le -3$. Hence the product of $\sum_{s\ge 0} p^{s(m+1-n)} A(m,s)$ on all p is convergent. If $p \in P(N)$, then N_p is unimodular and $d_p(M', N_p)$ is equal to $p^{m(m+1)/2-mn} r(U,V)$, putting $U = M'/pM'$, $V = N_p/pN_p$. Using Lemma 5.6.10, the product of $\beta_p(M_p, N_p)$ for $p \in P(N)$ is convergent. For $p \notin P(N)$, (c) of Theorem 5.6.5 implies the existence of a positive number c_p dependent only on N_p such that $\beta_p(M_p, N_p) \le c_p$, which completes the proof. □

The estimate from below in this corollary induces Theorem 6.6.2 via an analytic approach explained in the notes of the next chapter, although only the case of $m = 1, 2$ is proved. The estimate from below in the corollary does not hold in general when $n = 2m+2$, and the theorem is false when $n = 4, m = 1$. What is the global implication of the behaviour of $\prod_p \alpha_p(M_p, N_p)$?

Exercise 1.

Let $n = 2m+2$, let U, V be regular quadratic spaces over $\mathbb{Q}$ with $\dim U = m, \dim V = n$ and let M, N be lattices on U, V, i.e. finitely generated modules containing bases of U, V respectively. Suppose that there is a primitive submodule in N_p isometric to M_p for every prime p. Show

$$\prod_p \beta_p(M_p, N_p) > c \prod_{p \mid \mathrm{n}(M)} (1 - p^{-1}),$$

where c is a positive number dependent only on N.
(Hint: By (a) in Theorem 5.6.5 and Proposition 5.6.2, $\prod_p \beta_p(M_p, N_p) > c_1$ (> 0) where p ranges over

$$\{p \mid N_p \text{ is either not unimodular or both } N_p \text{ and } M_p \text{ are unimodular}\}.$$

If N_p is unimodular but M_p is not so, then use $\beta_p(M_p, N_p) \ge \mathrm{d}_p(M_p, N_p)$ and Lemma 5.6.10.)

Exercise 2.

Let $n = 2m+1, m > 1$, and let U, V be regular quadratic spaces over $\mathbb{Q}$ with $\dim U = m, \dim V = n$ and M, N be lattices on U, V respectively. Suppose that there is a primitive submodule in N_p isometric to M_p for every prime p. Show

$$\prod_{p \in P(N)} \beta_p(M_p, N_p) > c \prod_p (1 + \epsilon_p p^{-1}),$$

where c is a positive number dependent only on N, $\epsilon_p = 0, \pm 1$ and p in the right-hand side ranges over

$$\{p \mid N_p \text{ is unimodular but } M_p \text{ is not so}\}.$$

(Hint: Use Lemma 5.6.10 and Proposition 5.6.2.)

Exercise 3.

Let $n = 2m+2$ or $2m+1$ and let N be a regular quadratic lattice over R with $\operatorname{ind} FN = m-1$. For a regular submodule M of N with $\operatorname{rank} M = m$, show $\beta_p(p^t M, N) \to 0$ as $t \to \infty$.
(Hint: Use (c) in Theorem 5.6.5.)

Exercise 4.

Let F_q be a finite field with q elements. Let $U = U_0 \perp \operatorname{Rad} U$ be a quadratic space over F_q and H_k be the hyperbolic space of $\dim = 2k$ over F_q. Suppose that U is represented by H_k. Then show

$$\begin{aligned} & q^{u(u+1)/2-2ku} r(U, H_k) \\ & = (1-q^{-k})(1+\chi(U_0)q^{u-d/2-k}) \prod_{1 \le i < u-d/2} (1-q^{2i-2k}) \end{aligned}$$

where $d := \dim U_0, u := \dim U$ and

$$\chi(U_0) = \begin{cases} 0 & \text{if } d \text{ is odd} \\ 1 & \text{if } U_0 \text{ is a hyperbolic space,} \\ -1 & \text{otherwise;} \end{cases}$$

(q is not necessarily odd).

6

Quadratic forms over $\mathbb{Z}$

In this chapter and the next, we study quadratic forms over $\mathbb{Z}$, which is the main goal of this book. All previous chapters converge here. The next has another character; there are still many facts which we do not cover in this chapter.

6.1 Fundamentals

Here we give some basic notations and facts. First of all we give a correspondence between local and global lattices.

Let a field F and a ring R be either $\mathbb{Q}$ and $\mathbb{Z}$ or $\mathbb{Q}_p$ and $\mathbb{Z}_p$. A module L over R in a finite-dimensional vector space V over F is called a *lattice* on V if L is finitely generated and contains a basis of V. Then L has a basis over R and $\operatorname{rank} L = \dim V$.

Let V be a regular quadratic space over $\mathbb{Q}$. A lattice on it is called a *regular (quadratic) lattice.*

If V_∞ is positive definite, i.e. $Q(x) > 0$ for every non-zero $x \in V_\infty$, then V is called *positive definite* and a lattice on it is called a *positive (definite quadratic) lattice.* If V_∞ is negative definite, i.e. $Q(x) < 0$ for every non-zero $x \in V_\infty$, then V is called *negative definite* and a lattice on it is called a *negative (definite quadratic) lattice.*

If V_∞ is isotropic, then V is called *indefinite* and a lattice on it is called an *indefinite (quadratic) lattice* . We note that if V is indefinite and $\dim V \geq 5$, then V is isotropic by Theorem 4.4.1 and 3.5.1. However the indefiniteness of V in the case of $\dim V \leq 4$ does not always imply that V is isotropic.

For a vector space V over $\mathbb{Q}$, we denote by V_p the scalar extension $\mathbb{Q}_p \otimes_{\mathbb{Q}} V$ of V to the vector space over $\mathbb{Q}_p$ as in Chapter 4 and for a lattice L on V, we denote $\mathbb{Z}_p \otimes_{\mathbb{Z}} L$ by L_p. Through a basis $\{v_i\}$ of L over $\mathbb{Z}$, we can identify $L = \mathbb{Z}^n$, $V = \mathbb{Q}^n$, $L_p = \mathbb{Z}_p^n$, $V_p = \mathbb{Q}_p^n$ canonically. Thus L_p is the closure of L in V_p, and we have $L = \cap_p(V \cap L_p)$, where p runs over all primes, since $\mathbb{Z} = \cap_p(\mathbb{Q} \cap \mathbb{Z}_p)$.

Let us recall that a place is a prime number or the formal symbol ∞ as defined in Chapter 4.

Lemma 6.1.1. *Let S be a finite set of primes and v a place not in S. For a given $a_p \in \mathbb{Q}_p$ for $p \in S$, there is a rational number $a \in \mathbb{Q}$ such that a and a_p are sufficiently close in $\mathbb{Q}_p$ for $p \in S$, and for every prime $p \notin S \cup \{v\}$ a is contained in $\mathbb{Z}_p$. Moreover if $v = \infty$, then a can be positive and sufficiently large, and if v is a prime, then a is sufficiently close to a_∞, a real number given in advance.*

Proof. First we assume $v = \infty$. Write $a_p = b_p p^{e(p)}$ $(b_p \in \mathbb{Z}_p^\times, e(p) \in \mathbb{Z})$ for $p \in S$. Put $c = \prod_{p \in S} p^{e(p)}$ and take a sufficiently large integer t. We can choose a sufficiently large integer d such that $d \equiv b_p(cp^{-e(p)})^{-1} \bmod p^t\mathbb{Z}_p$ for $p \in S$. Then $dc \equiv b_p p^{e(p)} \bmod cp^t\mathbb{Z}_p$ for $p \in S$. Putting $a := dc$, we have $a \equiv a_p \bmod cp^t\mathbb{Z}_p$ for $p \in S$. This a has the required properties.

Next suppose that $v = q$ is a prime and $a_\infty \in \mathbb{R}$ is given. Let h be a natural number so that $q^h \equiv 1 \bmod (\prod_{p \in S} p)^t$. If $a_\infty = 0$, then the rational number aq^{-mh} is the required number for the integer as constructed above and a sufficiently large integer m.

Suppose $a_\infty \neq 0$; for $sgn(a_\infty)a_p$ instead of a_p we construct c, d as above, letting t satisfy $c(\prod_{p \in S})^t > |a_\infty|$. We put

$$a := sgn(a_\infty)c(d + x^n(\prod_{p \in S} p)^t)q^{-hmn}$$

for $x := q^{hm} - 1$ and integers m, n. We can take an integer n so that

$$\begin{aligned} 0 \leq \log(a/a_\infty) &= \log((\prod_{p \in S} p)^t + dx^{-n})|a_\infty|^{-1} - n\log(q^{hm}/x) \\ &< \log((q^{hm}/x), \end{aligned}$$

which implies $1 \leq a/a_\infty < q^{hm}/x$. By virtue of $c(\prod_{p \in S} p)^t > |a_\infty|$ and $q^{hm} > x$, we have $n \geq 0$ and then a is the desired number. □

The following basic correspondence between global and local lattices is due to Eichler.

Theorem 6.1.1. *Let V be a finite-dimensional vector space over $\mathbb{Q}$, and L be a lattice on it. For each prime p, let $M(p)$ be a lattice on V_p. Then there is a lattice M on V such that $M_p = M(p)$ for every prime p if and only if $L_p = M(p)$ for almost all primes. When this is the case, there is only one such lattice M, which is given by $M := \cap_p(V \cap M(p))$.*

Proof. Let $\{v_i\}_{i=1}^n$ be a basis of L. Suppose that there is a lattice M on V such that $M_p = M(p)$ for all primes. Let $\{u_i\}_{i=1}^n$ be a basis of M and $(u_1, \cdots, u_n) = (v_1, \cdots, v_n)A$ for $A \in GL_n(\mathbb{Q})$. Then for almost all primes p, $A, A^{-1} \in GL_n(\mathbb{Z}_p)$ holds and for such a prime p, $\{u_i\}$ and $\{v_i\}$ span the same lattice over $\mathbb{Z}_p$, i.e. $L_p = M_p = M(p)$.

Conversely suppose that $L_p = M(p)$ for almost all primes p. Let S be a finite set of primes p such that $M(p) \neq L_p$ and put $q = \prod_{p \in S} p$. We take a large natural number t such that $q^t L_p \subset M(p)$ for $p \in S$. Fix a set of the representatives $\{w_{p,j_p}\}$ of $M(p)/q^t L_p$ for $p \in S$ and for each collection $J = \{j_p\}_{p \in S}$ we can take $w_J \in V$ such that

$$w_J - w_{p,j_p} \in q^t L_p \text{ for every prime } p \in S, \text{ and } w_J \in L_p \text{ for } p \notin S$$

by using the last lemma, considering the coordinate representation by a basis of L. Denote by M the lattice spanned by $q^t L$ and w_J's. Then by the method of the construction, we have $L_p = M_p = M(p)$ for $p \notin S$, and moreover $M \subset M(p)$ in V_p and hence $M_p \subset M(p)$. Let us show $M_p = M(p)$ for $p \in S$. For $x \in M(p)$ $(p \in S)$, we can take a prescribed representative w_{p,j_p} such that $x \in w_{p,j_p} + q^t L_p$. Take any representative $j_{p'}$ for $p' \in S$ with $p' \neq p$; then for a collection J of $j_{p'}$ $(p' \in S)$ we have $x \in w_{p,j_p} + q^t L_p = w_J + q^t L_p \subset M_p$, since M contains w_J and $q^t L$. Thus $M(p) \subset M_p$ and hence $M_p = M(p)$ holds for a prime $p \in S$. Thus we have completed the proof of $M_p = M(p)$ for all primes. The uniqueness of M follows from $M = \cap_p(V \cap M_p) = \cap_p(V \cap M(p))$. □

Let us introduce some important notions about quadratic lattices. Let V be a quadratic space over $\mathbb{Q}$ and L, M be lattices on V. We say that L and M are in the same *class* if

$$L = \sigma(M) \quad \text{for some} \quad \sigma \in O(V).$$

The set of lattices on V in the same class as L is denoted by

$$cls(L).$$

In the above, when we replace $O(V)$ by $O^+(V)$, we write the *proper class* and $cls^+(L)$ instead of the class and $cls(L)$.

We define the *genus* $gen(L)$ of L to be the set of all lattices M on V such that for every prime p

$$M_p = \sigma_p(L_p) \quad \text{for some} \quad \sigma_p \in O(V_p).$$

By Proposition 5.3.1, we can replace $O(V_p)$ by $O^+(V_p)$ in the definition.

The *spinor genus* $spn(L)$ of L is defined by the set of lattices M such that for $\eta \in O(V)$ and $\sigma_p \in O'(V_p)$

$$\eta(M)_p = \sigma_p(L_p) \quad \text{for every prime } p.$$

Here $O'(V_p)$ is the kernel of the spinor norm from $O^+(V_p)$ to $\mathbb{Q}_p^\times/(\mathbb{Q}_p^\times)^2$. When we replace the condition $\eta \in O(V)$ by $\eta \in O^+(V)$, we say the *proper spinor genus* and use $spn^+(L)$.

Let us introduce some groups. We denote the set of all bijective linear mappings of a vector space V over $\mathbb{Q}$ into itself by $GL(V)$. Taking a basis of V, $GL(V)$ is isomorphic to $GL_n(\mathbb{Q})$ for $n := \dim V$. The *adele group* $GL(V)_A$ is the subset of $g = (g_v)$ in the product of $\prod_v GL(V_v)$ where v runs through all places such that

$$g_p L_p = L_p \quad \text{for almost all primes } p$$

for any fixed lattice L on V. This definition is independent of the choice of a lattice L since for another lattice M, $L_p = M_p$ holds for almost all primes p. $GL(V)$ is canonically embedded in $GL(V)_A$ by $g \mapsto (g_v)$ $(g_v = g)$. When $\dim V = n$, V is isomorphic to $\mathbb{Q}^n$ and $GL(V)_A$ is canonically identified with $g = (g_v)$ such that g_v is in $GL_n(\mathbb{Q}_v)$ for every place v and g_p is in $GL_n(\mathbb{Z}_p)$ for almost all primes p. When $n = 1$, $GL(V)_A$ is

$$\{(g_v) \mid g_v \in \mathbb{Q}_v^\times \text{ and } g_p \in \mathbb{Z}_p^\times \text{ for almost all primes } p\}$$

which is called the *idele group* of $\mathbb{Q}$ and denoted by $\mathbb{Q}_A^\times$. As above, $\mathbb{Q}^\times$ is canonically embedded in $\mathbb{Q}_A^\times$ by $x \mapsto (x_v)$ where $x_v = x$ for all places.

For a quadratic space V over $\mathbb{Q}$, the adelization groups

$$O_A(V),\ O_A^+(V),\ O'_A(V)$$

are defined by the set of $g = (g_v)$ in $GL(V)_A$ such that $g_v \in O(V_v)$, $O^+(V_v)$, and $O'(V_v)$ for all places v respectively. Since V_v is a topological space through $V_v \cong \mathbb{Q}_v^n$, $GL(V_v) \cong GL_n(\mathbb{Q}_v)$ is also a topological space. Then we take as bases of open sets of $GL(V)_A$ all the subsets which are of form

$$\prod_{v \in S} U_v \times \prod_{v \notin S} GL_n(\mathbb{Z}_p)$$

for which S is a finite set of places and U_v is an open set in $GL_n(\mathbb{Q}_v)$ for $v \in S$. This topology is called the restricted direct product topology. Now the topology on groups $O_A(V)$, $O_A^+(V)$, $O'_A(V)$ is induced by restriction.

Let us consider the action of $GL(V)_A$ on lattices. For $g = (g_v) \in GL(V)_A$ and a lattice L on V, we put

$$gL := \cap_p (V \cap g_p L_p).$$

By definition of $GL(V)_A$, $g_p L_p = L_p$ for almost all primes p, and by Theorem 6.1.1, $\cap_p(V \cap g_p L_p)$ becomes a lattice M on V with $M_p = g_p L_p$ for all p. We put

$$O_A^+(L) = \{g \in O_A^+(V) \mid gL = L\}.$$

Using the adele groups, we can write

$$cls^+(L) = O^+(V)L,\ cls(L) = O(V)L,\ gen(L) = O_A(V)L = O_A^+(V)L,$$
$$spn(L) = O(V)O'_A(V)L,\ spn^+(L) = O^+(V)O'_A(V)L$$

Then it is easy to see

$$cls^+(L) \subset cls(L) \subset spn^+(L) \subset spn(L) \subset gen(L).$$

Two symmetric regular matrices A and $B \in M_n(\mathbb{Q})$ are said to be in the same genus if there exists $g_p \in GL_n(\mathbb{Z}_p)$ such that $A = B[g_p]$ for every prime p and $A = B[g_\infty]$ for $g_\infty \in GL_n(\mathbb{R})$. This is compatible with our definition. For quadratic spaces $U \cong \langle A \rangle$, $V \cong \langle B \rangle$, the above assumptions mean $U_v \cong V_v$ for all places and then Corollary 4.4.3 yields $U \cong V$.

The fundamental result on the number of the classes in $gen(L)$ is the following.

Theorem 6.1.2. *Let V be a regular quadratic space over $\mathbb{Q}$, and L be a lattice on V. Then the number of the different proper classes in $gen(L)$ is finite.*

Proof. Take an integer a such that $\mathrm{s}(L^{(a)}) \subset \mathbb{Z}$. Since the number of the different proper classes in $gen(L)$ is the same as in $gen(L^{(a)})$, we may assume $\mathrm{s}(L) \subset \mathbb{Z}$. If $M \in gen(L)$, then $\mathrm{s}(M_p) = \mathrm{s}(L_p)$ for every prime p implies $\mathrm{s}(M) = \mathrm{s}(L) \subset \mathbb{Z}$. If $L \cong \langle A \rangle$, $M \cong \langle B \rangle$ for symmetric matrices A, B, then A, B are integral matrices and $M \in cls^+(L)$ if and only if $A = B[T]$ for some $T \in SL_n(\mathbb{Z})$. Noting $[GL_n(\mathbb{Z}) : SL_n(\mathbb{Z})] = 2$, Corollary 2.1.1 asserts the finiteness of the number of equivalence classes in question. □

Theorem 6.1.3. *Let U, V be quadratic spaces over $\mathbb{Q}$ and L, M be lattices on U, V respectively. If there is an isometry $\sigma_v : L_v \hookrightarrow M_v$ for every place v, then $L \hookrightarrow N$ holds for some $N \in gen(M)$. Here L_v, M_v mean U_v, V_v for $v = \infty$.*

Proof. The existence of σ_v induces $U_v \hookrightarrow V_v$ for every place v and so $U \hookrightarrow V$. Hence we may assume $U \subset V$ and $\sigma_v \in O(V_v)$, and so $\sigma_p(L_p) \subset M_p$. We define a lattice N on V by

$$N_p := \begin{cases} M_p & \text{if } L_p \subset M_p, \\ \sigma_p^{-1}(M_p) & \text{otherwise.} \end{cases}$$

This is well-defined because $L_p \subset M_p$ holds for almost all primes p. Then $L_p \subset N_p$ is clear for all primes p and hence $L \subset N \in gen(M)$ holds. □

6.2 Approximation theorems

In this section, we give two approximation theorems which are needed later.

Theorem 6.2.1. *Let V be a regular quadratic space over $\mathbb{Q}$ with $m := \dim V \geq 2$, and suppose that V is not a hyperbolic plane and that $V_\infty \cong (\perp_r \langle 1 \rangle) \perp (\perp_s \langle -1 \rangle)$ $(r + s = m)$. Suppose that the following data are given:*

(a) *a lattice M on V,*
(b) *a finite set S of primes p such that $S \ni 2$ and M_p is unimodular if $p \notin S$,*
(c) *integers r', s' with $0 \leq r' \leq r$, $0 \leq s' \leq s$ and $n := r' + s' < m$,*
(d) *$x_{1,p}, \cdots, x_{n,p} \in M_p$ for $p \in S$.*

Then there are vectors $x_1, \cdots, x_n \in M$ satisfying

(i) *x_i and $x_{i,p}$ are sufficiently close in V_p for $1 \leq i \leq n$ and $p \in S$,*
(ii) *$\det(B(x_i, x_j)) \in \mathbb{Z}_p^\times$ for $p \notin S$ with precisely one exception $p = q$, where $\det(B(x_i, x_j)) \in q\mathbb{Z}_q^\times$,*
(iii) *the subspace spanned by $x_1, \cdots, x_n$ in V_∞ is isometric to $(\perp_{r'} \langle 1 \rangle) \perp (\perp_{s'} \langle -1 \rangle)$.*

Proof. We use induction on n, m. First suppose $n = 1$ and $m = 2$. We begin by showing that the theorem is true for V if it holds for a scaling $V^{(a)}$. Suppose that the theorem is true for $V^{(a)}$ and that the data $M, S, x_{1,p}, r', s'$ in (a),...,(d) are given. Put $S(a) = S \cup \{p \mid a \notin \mathbb{Z}_p^\times\}$. Now V and $V^{(a)}$ have the same underlying vector space and we denote by M' the lattice M when M is considered in $V^{(a)}$. Then condition (b) for M' is satisfied for $S(a)$. For a prime $p \in S(a)\backslash S$, $p \neq 2$ and M_p is unimodular, hence there is an element $x_{1,p} \in M_p$ such that $Q(x_{1,p}) \in \mathbb{Z}_p^\times$. If $a > 0$, then we put $r'' = r', s'' = s'$. Otherwise, put $r'' = s'$ and $s'' = r'$. Applying the theorem to $V^{(a)}$, there is a vector $x \in M'$ for which

(i)$_1$ x and $x_{1,p}$ are sufficiently close in V_p for $p \in S(a)$,

(ii)$_1$ for $p \notin S(a)$, $aQ(x) \in \mathbb{Z}_p^\times$ with precisely one exception $p \neq q$, where $aQ(x) \in q\mathbb{Z}_q^\times$, and

(iii)$_1$ $aQ(x) > 0$ (resp. < 0) if $r'' = 1,\ s'' = 0$ (resp. $r'' = 0,\ s'' = 1$).

The conditions (i)$_1$, (iii)$_1$ imply immediately (i), (iii) for V. For $p \in S(a)\backslash S$, $Q(x_1)$ and $Q(x_{1,p})$ are sufficiently close by (i)$_1$. Since $Q(x_{1,p}) \in \mathbb{Z}_p^\times$, $Q(x_1)$ is also in $\mathbb{Z}_p^\times$. For $p = q$ ($\notin S(a)$), $aQ(x) \in q\mathbb{Z}_q^\times$ is nothing but $Q(x) \in q\mathbb{Z}_q^\times$. Thus the conditions (i), (ii), (iii) have been verified for V.

Let $\{v_1, v_2\}$ be an orthogonal basis of V; then

$$Q(a_1v_1 + a_2v_2) = a_1^2Q(v_1) + a_2^2Q(v_2) = Q(v_1)(a_1^2 + Q(v_1)^{-1}Q(v_2)a_2^2).$$

Since V is not a hyperbolic plane, $-Q(v_1)Q(v_2) \notin \mathbb{Q}^2$ holds and hence

$$F := \mathbb{Q}(\sqrt{-Q(v_1)Q(v_2)})$$

is a quadratic extension of $\mathbb{Q}$ and V is isometric to the scaling of F by $Q(v_1)$, where F is considered as a quadratic space over $\mathbb{Q}$ whose quadratic form is given by the norm N from F to $\mathbb{Q}$. By the above reduction, we may assume $V = (F, \mathrm{N})$. We denote by R_F the maximal order of F. Put

$$S' := S \cup \{p \mid M_p \neq (R_F)_p\}.$$

For $p \in S'\backslash S$, $p \neq 2$ and M_p is unimodular and hence we can choose $x_{1,p} \in M_p$ such that $Q(x_{1,p}) \in \mathbb{Z}_p^\times$. The assertions (i), (ii), (iii) for S follow obviously from that for S'. Thus we may assume that $M_p = (R_F)_p$ for $p \notin S$. Now, applying Lemma 6.1.1 to coordinates of $x_{1,p}$ which are expressed in terms of an orthogonal basis of F, we take $y \in F$ sufficiently close to $x_{1,p}$ for $p \in S$, $y \in M_p$ for $p \notin S$, and we can choose y so that $\mathrm{N}(y) = Q(y) > 0$ (resp. < 0) if $r' = 1,\ s' = 0$ (resp. $r' = 0,\ s' = 1$) since for an element u of an orthogonal basis $\{v_i\}$ such that $Q(u) > 0$ (resp. $Q(u) < 0$) for $r' = 1$ (resp. $r' = 0$) we can make a coefficient of y at u sufficiently large. Decompose the principal ideal $(y) = yR_F$ as

$$(y) = \tilde{m}\tilde{n},$$

for ideals $\tilde{m}$, $\tilde{n}$ of F such that the prime divisor $\tilde{p}$ appears in $\tilde{m}$ if and only if $\tilde{p}$ lies above a prime number in S. Since $y \in M_p = (R_F)_p$ for $p \notin S$, $\tilde{n}$ is an integral ideal. Put $P := \prod_{p \in S} p$ and take a sufficiently large integer t. Let us consider the ray class group mod P^t in F. It is known that the ray class group represented by $\tilde{n}$ contains infinitely many prime ideals $\tilde{q}$ of degree 1 (hence $\mathrm{N}(\tilde{q})$ is a prime $\notin S$). Hence there exists $z \in F$ such that

$$\tilde{q} = \tilde{n}z, \; z \equiv 1 \bmod P^t(R_F)_p \;\text{ for }\; p \in S \;\text{ and }\; \mathrm{N}(z) > 0.$$

Put $x := yz$. Then the conditions (i), (iii) are obviously satisfied. In particular $x \in M_p$ holds for $p \in S$, and $x = yz \in \tilde{m}\tilde{q}$ implies $x \in (R_F)_p = M_p$ for $p \notin S$. Thus x is in M. The second condition follows from $Q(x) = \mathrm{N}(x) = \mathrm{N}(yz) = \pm\mathrm{N}(\tilde{m}\tilde{q}) = \pm\mathrm{N}(\tilde{m})\,\mathrm{N}(\tilde{q})$, since $\mathrm{N}(\tilde{q})$ is a prime $\notin S$ and the primes in $\mathrm{N}(\tilde{m})$ are in S. Thus the case of $n = 1$ and $m = 2$ has been completed.

Suppose $n = 1$ and $m = \dim V \geq 3$. Take any prime $q \notin S$. Then $q \neq 2$ and M_q is unimodular. By Theorem 5.2.4, there is a basis $\{v_i\}$ of M_q such that $M_q = \mathbb{Z}_q[v_1, v_2] \perp \mathbb{Z}_q v_3 \perp \cdots$ with $B(v_1, v_2) = 1$, $Q(v_1) = Q(v_2) = 0$ and $Q(v_3) \in \mathbb{Z}_q^\times$. We choose $x_1 \in M$ for which x_1 and $x_{1,p}$ are sufficiently close in V_p for $p \in S$, and x_1 and $v_1 + qv_2$ are sufficiently close in V_q. Then $Q(x_1)$ and $Q(v_1 + qv_2) = 2q$ are close in $\mathbb{Q}_q$ and hence $Q(x_1) \in q\mathbb{Z}_q^\times$. Put

$$T := \{p : \text{prime} \mid p \notin S \text{ and } Q(x_1) \notin \mathbb{Z}_p^\times\}.$$

$Q(x_1) \in q\mathbb{Z}_q^\times$ implies $Q(x_1) \neq 0$ and then T is a finite set of primes and $q \in T$. If $p \in T$, then p is odd and M_p is unimodular and so $Q(M_p) \supset \mathbb{Z}_p^\times$. We take $x_2 \in V$ by Lemma 6.1.1 such that $Q(x_2) > 0$ (resp. < 0) if $r' = 1, s' = 0$ (resp. $r' = 0, s' = 1$), $x_2 \in M_p$ and $Q(x_2) \in \mathbb{Z}_p^\times$ for $p \in T$ and lastly x_2 and v_3 are sufficiently close in V_q. Taking a natural number a such that $ax_2 \in M$ and p does not divide a for $p \in T$, we put $M' := \mathbb{Z}[x_1, ax_2] \subset M$. Then $Q(x_1)Q(ax_2) - B(x_1, ax_2)$ is sufficiently close in $\mathbb{Q}_q$ to $2qa^2Q(v_3) - B(v_1 + qv_2, av_3) = 2qa^2Q(v_3) \in q\mathbb{Z}_q^\times$ and hence M' is regular and $\mathrm{d}\,M'_q \in q\mathbb{Z}_q^\times$. Thus $\mathbb{Q}M'$ is binary and not hyperbolic. Put

$$U := \{p \notin S \cup T \mid \mathrm{d}\,M' \notin \mathbb{Z}_p^\times\}$$

and then $S' := S \cup T \cup U$ is a finite set. Put

$$x'_{1,p} := \begin{cases} x_1 & \text{if } p \in S \cup U, \\ x_2 & \text{if } p \in T. \end{cases}$$

Since $M'_p \subset M_p$, $\mathrm{d}\,M' \in \mathbb{Z}_p^\times$ and M_p is unimodular for $p \notin S'$, M'_p is unimodular for $p \notin S'$. Applying the above case of $n = 1, m = 2$ to $x'_{1,p}$ and M'_p, there is an element $x \in M' \subset M$ such that

(i)$_2$ x and $x'_{1,p}$ are sufficiently close in V_p for $p \in S'$,

(ii)$_2$ $Q(x) \in \mathbb{Z}_p^\times$ for $p \notin S'$ with precisely one exception $p = h$, where $Q(x) \in h\mathbb{Z}_h^\times$, (iii)$_2$ $Q(x) > 0$ (resp. < 0) if $r' = 1, s' = 0$ (resp. $r' = 0, s' = 1$).

Let us show that x is what we want. If $p \in S$, then x and $x'_{1,p} = x_1$, and x_1 and $x_{1,p}$ are sufficiently close in V_p respectively and hence so are x and $x_{1,p}$. If $p \in T$, then (i)$_2$ yields that x and x_2 are sufficiently close in V_p, and $Q(x_2) \in \mathbb{Z}_p^\times$. Hence $Q(x) \in \mathbb{Z}_p^\times$ for $p \in T$. If $p \in U$, then x and x_1 are sufficiently close in V_p, and $p \notin T$ and the definition of T imply $Q(x_1) \in \mathbb{Z}_p^\times$, hence $Q(x) \in \mathbb{Z}_p^\times$. Thus we have proved that if $p \in S' \backslash S$, $Q(x) \in \mathbb{Z}_p^\times$ then (ii)$_2$ implies the condition (ii) and we have completed the proof of the case $n = 1$ and $m \geq 3$.

Now suppose $1 < n < m$. We use the induction hypothesis twice. First, we apply it to $x_{1,p}, \cdots, x_{n-1,p}$, S and M. Then there exist

$$(1) \qquad x_1, \cdots, x_{n-1} \in M$$

such that

(i)$_3$ x_i and $x_{i,p}$ are sufficiently close in V_p for $1 \leq i \leq n-1$, $p \in S$,

(ii)$_3$ $\det(B(x_i, x_j))_{1 \leq i,j \leq n-1} \in \mathbb{Z}_p^\times$ for $p \notin S$ with precisely one exception $p = h$, where

$$(2) \qquad \det(B(x_i, x_j)) \in h\mathbb{Z}_h^\times,$$

(iii)$_3$ the orthogonal sum of the subspace spanned by $x_1, \cdots, x_{n-1}$ in V_∞ and $\langle \delta \rangle$ is isometric to $(\perp_{r'} \langle 1 \rangle) \perp (\perp_{s'} \langle -1 \rangle)$ for $\delta = \pm 1$.

Put $U = \mathbb{Q}[x_1, \cdots, x_{n-1}]$ and $W = U^\perp$. By (ii)$_3$ or (iii)$_3$, U is regular and hence $V = U \perp W$. We decompose $x_{n,p}$ as

$$(3) \qquad x_{n,p} = y_{n,p} + z_{n,p} \quad (y_{n,p} \in U_p,\ z_{n,p} \in W_p)$$

for $p \in S$. We will construct $x_{n,h} \in M_h$, and decompose it as above. Put

$$N := \mathbb{Z}[x_1, \cdots, x_{n-1}].$$

Then $N \subset M$ and $\mathrm{d}\, N_h \in h\mathbb{Z}_h^\times$ are clear. The proof of Proposition 5.3.3 shows $[M_h : N_h \perp N_h^\perp]$ divides $[N_h^\sharp : N_h] = h$ and hence

$$[M_h : N_h \perp N_h^\perp] = 1 \text{ or } h.$$

If $[M_h : N_h \perp N_h^\perp] = 1$, then N_h is unimodular, since M_h is unimodular. This contradicts (2). Therefore $[M_h : N_h \perp N_h^\perp] = h$ and hence $\mathrm{d}\, N_h^\perp \in h\mathbb{Z}_h^\times$. Write $N_h = L_{1,h} \perp L_{2,h}$, $N_h^\perp = L_{3,h} \perp L_{4,h}$, where $L_{1,h}$, $L_{3,h}$ are unimodular or $\{0\}$ and $L_{2,h}, L_{4,h}$ are of rank 1 and $\mathrm{d}\, L_{2,h}, \mathrm{d}\, L_{4,h} \in h\mathbb{Z}_h^\times$. Since M_h is unimodular and $L_{1,h} \perp L_{3,h}$ is a unimodular submodule of rank $= m-2$, $(L_{1,h} \perp L_{3,h})^{\perp}$ in M_h is unimodular and contains $L_{2,h} \perp L_{4,h}$. Then $N_h = L_{1,h} \perp L_{2,h}$ is a submodule of a unimodular module

$$L_h := L_{1,h} \perp ((L_{1,h} \perp L_{3,h})^{\perp} \text{ in } M_h).$$

By $\mathrm{d}\, N_h \in h\mathbb{Z}_h^{\times}$, N_h is a primitive submodule in L_h. Hence from rank $N_h = n-1$ and rank $L_h = n$ it follows that there is an element $x_{n,h} \in L_h \subset M_h$ such that

$$N_h + \mathbb{Z}_h x_{n,h} = L_h = L_{1,h} \perp (L_{1,h} \perp L_{3,h})^{\perp}.$$

Now we write $x_{n,h} = y_{n,h} + z_{n,h}$ $(y_{n,h} \in U_h, z_{n,h} \in W_h)$ so that (3) holds for $p \in S \cup \{h\}$. We take $y_n \in U = \mathbb{Q}N$ such that y_n and $y_{n,p}$ are sufficiently close in U_p for $p \in S \cup \{h\}$ and $y_n \in N_p$ for $p \notin S \cup \{h\}$ by Lemma 6.1.1. We claim

($\sharp$) there exist an element z_n in the projection M' of M to W and a prime $q \notin S \cup \{h\}$ such that

(i)$_4$ z_n and $z_{n,p}$ are sufficiently close in W_p for $p \in S \cup \{h\}$,

(ii)$_4$ $Q(z_n) \in \mathbb{Z}_p^{\times}$ for $p \notin S \cup \{q,h\}$ and $Q(z_n) \in q\mathbb{Z}_q^{\times}$, and

(iii)$_4$ $Q(z_n)\delta > 0$.

Let us show the claim ($\sharp$). For $p \notin S \cup \{h\}$, M_p and N_p are unimodular by (b) and (ii)$_3$, and hence $M_p = N_p \perp N_p^{\perp}$ and then $N_p^{\perp} \subset W_p$ yields $M'_p = N_p^{\perp}$. Thus M'_p is unimodular for $p \notin S \cup \{h\}$. Let us recall that (3) is valid even for $p = h$, and then the given condition $x_{n,p} \in M_p$ implies $z_{n,p} \in M'_p$ for $p \in S \cup \{h\}$. (iii)$_3$ implies that $\mathbb{R}U \perp \langle\delta\rangle \hookrightarrow \mathbb{R}V$ and hence

$$\mathbb{R}W = \langle\delta\rangle \perp \cdots .$$

Now we can apply the induction hypothesis to $M', S \cup \{h\}, z_{n,p}$ unless W is a hyperbolic plane. $V_h \supset M_h$ and $U_h \supset N_h$ imply $\mathrm{d}\, V_h \in \mathbb{Z}_h^{\times}(\mathbb{Q}_h^{\times})^2$ and $\mathrm{d}\, U_h \in h\mathbb{Z}_h^{\times}(\mathbb{Q}_h^{\times})^2$ and then

$$\mathrm{d}\, W_h = \mathrm{d}\, V_h \,\mathrm{d}\, U_h \in h\mathbb{Z}_h^{\times}(\mathbb{Q}_h^{\times})^2,$$

which means that W is not a hyperbolic plane. Thus we get the claim ($\sharp$) by the induction hypothesis.

Putting

$$x_n := y_n + z_n,$$

we will show $\{x_1, \cdots, x_n\}$ is what we want. To verify the condition (i), we have only to check that x_n and $x_{n,p}$ are sufficiently close for $p \in S$ by (i)$_3$. By the choice of y_n, y_n and $y_{n,p}$ are sufficiently close for $p \in S \cup \{h\}$ and then (3) and (i)$_4$ yield that x_n and $x_{n,p}$ are sufficiently close for $p \in S \cup \{h\}$, which implies the condition (i).

Let us verify $x_i \in M$. By (1), we have only to consider x_n. $x_{n,p} \in M_p$ holds for $p \in S$ by the original assumption and for $p = h$ by choice. Since x_n and $x_{n,p}$ are sufficiently close for $p \in S \cup \{h\}$ as above, $x_n \in M_p$ holds for $p \in S \cup \{h\}$. Suppose $p \notin S \cup \{h\}$. Then we get $x_n = y_n + z_n \in N_p \perp M'_p = M_p$ as above and then $x_n \in M_p$ for every prime p, i.e. $x_n \in M$.

Let us verify the condition (ii). Since $N = \mathbb{Z}[x_1, \cdots, x_{n-1}]$, x_n and $x_{n,h}$ are close for h, and $N_h + \mathbb{Z}_h x_{n,h} = L_h$ is unimodular,

$$\det(B(x_i, x_j)) \in \mathbb{Z}_h^\times.$$

For $p \notin S \cup \{h\}$,

$$\begin{aligned} \det(B(x_i, x_j))(\mathbb{Z}_p^\times)^2 &= \mathrm{d}(N_p + \mathbb{Z}_p[x_n]) \\ &= \mathrm{d}(N_p + \mathbb{Z}_p[y_n + z_n]) = \mathrm{d}(N_p + \mathbb{Z}_p[z_n]) \end{aligned}$$

since $y_n \in N_p$. Moreover $z_n \in W$ and $\mathbb{Q}N = U = W^\perp$ imply

$$\det(B(x_i, x_j)) = \mathrm{d}\, N_p \cdot Q(z_n)$$

for $p \notin S \cup \{h\}$ and then (ii)$_3$,(ii)$_4$ imply the condition (ii).

(iii) follows from $\mathbb{R}[x_1, \cdots, x_n] = \mathbb{R}N + \mathbb{R}x_n = \mathbb{R}U \perp \mathbb{R}z_n$, (iii)$_3$ and (iii)$_4$. Thus we have completed the proof of the theorem. □

The next theorem is a strong approximation theorem for $O'(V)$. To state the theorem, we recall the topology of $End(V)$ for an n-dimensional vector space V over a local field F. Fix any basis $\{v_i\}$ of V, and then we can identify $End(V)$ with $M_n(F)$, and the topology of $M_n(F)$ is transferred to $End(V)$. It is independent of the choice of a basis once it has been fixed, because $f \in End(V)$ is close to the zero mapping if and only if the coordinates $f_{i,j}$ defined by $(f(v_i), \cdots, f(v_n)) = (v_1, \cdots, v_n)(f_{i,j})$ are close to zero, and the base change induces $(f_{i,j}) \to g(f_{i,j})g^{-1}$ for some $g \in GL_n(F)$. The following strong approximation theorem is due to Eichler.

Theorem 6.2.2. *Let V be a regular quadratic space over $\mathbb{Q}$ with $\dim V \geq 3$ and suppose that V_v is isotropic for some place v; (v may be finite or infinite). Let L be a lattice on V and S be a finite set of primes with $v \notin S$. Given $\sigma_p \in O'(V_p)$ for $p \in S$, there is an isometry $\sigma \in O'(V)$ such that*

$$\sigma(L_p) = L_p \text{ for every prime } p \notin S \cup \{v\} \text{ and}$$
$$\sigma \text{ and } \sigma_p \text{ are sufficiently close in } End(V_p) \text{ for } p \in S.$$

To prove the theorem, we need several preparatory lemmas.

Lemma 6.2.1. *Let V be a regular quadratic space over $\mathbb{Q}$ and S a finite set of places including the infinite place ∞. Given isometries $\sigma_v \in O^+(V_v)$ for $v \in S$, there are vectors $x_1, \cdots, x_{2n} \in V$ such that σ_v and $\tau_{x_1} \cdots \tau_{x_{2n}}$ are sufficiently close for $v \in S$, where τ_x denotes the symmetry with respect to x.*

Proof. Put $\sigma_v = \tau_{x_1(v)} \cdots \tau_{x_{2n}(v)}$ for $x_i(v) \in V_v$. Since the order of a symmetry is 2, we may suppose that n is independent of $v \in S$. By Lemma 6.1.1 we can choose $x_i \in V$ such that x_i and $x_i(v)$ are sufficiently close in V_v for $v \in S$. Then τ_{x_i} and $\tau_{x_i(v)}$ are sufficiently close, and hence σ_v and $\tau_{x_1(v)} \cdots \tau_{x_{2n}(v)}$ are, for $v \in S$. □

Lemma 6.2.2. *Let W be a regular quadratic space over $\mathbb{Q}$ with* $\dim W \geq 3$, *S a finite set of places and v a place not in S. For a lattice K on W, there is an integer r dependent on K such that*
(i) $r \in \mathbb{Z}_p^\times$ for every prime $p \in S$ and
(ii) if a rational number c is represented by W and $c \in r\mathbb{Z}_p \cap Q(K_p)$ for every prime $p \neq v$, then there is an element $w \in W$ for which $Q(w) = c$ and $w \in K_p$ for every prime $p \neq v$.

Proof. Enlarging S, we may assume that a prime p different from v is in S if $p = 2$ or K_p is not unimodular. Let $\{K_i\}_{i=1}^h$ be a complete set of representatives of the classes in $gen(K)$. We show first that K_i can be chosen so that $(K_i)_p = K_p$ for every prime $p \in S$. By the definition of the genus and Proposition 5.3.1 there is an isometry $\sigma_{i,p} \in O^+(W_p)$ such that $\sigma_{i,p}((K_i)_p) = K_p$. By the last lemma, there is $\sigma_i \in O^+(W)$ such that σ_i and $\sigma_{i,p}$ are sufficiently close for $p \in S$. As representatives we have only to take $\sigma_i(K_i)$ again. Thus we may assume $(K_i)_p = K_p$ for $1 \leq i \leq h, \quad p \in S$. Now we take an integer s such that

$$sK_i \subset K \quad \text{for every } i \text{ and}$$
$$s \in \mathbb{Z}_p^\times \quad \text{for } p \in S,$$

and put $r := s^2$. The condition (i) is satisfied. Suppose that c is a rational number as in the condition (ii). If $p \in S$, then $p \neq v$ and hence $c/r = c/s^2 \in \mathbb{Q}(K_p)$. If $p \notin S$ and $p \neq v$, then $p \neq 2$ and K_p is unimodular, and then $K_p \cong \langle\begin{pmatrix} 0 & 1 \\ 1 & 0 \end{pmatrix}\rangle \perp$ (some lattice). Since $c/r \in \mathbb{Z}_p$, c/r is represented by $\langle\begin{pmatrix} 0 & 1 \\ 1 & 0 \end{pmatrix}\rangle$ and hence by K_p. Thus $c/r \in Q(K_p)$ if a prime p is not the place v. If v is a finite prime q, then $c/r \cdot q^{2t} \in Q(K_q)$ for a large integer t since c is represented by W. Thus c/r or $c/r \cdot q^{2t}$ is represented by K_p for every prime p if $v = \infty$ or q respectively. By Theorem 6.1.3, there exists an element x in K_i for some i such that $Q(x) = c/r$ or $c/r \cdot q^{2t}$ for $v = \infty$ or q, respectively. Put $w = sx$ or $sq^{-t}x$ for $v = \infty$ or q, respectively; then $Q(w) = c$ holds. If $v = \infty$, then $w = sx \in sK_i \subset K$. If $v = q$, then $w = sq^{-t}x \in sq^{-t}K_i \subset q^{-t}K$. Thus w is what we want. □

Lemma 6.2.3. *Let V be a regular quadratic space over $\mathbb{Q}$ with* $\dim V \geq 4$ *for which V_v is isotropic for some place v, L a lattice on V and T a finite set of prime numbers with $v \notin T$. Suppose that a non-zero rational number $a \in Q(V)$ and $z_p \in V_p$ $(p \in T)$ satisfy the conditions*

(a) $a = Q(z_p)$ *for every prime* $p \in T$,

(b) $a \in Q(L_p)$ *for every prime* $p \notin T$.

Then there is an element $z \in V$ satisfying the conditions

(i) *z and z_p are sufficiently close in V_p for every prime $p \in T$,*

(ii) $z \in L_p$ *for every prime* $p \notin T \cup \{v\}$,

(iii) $Q(z) = a$.

Proof. Scaling by a^{-1}, we may suppose $a = 1$ without loss of generality. We take a finite set T_1 of primes such that $T_1 \supset T$, $T_1 \not\ni v$, and $T_1 \cup \{v\} \not\ni p$ implies $p \neq 2$ and that L_p is unimodular. For a prime $p \in T_1 \setminus T$, we take $z_p \in L_p$ for which $Q(z_p) = 1$ by (b). Hence the conditions (a), (b) are satisfied for T_1. If the conclusions are true for T_1 instead of T, then they imply the assertion for T. So we may assume

(c) if $p \notin T \cup \{v\}$, then $p \neq 2$ and L_p is unimodular.

If V_∞ is isotropic and v is a prime number q, then we put $T' := T \cup \{q\}$; by the condition (b) for T, there exists an element $z_q \in L_q$ with $Q(z_q) = 1$ and hence (a) holds for T' instead of T. If the lemma has been verified for T' and $v = \infty$, then the conclusion for T' implies it for T. Thus we may assume

(d) if V_∞ is isotropic, then $v = \infty$.

Moreover, taking $(\prod_{p \in T} p)^{-t} L$ for a large integer t instead of L, we may assume

(e) $z_p \in L_p$ for $p \in T$.

Fix any element $x \in V$ with $Q(x) = 1$, and take an isometry $\sigma_p \in O^+(V_p)$ such that $\sigma_p(x) = z_p$ for $p \in T$. By Lemma 6.2.1, there is an isometry $\sigma \in O^+(V)$ for which σ and σ_p are sufficiently close for $p \in T$. Put $y = \sigma(x)$; then $Q(y) = 1$, and y $(= \sigma(x))$ and z_p $(= \sigma_p(x))$ are sufficiently close for $p \in T$. Then we have by (e)

(f) $y \in L_p$ for $p \in T$.

We choose an integer b by Lemma 6.1.1 such that

(g) b and 1 are sufficiently close in $\mathbb{Q}_p$ for $p \in T$ and $u := by \in L_p$ for $p \notin T \cup \{v\}$.

Then we have, with (f) and (g)

(h) $u \in L_p$ for $p \neq v$ and $Q(u) = b^2$.

Put $W := u^\perp$ and determine a lattice K on W by

(j) $$\begin{cases} K_p = (L \cap W)_p = L_p \cap W_p & \text{if } p \notin T, \\ K_p \subset p^t L_p & \text{for a sufficiently large } t \text{ if } p \in T. \end{cases}$$

Obviously $K \subset L$ holds. Put

$$T_b := \{p \mid b \notin \mathbb{Z}_p^\times \text{ and } p \neq v\}.$$

Then T_b is a finite set and $T \cap T_b = \emptyset$ since $b \in \mathbb{Z}_p^\times$ for $p \in T$ by (g). If $p \notin T_b \cup \{v\}$, then (h) implies $Q(u) = b^2 \in \mathbb{Z}_p^\times$, and if moreover $p \notin T$, then $K_p = L_p \cap (u^\perp)_p$ is unimodular since $u \in L_p$ by (h) and L_p is unimodular by (c). Therefore Corollary 5.2.2 yields

(k) $Q(K_p) = \mathbb{Z}_p$ for $p \notin T \cup T_b \cup \{v\}$.

Applying the last lemma to $W, S := T \cup T_b, v$, there is an integer r such that

(l) $r \in \mathbb{Z}_p^\times$ for $p \in T \cup T_b$, and for a rational number $c \in Q(W)$ with $c \in r\mathbb{Z}_p \cap Q(K_p)$ for every $p \neq v$, there is an element $w \in W$ for which $Q(w) = c$ and $w \in K_p$ for every $p \neq v$.

Put

$$T_r := \{p \mid r \notin \mathbb{Z}_p^\times \text{ and } p \neq v\}.$$

Then the first part of (l) implies

(m) T, T_b, T_r and $\{v\}$ are disjoint.

We claim

(n) there is a rational number d such that $d \in \mathbb{Z}_p$ if $p \neq v$, and d and 1 are sufficiently close in $\mathbb{Q}_p$ for $p \in T$, and moreover $1 - (bd)^2 \in Q(W)$ and

$$1 - (bd)^2 \in r\mathbb{Z}_p \cap Q(K_p) \quad \text{if } p \neq v.$$

We will come back to the proof of this later; but first we complete the proof of the lemma with its help. Applying (l) to $c := 1 - (bd)^2$, there is an element $w \in W$ such that $Q(w) = 1 - (bd)^2$ and $w \in K_p$ for $p \neq v$. We show that

$$z := du + w$$

is what we want. For $p \in T$, $z - z_p = (dby - z_p) + w$ is sufficiently close to 0, since b, d and 1 are sufficiently close, by (g) and (n), y and z_p are sufficiently close and $w \in K_p \subset p^t L_p$, by (j). Thus condition (i) holds. If $p \notin T \cup \{v\}$, then (h) and $K \subset L$ imply $z = du + w \in dL_p + K_p \subset L_p$, which is condition (ii). Lastly, $Q(z) = d^2 Q(u) + Q(w) = (bd)^2 + (1 - (bd)^2) = 1$ is condition (iii).

It remains to verify the claim (n). First we construct $d_p \in \mathbb{Z}_p$ for $p \in T \cup T_b \cup T_r$ which satisfies the local version of (n):

(n′) $0 \neq 1-(bd_p)^2 \in r\mathbb{Z}_p \cap Q(K_p)$ and $d_p \in \mathbb{Z}_p$ hold for $p \in T \cup T_b \cup T_r$, and d_p and 1 are sufficiently close for $p \in T$.

Let $p \in T$; taking a non-zero number $e_p \in Q(K_p)$ that is sufficiently close to 0, we put $d_p = b^{-1}(1-e_p/2)$. By (g), d_p is sufficiently close to 1 in $\mathbb{Q}_p$. Clearly

$$1-(bd_p)^2 = 1-(1-e_p/2)^2 = e_p(1-e_p/4) \in e_p(\mathbb{Z}_p^\times)^2.$$

It implies $(bd_p)^2 \neq 1$ and $1-(bd_p)^2 \in Q(K_p)$. By (l) we have $r \in \mathbb{Z}_p^\times$ and hence $1-(bd_p)^2 \in r\mathbb{Z}_p$. Thus (n′) has been proved in this case.

Let $p \in T_r$; then $p \notin T_b$ and so $b \in \mathbb{Z}_p^\times$. Choose $d_p \in \mathbb{Z}_p$ such that d_p $(\neq b^{-1})$ and b^{-1} are sufficiently close in $\mathbb{Q}_p$. Then $1-(bd_p)^2$ is sufficiently close to 0 and hence it is in $r\mathbb{Z}_p$. By (k) and (m), $Q(K_p) = \mathbb{Z}_p$ holds and then the assertion (n′) is true in this case.

Let $p \in T_b$; we put $d_p := 0$. Since (l) implies $r \in \mathbb{Z}_p^\times$,

$$1 = 1-(bd_p)^2 \in r\mathbb{Z}_p.$$

It remains to show $1 \in Q(K_p)$. Suppose that $v_1 = fy$ for $f \in \mathbb{Q}_p^\times$ is primitive in L_p. By (c), L_p is unimodular and hence $Q(v_1) = f^2 \in \mathbb{Z}_p$. Suppose $f \in \mathbb{Z}_p^\times$. Since $v_1 = fb^{-1}u$ and (j) imply that $K_p = v_1^\perp$ in L_p, K_p is unimodular and then $Q(K_p) \ni 1$, since $\operatorname{rank} K_p = \dim V - 1 \geq 3$ and $p \neq 2$. If $f \in p\mathbb{Z}_p$, then $Q(v_1) \in p\mathbb{Z}_p$ holds. Since L_p is unimodular, $B(v_1, L_p) = \mathbb{Z}_p$ and hence there is an element $v_2 \in L_p$ such that $B(v_1, v_2) = 1$. Then $\mathbb{Z}_p[v_1, v_2]$ is unimodular and so is $\mathbb{Z}_p[v_1, v_2]^\perp$ in L_p which is contained in $K_p = (u^\perp)_p$ in L_p. Thus $Q(K_p) \supset Q(\mathbb{Z}_p[v_1, v_2]^\perp) \ni 1$ holds since $\operatorname{rank} \mathbb{Z}_p[v_1, v_2]^\perp \geq 2$ and $p \neq 2$, and we have verified the assertion (n′).

Let us verify assertion (n). First, suppose $v = \infty$; then by Lemma 6.1.1, we can choose a large number $d \in \mathbb{Q}$ such that $1-(bd)^2$ is negative, and d and d_p are sufficiently close in $\mathbb{Q}_p$ for $p \in T \cup T_b \cup T_r$ and $d \in \mathbb{Z}_p$ otherwise. Since $d_p \in \mathbb{Z}_p$ for $p \in T \cup T_b \cup T_r$ by (n′), $d \in \mathbb{Z}_p$ there; but $d \in \mathbb{Z}_p$ for $p \notin T \cup T_b \cup T_r$ implies that d is an integer. By (n′), d and 1 are sufficiently close for $p \in T$. For $p \in T \cup T_b \cup T_r$,

$$1-(bd_p)^2 \in (r\mathbb{Z}_p \cap Q(K_p)) \backslash \{0\}$$

which is open, and d and d_p are sufficiently close. Therefore

$$1-(bd)^2 \in r\mathbb{Z}_p \cap Q(K_p)$$

holds for $p \in T \cup T_b \cup T_r$. If $p \notin T \cup T_b \cup T_r$, then $b, d \in \mathbb{Z}_p$ and so by (m) and (k)

$$1-(bd)^2 \in \mathbb{Z}_p = r\mathbb{Z}_p \cap Q(K_p).$$

It remains to show $1-(bd)^2 \in Q(W_\infty)$. By the way of the construction, $V_\infty = \langle 1\rangle \perp W_\infty$ and V_∞ is isotropic since $v=\infty$. Thus $Q(W_\infty)$ contains a negative number and hence $1-(bd)^2 \in Q(W_\infty)$.

Next suppose $v := q \neq \infty$; for a sufficiently large integer t, we put $d_q := q^{-t}$ and let d_p be the integer in $\mathbb{Z}_p$ in (n′). We choose a rational number d' such that d' and d_p are sufficiently close for $p \in T\cup T_b \cup T_r \cup \{q\}$ and $d' \in \mathbb{Z}_p$ otherwise. Take a sufficiently large integer m for which q^m and 1 are sufficiently close in $\mathbb{Q}_p$ for $p \in T\cup T_b\cup T_r$ and $1-(bd)^2>0$ for $d := d'q^{-m}$. Then d and d_p are sufficiently close for $p \in T\cup T_b\cup T_r$. (n′) implies that $d' \in \mathbb{Z}_p$ if $p \in T\cup T_b \cup T_r$, and hence $d' \in \mathbb{Z}_p$ if $p \neq q$. Thus $d \in \mathbb{Z}_p$ if $p \neq q$, which is the first condition in (n). Since d_p and 1 are sufficiently close for $p \in T$ by (n′), d and 1 are too, which is the second part of (n). For $p \in T\cup T_b \cup T_r$

$$1-(bd)^2 \in (r\mathbb{Z}_p \cap Q(K_p)) \setminus \{0\}$$

follows from (n′). If $p \notin T\cup T_b\cup T_r\cup\{q\}$, then

$$r\mathbb{Z}_p \cap Q(K_p) = \mathbb{Z}_p \ni 1-(bd)^2$$

follows from (k), $b \in \mathbb{Z}_p$ and $d \in \mathbb{Z}_p$. Thus $1-(bd)^2 \in r\mathbb{Z}_p \cap Q(K_p)$ if $p \neq v = q$ which is the last condition in (n). Since $V_q = \langle 1\rangle \perp W_q$ is isotropic, $Q(W_q) \ni -1$ holds and then

$$1-(bd)^2 = -(bd)^2(1-(bd')^{-2}q^{2m}) \in Q(W_q).$$

By (d), $V_\infty = \langle 1\rangle \perp W_\infty$ is anisotropic and so $Q(W_\infty) = \{x>0\} \ni 1-(bd)^2$. Thus $1-(bd)^2$ is represented locally by W and so $1-(bd)^2 \in Q(W)$ which proves (n) and so the lemma. □

Proof of Theorem 6.2.2 with $\dim V \geq 4$. We keep the notions in the theorem. We divide the proof into three steps.

(i) Suppose $\sigma_p = \tau_{x_p}\tau_{y_p}, Q(x_p) = Q(y_p)$ $(x_p, y_p \in V_p)$ for every $p \in S$.

We take any vector x such that x and x_p are sufficiently close for $p \in S$, and $x \in L_p$ otherwise, using Lemma 6.1.1, and take $\eta_p \in O(V_p)$ such that $y_p = \eta_p(x_p)$. Since x and x_p are sufficiently close for $p \in S$, $Q(x) \neq 0$ follows from $Q(x_p) \neq 0$. Choose a finite set S' of primes such that $S'\cap(S\cup\{v\}) = \emptyset$, and that $p \notin S'\cup S\cup\{v\}$ implies that $p \neq 2$, $Q(x) \in \mathbb{Z}_p^\times$, $\tau_x L_p = L_p$ and L_p is unimodular. Since $Q(x)$ and $Q(y_p)$ $(\neq 0)$ are sufficiently close for $p \in S$, we have $Q(x) = c_p^2 Q(y_p)$ for some $c_p \in \mathbb{Z}_p^\times$. Put

$$z_p := \begin{cases} c_p y_p & \text{for } p \in S, \\ x & \text{for } p \in S'. \end{cases} \tag{5}$$

If $p \notin S' \cup S \cup \{v\}$, then $p \neq 2$ and L_p is unimodular and hence $Q(x) \in \mathbb{Z}_p^\times \subset Q(L_p)$. Applying the last lemma to $z_p, T = S' \cup S$ and $0 \neq a := Q(x) \in Q(V)$, there is a vector $z \in V$ with $Q(z) = Q(x)$ such that

$$(6) \qquad \begin{aligned} &z \text{ and } z_p \text{ are sufficiently close for } p \in S' \cup S, \\ &z \in L_p \text{ for } p \notin S' \cup S \cup \{v\}. \end{aligned}$$

Put

$$\sigma := \tau_x \tau_z.$$

If $p \in S$, then $\sigma = \tau_x\tau_z$ and $\sigma_p = \tau_{x_p}\tau_{y_p}$ are sufficiently close since x and x_p, z and $z_p = c_p y_p$ are sufficiently close. If $p \in S'$, then $\sigma = \tau_x\tau_z$ and $\tau_x^2 = \mathrm{id}$ are sufficiently close since by (5), (6) x and z are sufficiently close, and hence $\sigma(L_p) = L_p$. If $p \notin S' \cup S \cup \{v\}$, then $\tau_x L_p = L_p$ follows from the definition of S'. On the other hand, $z \in L_p$ and $Q(z) = Q(x) \in \mathbb{Z}_p^\times$ imply $\tau_z \in O(L_p)$ since L_p is unimodular. Thus $\sigma(L_p) = L_p$ holds for $p \notin S' \cup S \cup \{v\}$, and σ is what we want.

(ii) Suppose that $\sigma_p = \tau_{x_{1,p}}\tau_{y_{1,p}} \cdots \tau_{x_{r,p}}\tau_{y_{r,p}}$, $Q(x_{i,p}) = Q(y_{i,p})$ for every $p \in S$.

In this case, we may assume that r is independent of $p \in S$, since the order of symmetries is 2. Applying (i) to $\tau_{x_{i,p}}\tau_{y_{i,p}}$, we complete this case.

(iii) General case.

Since V_p is isotropic for almost all primes, we may enlarge S by adding anisotropic primes where we make σ_p the identity mapping. So we may assume that V_p is isotropic if $p \notin S$. Put

$$\sigma_p := \tau_{x_{1,p}} \cdots \tau_{x_{2r,p}} \qquad \text{with} \qquad \prod_{i=1}^{2r} Q(x_{i,p}) = 1,$$

and assume that r is independent of $p \in S$. Take $x_1, \cdots, x_{2r-1} \in V$ satisfying

(a) x_i and $x_{i,p}$ are sufficiently close for $p \in S$ and $1 \leq i \leq 2r-1$,

and so are $\prod_{i=1}^{2r-1} Q(x_i)$ and $\prod_{i=1}^{2r-1} Q(x_{i,p})$ $(\neq 0)$ for $p \in S$. Hence there is a unit ϵ'_p such that

$$Q(x_{2r,p})^{-1} = \prod_{i=1}^{2r-1} Q(x_{i,p}) = \epsilon'_p \prod_{i=1}^{2r-1} Q(x_i),$$

and ϵ'_p is sufficiently near 1. Therefore there is a unit ϵ_p such that $\epsilon'_p = \epsilon_p^2$, and ϵ_p is sufficiently close to 1 for $p \in S$.

We claim the existence of a vector $x_{2r} \in V$ which satisfies

$$Q(x_{2r}) = \prod_{i=1}^{2r-1} Q(x_i)^{-1}$$

and
(b) $x_{2r}, x_{2r,p}$ are sufficiently close for $p \in S$.
Put

$$a := \prod_{i=1}^{2r-1} Q(x_i)^{-1}. \tag{7}$$

First let us verify that a is locally represented by V. For $p \in S$, $a = Q(\epsilon_p x_{2r,p}) \in Q(V_p)$ is clear. Since V_p is isotropic for $p \notin S$, a is represented by V_p for $p \notin S$. If V_∞ is isotropic, then a is also represented by V_∞. If V_∞ is anisotropic, then $aQ(x_{2r-1}) = \prod_{i=1}^{2r-2} Q(x_i)^{-1} > 0$ implies $a \in Q(V_\infty)$. Thus a is locally represented by V and hence a is represented by V, that is there is a vector $w \in V$ with

$$Q(w) = a. \tag{8}$$

Take $\eta_p \in O^+(V_p)$ such that $\eta_p(w) = \epsilon_p x_{2r,p}$ for $p \in S$ and by Lemma 6.2.1 there is an isometry $\eta \in O^+(V)$ such that η and η_p are sufficiently close for $p \in S$. Put $x_{2r} := \eta(w)$; then $Q(x_{2r}) = \prod_{i=1}^{2r-1} Q(x_i)^{-1}$ by (7) and (8), and $x_{2r} = \eta(w)$, $\epsilon_p x_{2r,p} = \eta_p(w)$ are sufficiently close for $p \in S$. Since ϵ_p and 1 are sufficiently close, x_{2r} and $x_{2r,p}$ are so. Thus we have verified the claim.

Put

$$S' := \{p \notin S \cup \{v\} \mid \tau_{x_1} \cdots \tau_{x_{2r}} L_p \neq L_p\}.$$

Since V_p is isotropic for $p \in S'$, $\tau_{x_1} \cdots \tau_{x_{2r}}$ is a product of $\tau_x \tau_y$ (where $x, y \in V_p, Q(x) = Q(y) \neq 0$) by Proposition 1.6.3. Hence by virtue of (ii) there exists an isometry $\sigma_1 \in O'(V)$ such that σ_1 and the identity are sufficiently close for $p \in S$, and σ_1 and $\tau_{x_1} \cdots \tau_{x_{2r}}$ are sufficiently close for $p \in S'$ and $\sigma_1(L_p) = L_p$ for $p \notin S \cup S' \cup \{v\}$. Now $\sigma := \sigma_1^{-1} \tau_{x_1} \cdots \tau_{x_{2r}}$ is what we want, since x_i and $x_{i,p}$ are sufficiently close for $p \in S$ and $1 \leq i \leq 2r$. Thus we have completed the proof of the theorem in the case of $\dim \geq 4$.

Suppose now $\dim V = 3$.

By scaling, we may assume $\mathrm{d}V = 1$, and then V is identified with the set of all pure elements of some quaternion algebra C as in 1.5. Then $C = \mathbb{Q} \oplus V$ and for $x = a + w \in C$ $(a \in \mathbb{Q}, w \in V)$ we put $\bar{x} := a - w$

and $N(x) = x\bar{x}$. N is a quadratic form on C and the restriction on V is the original one on V. Put $L' = \mathbb{Z} \perp L$ and take a finite set T ($\not\ni v$) of primes such that $T \supset S$ and if $p \notin T \cup \{v\}$, then $\bar{L'_p} = L'_p$ is unimodular and a subring of C_p. By Proposition 1.6.4 there exists $z_p \in C_p$ such that

$$\sigma_p(x) = z_p x z_p^{-1} \quad \text{and} \quad \theta(\sigma_p) = N(z_p)$$

for the given σ_p for $p \in S$. By $N(z_p) = \theta(\sigma_p) \in (\mathbb{Q}_p^\times)^2$, we put $N(z_p) = a_p^2$, $a_p \in \mathbb{Q}_p^\times$. Taking $a_p^{-1} z_p$ instead of z_p, we may assume

$$\sigma_p(x) = z_p x z_p^{-1}, \quad N(z_p) = 1.$$

For $p \in T \setminus S$, we put $z_p = 1$. Applying Lemma 6.2.3 to a lattice L' on C with $a = 1$, there is a vector $z \in C$ such that $N(z) = 1$, z and z_p are sufficiently close for $p \in T$ and $z \in L'_p$ for $p \notin T \cup \{v\}$. We define an isometry $\sigma \in O(C)$ by

$$\sigma(y) = zyz^{-1}.$$

By Proposition 1.5.4, $\sigma \in O^+(C)$ holds and $\sigma(1) = 1$ implies $\sigma|_V \in O^+(V)$, which then by Proposition 1.6.4 yields $\sigma(x) = vxv^{-1}$ for $x \in V$, $v \in C$ and $\theta(\sigma|_V) = N(v)$. Then $\sigma(x) = vxv^{-1} = zxz^{-1}$ for $x \in C$. Thus $z^{-1}v$ is in the centre of C, that is $v = az$ for some $a \in \mathbb{Q}$. Thus $\theta(\sigma|_V) = N(v) = N(az) = a^2$ and so $\sigma|_V \in O'(V)$. Now we regard σ_p as an isometry of C_p. If $p \in S$, then z and z_p are sufficiently close. Hence σ and σ_p are sufficiently close, as are $\sigma|_{V_p}$ and $\sigma_p|_{V_p}$ for $p \in S$, using $\sigma(1) = \sigma_p(1) = 1$. If $p \in T \setminus S$, then σ and id are sufficiently close since z and $z_p = 1$ are so and hence $\sigma|_{V_p}$ is sufficiently close to id, which yields $\sigma|_{V_p}(L_p) = L_p$. Suppose $p \notin T \cup \{v\}$; then by the assumption on T, $\bar{L'_p} = L'_p$ is unimodular and a ring. Then $z \in L'_p$ and $N(z) = 1$ imply $L'_p = \mathbb{Z}_p z \perp$ (some lattice); hence $\tau_z(L'_p) = L'_p$ and $\bar{z}^{-1} = z \in L'_p = \bar{L'_p}$ and so $z^{-1} \in L'_p$. Thus $\tau_z(x) = -z\bar{x}\bar{z}^{-1} = -z\bar{x}z$ yields

$$\sigma(L'_p) = zL'_p z^{-1} = zL'_p z^{-2} \cdot z \subset zL'_p z = -z\bar{L'_p}z = \tau_z(L'_p) = L'_p.$$

Therefore $\sigma|_{V_p}(L_p) \subset L_p$ holds since L_p is the orthogonal complement of 1, and the theorem has been proved. □

6.3 Genus, spinor genus and class

In this section, we generalize the concepts of genus, spinor genus and class given in 6.1. Let V be a regular quadratic space over $\mathbb{Q}$ and S a finite set of places with $\infty \in S$. Put $\mathbb{Z}_S := \{a \in \mathbb{Q} \mid a \in \mathbb{Z}_p \text{ for } p \notin S\}$. For a lattice L on V, we can associate an S-lattice L_S by

$$\bigcap_{p \notin S} (V \cap L_p)$$

which is equal to $\cup_{t \geq 0} (\prod_{p \in S} p)^{-t} L$ and is a finitely generated module over $\mathbb{Z}_S$. In other words, an S-lattice L is a submodule of V such that $L_p = V_p$ if $p \in S$, and L_p is a lattice on V_p if $p \notin S$. If $S = \{\infty\}$, then an S-lattice is nothing but a lattice defined as before. We define genus, spinor genus and class for S-lattices in the same way as for the standard lattice:

$$gen_S(L) := O_A(V)L = O_A^+(V)L, \quad spn_S(L) := O(V)O'_A(V)L,$$
$$spn_S^+(L) := O^+(V)O'_A(V)L, \quad cls_S(L) := O(V)L, \quad cls_S^+(L) := O^+(V)L,$$

where $g = (g_v) \in O_A(V)$ acts on the set of S-lattices L by

$$gL := \bigcap_{p \notin S} (V \cap g_p L_p).$$

We define the stabilizer group of L by

$$\begin{aligned} O_A^+(L) &:= \{g \in O_A^+(V) \mid g(L) = L\} \\ &= \{g \in O_A^+(V) \mid g_p(L_p) = L_p \text{ for } p \notin S\} \end{aligned}$$

If $S = \{\infty\}$, then the above definitions are the same as those given in 6.1. The following is due to Kneser.

Theorem 6.3.1. *Let L be an S-lattice on a regular quadratic space V over $\mathbb{Q}$; then the number $g_S^+(L)$ of the different proper spinor genera spn_S^+ in $gen_S(L)$ is given by*

$$[\mathbb{Q}_A^\times : \prod_{v \in S} \mathbb{Q}_v^\times \cdot \theta(O^+(V))\theta(O_A^+(L))] \ \text{ if } \ \dim V \geq 3.$$

Proof. First we note that $O'_A(V)$ contains all commutators in $O_A^+(V)$; hence $O'_A(V)$ is a normal subgroup of $O_A^+(V)$ and $O_A^+(V)/O'_A(V)$ is commutative. Suppose $spn_S^+(M) = spn_S^+(N)$ for $M, N \in gen_S(L)$. Write $M = gL, N = hL$ for $g, h \in O_A^+(V)$. By definition, we have

$$O^+(V)O'_A(V)gL = O^+(V)O'_A(V)hL.$$

Thus $g^{-1}h \in O^+(V)O'_A(V)O^+_A(L)$ and

$$g^+_S(L) = [O^+_A(V) : O^+(V)O'_A(V)O^+_A(L)].$$

Now we consider the spinor norm $\theta : O^+_A(V) \to \mathbb{Q}_A$, where we identify the image with its inverse image by $\mathbb{Q}^\times_A \to \mathbb{Q}^\times_A/(\mathbb{Q}^\times_A)^2$. This norm induces (1)

$$\theta : O^+_A(V)/O^+(V)O^+_A(L)O'_A(V) \to \mathbb{Q}^\times_A/(\prod_{v\in S}\mathbb{Q}^\times_v \cdot \theta(O^+(V))\theta(O^+_A(L))).$$

Note that the v-components of the above factor groups are trivial for $v \in S$. Let us show the surjectivity. Let $x = (x_v) \in \mathbb{Q}^\times_A$ with $x_v = 1$ for $v \in S$. Since $\dim V \geq 3$, Proposition 5.5.1 guarantees the existence of $g_p \in O^+(V_p)$ such that $\theta(g_p) \ni x_p$ and moreover Proposition 5.5.4 implies the existence of $g_p \in O^+(L_p)$ such that $\theta(g_p) \ni x_p$ if $x_p \in \mathbb{Z}^\times_p$ and L_p is unimodular. Hence $g := (g_v)$ with $g_v = \mathrm{id}$ for $v \in S$ is in $GL(V)_A$, i.e. $g \in O^+_A(V)$. Thus the surjectivity has been shown.

Next suppose

$$\theta(g) \in \prod_{v\in S}\mathbb{Q}^\times_v \cdot \theta(O^+(V))\theta(O^+_A(L)).$$

Taking $h \in O^+(V)$, $k \in O^+_A(L)$, we may suppose

$$\theta(ghk) \subset \prod_{v\in S}\mathbb{Q}^\times_v \times \prod_{p\notin S}(\mathbb{Q}^\times_p)^2.$$

Moreover we may assume $(ghk)_v = 1$ if $v \in S$, since the v-component of $O^+_A(L)$ is $O^+(V_v)$ for $v \in S$. Then $\theta(ghk) \subset (\mathbb{Q}^\times_A)^2$, i.e. $ghk \in O'_A(V)$. Thus the injectivity has been proved. □

Corollary 6.3.1. *$g^+_S(L)$ is a power of 2.*

Proof. First, we note that the mapping θ at (1) in the proof of the theorem is injective even for dim 2. Since $\theta(O^+_A(L)) \supset (\mathbb{Q}^\times_A)^2$, the injectivity of the mapping θ at (1) in the proof of the theorem yields $g^+_S(L)$ is a power of 2 if it is finite. If $S = \{\infty\}$, then $g^+(L)$ is finite by Theorem 6.1.2. Suppose $S \neq \{\infty\}$. Take any lattice L_p on V_p for every prime $p \in S$, and put $L' := \cap_{p\neq\infty}(V \cap L_p)$. For a lattice M on V, we associate an S-lattice M_S by $M_S := \cap_{p\notin S}(V \cap M_p)$. Then

$$cls(M) \to cls_S(M_S), spn^+(M) \to spn^+_S(M_S), gen(M) \to gen_S(M_S)$$

are well-defined and we have a surjection from cls in $gen(L')$ to cls_S in $gen_S(L)$, since for $N \in gen_S(L)$,

$$N' = \cap_{p\in S}(V \cap L_p) \cap_{p\notin S} (V \cap N_p) \in gen(L')$$

corresponds to N. Hence the finiteness of the number of cls in $gen(L')$ implies that the number g^+_S of spn^+_S in $gen_S(L)$ is finite. □

Let us determine $\theta(O^+(V))$.

Proposition 6.3.1. *Let V be a regular quadratic space over $\mathbb{Q}$ with $\dim V \geq 3$. Then we have*

$$\theta(O^+(V)) = \begin{cases} \{a \in \mathbb{Q}^\times \mid a > 0\} & \text{if } V_\infty \text{ is anisotropic,} \\ \mathbb{Q}^\times & \text{otherwise.} \end{cases}$$

Proof. Let $a \in \mathbb{Q}^\times$ and suppose $a > 0$ if V_∞ is anisotropic. Put

$$\delta = \begin{cases} -1 & \text{if } V_\infty \text{ is positive definite,} \\ 1 & \text{otherwise.} \end{cases}$$

If V_p is anisotropic for a prime p, then we take $b_p \in \mathbb{Q}_p^\times$ such that $\delta a b_p \, \mathrm{d}V \not\subset (\mathbb{Q}_p^\times)^2$ and $\delta b_p \, \mathrm{d}V \not\subset (\mathbb{Q}_p^\times)^2$. Take $0 < b \in \mathbb{Q}$ such that b and b_p are sufficiently close in $\mathbb{Q}_p$ for a prime p for which V_p is anisotropic. Note that the number of primes for which V_p is anisotropic is finite by Theorem 3.5.1. If there is no anisotropic prime, then we put $b = 1$. Then $V \perp \langle \delta b \rangle$ and $V \perp \langle \delta ab \rangle$ are isotropic at every place by the same theorem. Thus they are isotropic and hence $-\delta b, -\delta ab \in Q(V)$. Taking $x, y \in V$ such that $Q(x) = -\delta b$, $Q(y) = -\delta ab$, $\theta(\tau_x \tau_y) = a(\mathbb{Q}^\times)^2$. Thus

$$\theta(O^+(V)) \supset \begin{cases} \{a \in \mathbb{Q}^\times \mid a > 0\} & \text{if } V_\infty \text{ is anisotropic,} \\ \mathbb{Q}^\times & \text{otherwise.} \end{cases}$$

The converse inclusion is clear. □

Corollary 6.3.2. *Let L be an S-lattice on a regular quadratic space V over $\mathbb{Q}$ with* $\dim \geq 3$. *If*

$$\theta(O^+(L_p)) \supset \mathbb{Z}_p^\times$$

for every prime $p \notin S$, then $gen_S(L) = spn_S^+(L)$.

Proof. By the assumption $\theta(O_A^+(L))$ contains $\prod_{p \notin S} \mathbb{Z}_p^\times$. Putting $\mathbb{Q}_+^\times = \{a \in \mathbb{Q} \mid a > 0\}$, we know $\mathbb{Q}_A^\times = \mathbb{R}^\times \times \mathbb{Q}_+^\times \times \prod \mathbb{Z}_p^\times$ and then

$$\mathbb{Q}_A^\times \subset \prod_{v \in S} \mathbb{Q}_v^\times \times \mathbb{Q}_+^\times \times \theta(O_A^+(L)) \subset \prod_{v \in S} \mathbb{Q}_v^\times \times \theta(O^+(V))\theta(O_A^+(L)) \subset \mathbb{Q}_A^\times.$$

Thus Theorem 6.3.1 implies the corollary. □

Remark.

Let $L_p = L_1 \perp \cdots \perp L_t$ be a Jordan splitting. If either $\operatorname{rank} L_i \geq 2$ (resp. 3) for $p \neq 2$ (resp. $p = 2$) for some i, or L_p is maximal and $\operatorname{rank} L_p \geq 3$, then $\theta(L_p) \supset \mathbb{Z}_p^\times$ by Proposition 5.5.4 and 5.5.3. Thus if $\mathrm{s}(L) \subset \mathbb{Z}$ and $\mathrm{d}L$ is odd and square-free for a $(\{\infty\}$-)lattice L, then $gen(L) = spn^+(L)$ if $\operatorname{rank} L \geq 3$.

Next let us give a sufficient condition for $spn^+ = cls^+, spn = cls$.

Theorem 6.3.2. *Let V be a regular quadratic space over $\mathbb{Q}$ with* $\dim V \geq 3$, *and S be a finite set of places with $S \ni \{\infty\}$. Suppose that there is a place v in S such that V_v is isotropic. Then for every S-lattice L, we have $spn_S(L) = cls_S(L)$ and $spn_S^+(L) = cls_S^+(L)$.*

Proof. Suppose $K \in spn_S(L)$; then there exist isometries $\mu \in O(V), \sigma_p \in O'(V_p)$ such that

$$\mu(K_p) = \sigma_p(L_p) \quad \text{for } p \notin S.$$

Put $T := \{p \notin S \mid \mu(K_p) \neq L_p\}$, which is a finite set. Then by Theorem 6.2.2, there is an isometry $\sigma \in O'(V)$ such that $\sigma(L_p) = L_p$ for $p \notin S \cup T$, and σ and σ_p are sufficiently close for $p \in T$. Therefore $\sigma(L_p) = L_p = \mu(K_p)$ for $p \notin S \cup T$ and $\sigma(L_p) = \sigma_p(L_p) = \mu(K_p)$ for $p \in T$. Thus we have $\sigma(L_p) = \mu(K_p)$ for $p \notin S$ and so $K \in cls_S(L)$. Therefore $spn_S(L) = cls_S(L)$ holds. If we take $\mu \in O^+(V)$ in the above, then we obtain $spn_S^+(L) = cls_S^+(L)$. □

Remark.

The most important case is $S = \{\infty\}$. For example, if L is a lattice on an indefinite regular quadratic space V over $\mathbb{Q}$ with $\dim V \geq 3$ and $\mathrm{s}(L) \subset \mathbb{Z}$, and $\mathrm{d}\,L$ is odd and square-free, then we have $gen(L) = cls^+(L)$ by the above two theorems.

Exercise 1.

Let V be a regular quadratic space over $\mathbb{Q}$ of $\dim \geq 3$ and L a lattice on V and suppose that V_q is isotropic for a prime q and $\theta(O^+(L_p)) \supset \mathbb{Z}_p^\times$ for $p \neq q$. Prove for every $K \in gen(L)$, there is an isometry $\mu \in O^+(V)$ such that $\mu(K_p) = L_p$ for $p \neq q$.
(Hint: Put $S = \{\infty, q\}$ and apply Corollary 6.3.2 and Theorem 6.3.2.)

Exercise 2.

Let L be a unimodular lattice on an indefinite quadratic space V over $\mathbb{Q}$, i.e. $\mathrm{s}(L) \subset \mathbb{Z}$ and $\mathrm{d}\,L = \pm 1$. Prove:

(i) If $\mathrm{s}(L) = \mathbb{Z}$, then L is isometric to $(\perp \langle 1 \rangle) \perp (\perp \langle -1 \rangle)$.

(ii) If $\mathrm{s}(L) = 2\mathbb{Z}$, then L is isometric to an orthogonal sum of E_8, $E_8^{(-1)}$ and $\langle \begin{pmatrix} 0 & 1 \\ 1 & 0 \end{pmatrix} \rangle$, where E_8 is the unique lattice which is positive definite, with rank 8 and norm $2\mathbb{Z}$. (The explicit form of E_8 is given in 7.1 of the next chapter.)

(Hint: Use the above remark if $\operatorname{rank} L \geq 3$.)

6.4 Representation of codimension 1

In this section we consider the representation problem of codimension 1.

Let $L = L_1 \perp \cdots \perp L_k$ be a Jordan splitting of a regular quadratic lattice L over $\mathbb{Z}_p$, that is L_i is modular and $\operatorname{ord}_p \mathrm{s}(L_1) < \cdots < \operatorname{ord}_p \mathrm{s}(L_k)$. Then we put

$$t_p(L) = (a_1, \cdots, a_1, a_2, \cdots, a_2, \cdots, a_k, \cdots, a_k),$$

where a_i is defined by $p^{a_i}\mathbb{Z}_p = \mathrm{s}(L_i)$ hence $a_1 < \cdots < a_k$ and the number of a_i in the above sequence is $\operatorname{rank} L_i$. Note $t_p(L)$ is independent of the particular Jordan decomposition. If L is unimodular, then $t_p(L) = (0, \cdots, 0)$. For two ordered sets $a = (a_1, \cdots, a_n), b = (b_1, \cdots, b_n)$, we define the ordering $a \leq b$ by either $a_i = b_i$ for $i < k$ and $a_k < b_k$ for some $k \leq n$ or $a_i = b_i$ for all i.

Lemma 6.4.1. *Let U be a regular quadratic space over $\mathbb{Q}_p$ and L a lattice on it. Then L contains a submodule M such that*
(i) $\operatorname{rank} M = \operatorname{rank} L - 1$, $\mathrm{d}\, M \neq 0$ *and M is a direct summand of L as a module, and*
(ii) *any lattice L' on U containing M with $\mathrm{d}\, L' = \mathrm{d}\, L$ and $t_p(L') \geq t_p(L)$ is equal to L.*

Proof. We use induction on $\dim U$. First, we assume that L is unimodular and has an orthogonal basis $\{v_i\}_{i=1}^n$ ($n := \dim U$); then put

$$M = \mathbb{Z}_p[v_1, \cdots, v_{n-1}].$$

The condition (i) is clearly satisfied. Let L' be the lattice in (ii). Since L is unimodular, $t_p(L) = (0, \cdots, 0)$, and $t_p(L') \geq t_p(L)$ implies that every coordinate of $t_p(L')$ is non-negative and hence $\mathrm{s}(L') \subset \mathbb{Z}_p$. Since M is unimodular and $L' \supset M$, $L' = M \perp \mathbb{Z}_p a v_n$ for some $a \in \mathbb{Q}_p$ by Proposition 5.2.2. Then $\mathrm{d}\, L = \mathrm{d}\, L' = Q(av_n)\, \mathrm{d}\, M$ implies $a^2 Q(v_n) \in \mathbb{Z}_p^\times$ and so $a \in \mathbb{Z}_p^\times$. Thus $L' = L$.

Next, we assume that L is unimodular but has no orthogonal basis; then we have $p = 2$ and

$$L = \perp_{i<k} \mathbb{Z}_2[u_i, v_i] \perp \mathbb{Z}_2[u_k, v_k]$$

with $Q(u_i) = Q(v_i) = 0$ and $B(u_i, v_i) = 1$ for $i < k$, and $Q(u_k) = Q(v_k) = 2c$, $B(u_k, v_k) = 1$ for $c = 0$ or 1. Put

$$M = \perp_{i<k} \mathbb{Z}_2[u_i, v_i] + \mathbb{Z}_2[u_k + v_k];$$

then the condition (i) is satisfied. Let L' be the lattice in (ii); then as above $\mathrm{s}(L') \subset \mathbb{Z}_2$, and $L' \supset \perp_{i<k} \mathbb{Z}_2[u_i, v_i]$ yields

$$L' = \perp_{i<k} \mathbb{Z}_2[u_i, v_i] \perp L''.$$

Since $s(L') \subset \mathbb{Z}_2, \mathrm{d}\,L' = \mathrm{d}\,L$, L' is unimodular; then so is L'' and $L'' \ni u_k + v_k$. Since $Q(u_k + v_k) = 2(2c+1)$, $u_k + v_k$ is primitive in L'', and hence $L'' = \mathbb{Z}_2[u_k + v_k, au_k + bv_k]$ for some $a, b \in \mathbb{Q}_2$. Since L'' is unimodular and $Q(u_k + v_k) \in 2\mathbb{Z}_2$, we have $Q(au_k + bv_k) \in \mathbb{Z}_2$ and

$$B(u_k + v_k, au_k + bv_k) \in \mathbb{Z}_2^\times.$$

Now

$$B(u_k + v_k, au_k + bv_k) = (a + b)(2c + 1) \in \mathbb{Z}_2^\times$$

and

$$Q(au_k + bv_k) = 2ca^2 + 2ab + 2cb^2 = 2(1 - 2c)ab + 2cx^2 \in \mathbb{Z}_2 \text{ for } x := a + b$$

imply $x \in \mathbb{Z}_2^\times$ and $2ab \in \mathbb{Z}_2$. These imply $a, b \in \mathbb{Z}_2$ and hence $L' \subset L$. Then $\mathrm{d}\,L' = \mathrm{d}\,L$ yields $L' = L$. Thus we have proved the lemma when L is unimodular.

By scaling, the case when L is modular follows from the unimodular case. Coming back to the general cases, let $L = L_1 \perp \cdots \perp L_k$ where L_i is (p^{a_i})-modular and $a_1 < \cdots < a_k$. Denote by M_k the submodule of L_k which satisfies (i) and (ii), and put $M = \perp_{i=1}^{k-1} L_i \perp M_k$. Then condition (i) is clearly satisfied. For the lattice L' in (ii), $L' \supset M \supset L_1$, and $t_p(L') \geq t_p(L)$ yields $s(L') \subset s(L_1)$; hence $L' = L_1 \perp \tilde{L_1'}$ for some $\tilde{L_1'}$ by Proposition 5.2.2. Now $t_p(L') \geq t_p(L)$ implies $t_p(\tilde{L_1'}) \geq t_p(\perp_{i\geq 2} L_i)$. Then applying the induction hypothesis to $\tilde{L_1} := L_2 \perp \cdots \perp L_k$, $\tilde{M} := L_2 \perp \cdots \perp L_{k-1} \perp M_k$ and $\tilde{L_1'}$, we have $\tilde{L_1} = \tilde{L_1'}$ and hence $L = L'$. □

We call a submodule M in the lemma a *characteristic submodule* of L. The images of a characteristic submodule by $O(L)$ are obviously characteristic. Also, if L is a unimodular lattice over $\mathbb{Z}_p$ $(p \neq 2)$ and M is a unimodular submodule of L of codimension 1, then M is characteristic as in the proof.

Theorem 6.4.1. *Let U be a regular quadratic space over $\mathbb{Q}$ with $\dim U = n$. Let L be a lattice on it and S be a finite set of primes such that $S \ni 2$ and L_p is unimodular if $p \notin S$. Suppose that a submodule M of L satisfies*
(i) rank $M = n - 1$,
(ii) *M_p is a characteristic submodule of L_p for $p \in S$,*
(iii) *M_p is unimodular for $p \notin S$ with precisely one exception $p = q$, where $s(M_q) \subset \mathbb{Z}_q$ and $\mathrm{d}\,M_q \in q\mathbb{Z}_q^\times$.*

Assume that L' is a lattice on a quadratic space U' over $\mathbb{Q}$ with $\dim U' = n$ such that $\mathrm{d}\,L' = \mathrm{d}\,L$, $t_p(L'_p) \geq t_p(L_p)$ for every prime p. If there is an isometry u from M to L', then $U' \cong U$ and extending u to an isometry *from U to U', we have $L' = u(L)$ or $u\tau_x(L)$ where x is an element such that $U = \mathbb{Q}M \perp \mathbb{Q}x$, and $u(L) \neq u\tau_x(L)$.*

Proof. Let L' be the given lattice in the theorem. Since $M \hookrightarrow L'$,

$$U' = \mathbb{Q}L' \cong \mathbb{Q}M \perp \langle \mathrm{d}\, M\, \mathrm{d}\, L' \rangle = \mathbb{Q}M \perp \langle \mathrm{d}\, M\, \mathrm{d}\, L \rangle \cong U.$$

Hence we may suppose that $U' = U$ and $L' \supset M$, taking $u^{-1}(L')$ instead of L'. We will show $L' = L$ or $\tau_x(L)$. Take an orthogonal basis $\{w_i\}_{i=1}^{n-1}$ of M_q such that $Q(w_i) \in \mathbb{Z}_q^\times$ for $i < n-1$ and $Q(w_{n-1}) \in q\mathbb{Z}_q^\times$, by virtue of (iii). Since L_q is unimodular, $N := \mathbb{Z}_q[w_1, \cdots, w_{n-2}]^\perp$ is also unimodular and w_{n-1} is primitive in N. We take an element $w_n \in N$ such that $\{w_{n-1}, w_n\}$ is a basis of N. Let us see that $\mathbb{Q}_qN$ is hyperbolic. Since N is unimodular and $Q(w_{n-1}) \in q\mathbb{Z}_q$, we have $B(w_{n-1}, w_n) \in \mathbb{Z}_q^\times$, then

$$\begin{aligned} \mathrm{d}\, N &= Q(w_{n-1})Q(w_n) - B(w_{n-1}, w_n)^2 \\ &= -B(w_{n-1}, w_n)^2(1 - Q(w_{n-1})Q(w_n)B(w_{n-1}, w_n)^{-2}) \end{aligned}$$

and

$$Q(w_{n-1})Q(w_n)B(w_{n-1}, w_n)^{-2} \in q\mathbb{Z}_q$$

imply that $\mathrm{d}\, N \in -(\mathbb{Q}_q^\times)^2$ and $\mathbb{Q}_qN$ is hyperbolic. By Lemma 5.2.2, there is a basis $\{e_1, e_2\}$ of N such that $Q(e_1) = Q(e_2) = 0, B(e_1, e_2) = 1$. Put $w_{n-1} = a_1e_1 + a_2e_2$ $(a_i \in \mathbb{Z}_q)$; then $Q(w_{n-1}) = 2a_1a_2 \in q\mathbb{Z}_q^\times$. Multiplying e_i by a unit and renumbering, we may suppose $w_{n-1} = e_1 + \epsilon q e_2$ for some $\epsilon \in \mathbb{Z}_q^\times$. Now $t_q(L'_q) \geq t_q(L_q)$ implies $\mathrm{s}(L'_q) \subset \mathbb{Z}_q$ and then $\mathrm{d}\, L' = \mathrm{d}\, L$ implies that L'_q is also unimodular. Since L'_q contains M_q, there is a unimodular submodule K_q such that $L'_q = \perp_{i \leq n-2} \mathbb{Z}_q w_i \perp K_q$. Then K_q is unimodular and $K_q \ni w_{n-1}$. Let us claim $K_q = \mathbb{Z}_q[e_1, e_2]$ or $\mathbb{Z}_q[q^{-1}e_1, qe_2]$. Since w_{n-1} is primitive in K_q, let $K_q = \mathbb{Z}_q[w_{n-1}, ce_1 + de_2]$ for $c, d \in \mathbb{Q}_q$. Since K_q is unimodular and $Q(w_{n-1}) \in q\mathbb{Z}_q$,

$$(1) \qquad Q(ce_1 + de_2) = 2cd \in \mathbb{Z}_q$$

and

$$(2) \qquad B(w_{n-1}, ce_1 + de_2) = d + c\epsilon q \in \mathbb{Z}_q^\times.$$

If $c \in \mathbb{Z}_q$, then (2) implies $d \in \mathbb{Z}_q^\times$. Hence we have

$$\begin{aligned} K_q &= \mathbb{Z}_q[e_1 + \epsilon q e_2, ce_1 + de_2 - c(e_1 + \epsilon q e_2)] \\ &= \mathbb{Z}_q[e_1 + \epsilon q e_2, (d - c\epsilon q)e_2] = \mathbb{Z}_q[e_1, e_2]. \end{aligned}$$

If $\mathrm{ord}_q c = -1$, then (1) implies $d \in q\mathbb{Z}_q$ and so

$$\begin{aligned} K_q &= \mathbb{Z}_q[e_1 + \epsilon q e_2, c e_1 + d e_2 - d(\epsilon q)^{-1}(e_1 + \epsilon q e_2)] \\ &= \mathbb{Z}_q[e_1 + \epsilon q e_2, (c - d(\epsilon q)^{-1})e_1] = \mathbb{Z}_q[q^{-1}e_1, q e_2]. \end{aligned}$$

If $\mathrm{ord}_q c \le -2$, then (2) implies $\mathrm{ord}_q d = \mathrm{ord}_q c + 1$, which yields $\mathrm{ord}_q c + \mathrm{ord}_q d = 2\,\mathrm{ord}_q c + 1 < 0$. This contradicts (1). Thus we have verified the claim. Since $x^\perp = \mathbb{Q}M$,

$$\mathbb{Q}_q x = \mathbb{Q}_q M^\perp = (e_1 + \epsilon q e_2)^\perp \text{ in } \mathbb{Q}_q N = \mathbb{Q}_q[e_1, e_2].$$

So $\mathbb{Q}_q x = \mathbb{Q}_q(e_1 - \epsilon q e_2)$. Hence $\tau_x(e_1) = \epsilon q e_2$ and $\tau_x(e_2) = (\epsilon q)^{-1} e_1$ hold and $\tau_x(\mathbb{Z}_q[e_1, e_2]) = \mathbb{Z}_q[q^{-1}e_1, q e_2]$. Thus we have $L'_q = L_q$ or $\tau_x L_q$. If $p \ne q$, then $B(x, M_p) = 0$ implies $\tau_x M_p = M_p$.

Noting that M_p is characteristic of L_p for $p \ne q$ by the assumption and the remark preceeding the theorem, $L'_p \supset M_p$, $\tau_x(L_p) \supset M_p$ imply $L'_p = L_p = \tau_x(L_p)$ for $p \ne q$. Thus we have $L' = L$ or $\tau_x L$. Since $\mathbb{Z}_q[e_1, e_2] \ne \mathbb{Z}_q[q^{-1}e_1, q e_2]$, $L \ne \tau_x L$. □

Remark.

Applying the theorem to $L' = L$, for a submodule M and $u : M \hookrightarrow L$ in the theorem, u is uniquely extended to an element of $O(L)$, and hence if U is positive definite then we have $\sharp\{u : M \hookrightarrow L\} = \sharp O(L)$.

Next we show the existence of a submodule M in the theorem.

Proposition 6.4.1. *Let U, L and S be those in Theorem 6.4.1. Then there exists a submodule M which satisfies the three conditions there.*

Proof. First we suppose that U is a hyperbolic plane and then there is a basis $\{u_1, u_2\}$ such that $L = \mathbb{Z}[u_1, u_2] \cong \langle a \begin{pmatrix} 0 & b' \\ b' & c' \end{pmatrix} \rangle$, where $a \in \mathbb{Q}$, $b', c' \in \mathbb{Z}$ and we may suppose that $2|c'$ and $(b', c'/2) = 1$ without loss of generality. We note that $a \in \mathbb{Z}_p^\times$ for every $p \notin S$ and $\mathrm{n}(L_2) \subset 2a\mathbb{Z}_2$. Since $(b', c'/2) = 1$, there is an integer x such that $q := xb' + c'/2$ is a prime not in S. Then $Q(xu_1 + u_2) = 2aq$ and $L \cong \langle a \begin{pmatrix} 2q & b \\ b & c \end{pmatrix} \rangle$, where b, c are integers and moreover c is even by $\mathrm{n}(L_2) \subset 2a\mathbb{Z}_2$ and we may assume $0 \le b \le q$ without loss of generality. If $b = 0$ or q, then $\mathrm{d}\,L \in q\mathbb{Z}_q$, which contradicts the definition of S. Thus we have $0 < b < q$. Taking a basis $\{v_1, v_2\}$ of L such that $(B(v_i, v_j)) = a \begin{pmatrix} 2q & b \\ b & c \end{pmatrix}$, we put $M := \mathbb{Z}v_1$. Let us verify the conditions in the theorem. Since $a \in \mathbb{Z}_p^\times$ for $p \notin S$, we may suppose $a = 1$

by scaling by a^{-1}. (i) is clear. Let us examine the second condition. If $p \in S$ and $p \neq 2$, then $2q \in \mathbb{Z}_p^\times$ and so $t_p(L_p) = (0, \mathrm{ord}_p(2qc - b^2))$. If another lattice L'_p satisfies $t_p(L'_p) \geq t_p(L_p)$, $\mathrm{d}\, L'_p = \mathrm{d}\, L_p$ and $2q = Q(v_1) \in Q(L'_p)$, then $t_p(L'_p) = t_p(L)$ and we have $L'_p \cong L_p$. Thus M is characteristic. Suppose that $p = 2$. If b is even, then

$$L_2 \cong \langle \mathrm{diag}(2q, (2qc - b^2)/2q) \rangle.$$

Similarly to the above, M is characteristic. Suppose that b is odd. Then $t_2(L_2) = (0, 0)$. If L'_2 satisfies $t_2(L'_2) \geq t_2(L_2)$, $\mathrm{d}\, L'_2 = \mathrm{d}\, L_2$ and $2q \in Q(L'_2)$, then $t_2(L'_2) = t_2(L_2)$, $\mathrm{s}(L_2) = \mathrm{s}(L'_2)$ and L'_2 is unimodular. If L'_2 has no orthogonal basis, then $\mathrm{n}(L'_2) = 2\,\mathrm{s}(L'_2)$, $\mathrm{d}\, L'_2 = \mathrm{d}\, L_2$ yield $L'_2 \cong L_2$. Suppose L'_2 has an orthogonal basis. Then $L'_2 \cong \langle \epsilon_1 \rangle \perp \langle \epsilon_2 \rangle$ for $\epsilon_i \in \mathbb{Z}_2^\times$. Now $\mathrm{d}\, L_2 = \mathrm{d}\, L'_2$ implies $\epsilon_1\epsilon_2 \equiv 2qc - b^2 \bmod 8$ and so $\epsilon_1\epsilon_2 \equiv -1 \bmod 4$. On the other hand, $2q \in Q(L'_2)$ implies that $\epsilon_1 x_1^2 + \epsilon_2 x_2^2 = 2q$ has a solution x_i in $\mathbb{Z}_2$. This is impossible because $\epsilon_1\epsilon_2 \equiv -1 \bmod 4$. Thus the second condition has been verified. Clearly $M_p \cong \langle Q(v_1) \rangle = \langle 2q \rangle$ satisfies the third condition and we have completed the case that U is a hyperbolic plane.

Suppose that U is not a hyperbolic plane. Let M'_p be a characteristic submodule of L_p for $p \in S$, and $\{x_{1,p}, \cdots, x_{n-1,p}\}$ be a basis of M'_p. Applying Theorem 6.2.1, there are vectors $x_1, \cdots, x_{n-1} \in L$ such that

(i) x_i and $x_{i,p}$ are sufficiently close for $1 \leq i \leq n-1$, $p \in S$ and
(ii) $\det(B(x_i, x_j)) \in \mathbb{Z}_p^\times$ for $p \notin S$ with precisely one exception $p = q$, where $\det(B(x_i, x_j)) \in q\mathbb{Z}_q^\times$.

Then put $M := \mathbb{Z}[x_1, \cdots, x_{n-1}]$. By Corollary 5.4.5, there is an isometry $\sigma_p \in O(L_p)$ such that $M_p = \sigma_p(M'_p)$ for $p \in S$, and hence M_p is also characteristic of L_p. The other conditions are clear. □

Remark.

In general, the codimension two analogue of a global characteristic sublattice can not exist. The intuitive reason is this: the orthogonal complement of a sublattice of codimension two is two-dimensional, and the class number of a binary quadratic lattice is in general greater than 1.

Corollary 6.4.1. *Let $\{L_i\}_{i=1}^m$ be a set of lattices on quadratic spaces V_i such that* $\mathrm{rank}\, L_i = n$, $\mathrm{d}\, L_i = d$ $(1 \leq i \leq m)$ *and they are not mutually isometric. Then there is a submodule M of L_j for some j such that* $\mathrm{rank}\, M = n - 1$ *and M is not represented by L_i for every $i \neq j$.*

Proof. Let S be a finite set of primes such that $2 \in S$ and $(L_i)_p$ is unimodular for every i and $p \notin S$. Put $S = \{p_1, \cdots, p_r\}$ and define a subset A_j of $\{L_i\}_{i=1}^m$ inductively as follows:

$$A_1 := \{L_i \mid t_{p_1}(L_i) \leq t_{p_1}(L_k) \ \text{ for } \ k = 1, 2, \cdots, m\},$$
$$A_{j+1} := \{L_i \in A_j \mid t_{p_{j+1}}(L_i) \leq t_{p_{j+1}}(L_k) \ \text{ for every } \ L_k \in A_j\}.$$

Then $A_1 \supset \cdots \supset A_r$ is obvious. Take a lattice L_j in A_r, and let M be a submodule satisfying the conditions in Theorem 6.4.1 as constructed by Proposition 6.4.1. Assume that M is represented by L_i. Then $V_i = \mathbb{Q}L_i = \mathbb{Q}M \perp \langle d \cdot \mathrm{d}\, M\rangle \cong V_j$. Since $L_j \in A_r \subset A_1$, $t_{p_1}(L_j) \leq t_{p_1}(L_i)$. Furthermore since M_{p_1} is characteristic to $(L_j)_{p_1}$, $(L_j)_{p_1} \cong (L_i)_{p_1}$. Thus $t_{p_1}(L_i) = t_{p_1}(L_j)$ and L_i is in A_1, and $L_j \in A_r \subset A_2$ implies $t_{p_2}(L_j) \leq t_{p_2}(L_i)$ since $L_i \in A_1$. Thus we have $t_{p_2}(L_j) = t_{p_2}(L_i)$ similarly to the above. Repeating this argument, we have $L_i \in A_r$ which means $t_p(L_i) = t_p(L_j)$ for $p \in S$. Thus Theorem 6.4.1 yields $L_i \cong L_j$. □

6.5 Representation of codimension 2

In this section, we show the following fact due to Kneser, Hsia and Schulze-Pillot whose proof is group-theoretic. As an abbreviation, we say that K is represented by *gen*(L) (resp. *spn*$^+(L)$) if K is represented by a lattice in *gen*(L) (resp. *spn*$^+(L)$).

Theorem 6.5.1. *Let $V = W \perp U$ be a regular quadratic space over $\mathbb{Q}$ and suppose* $\dim U = 2$. *Let L, K be lattices on V, W respectively and suppose $L \supset K$. Then K is represented by either every spn$^+$ or exactly a half of the spn$^+$ in gen(L). The latter case occurs if and only if the following three conditions hold:*

(i) $-\mathrm{d}\, U \notin (\mathbb{Q}^\times)^2$;

(ii) $\theta(O_A^+(L)) \subset \tilde{N} := \prod_v N_v \cap \mathbb{Q}_A^\times$, *where*

$$N_v := \{x \in \mathbb{Q}_v^\times \mid (x, -\mathrm{d}\, U)_{\mathbb{Q}_v} = 1\},$$

(iii) *putting* $\theta(L_p; K_p) := \{\theta(\varphi) \subset \mathbb{Q}_p^\times \mid \varphi \in O^+(V_p) \text{ with } K_p \subset \varphi(L_p)\}$,

$$\theta(L_p; K_p) \subset N_p$$

for every prime p.

We need several lemmas. We take $\delta \in \mathbb{Q}^\times$ such that $\delta \in \mathrm{d}\, U$.

Lemma 6.5.1. *Suppose $-\delta \notin (\mathbb{Q}^\times)^2$; then we have*

$$[\mathbb{Q}_A^\times : \mathbb{Q}^\times \tilde{N}] = 2.$$

Proof. Putting $\tilde{\mathbb{Z}} := \prod_p \mathbb{Z}_p^\times$, we have $\mathbb{Q}_A^\times = \mathbb{Q}^\times \cdot \tilde{\mathbb{Z}} \cdot \mathbb{R}_+^\times$, where

$$\mathbb{R}_+^\times := \{a \in \mathbb{R}^\times \mid a > 0\}$$

is identified with elements in $\mathbb{Q}_A^\times$ with all finite components being 1. We define the mapping $\varphi : \mathbb{Q}_A^\times \to \{\pm 1\}$ by $\varphi(x) = \prod_v (x_v, -\delta)_{\mathbb{Q}_v}$ for $x =$

$(x_v) \in \mathbb{Q}_A^\times$. It is well-defined since for almost all odd primes p, $x_p, -\delta \in \mathbb{Z}_p^\times$ and hence $(x_p, -\delta)_{\mathbb{Q}_p} = 1$. Since $-\delta \notin (\mathbb{Q}^\times)^2$, there is a place v such that $-\delta \notin (\mathbb{Q}_v^\times)^2$. By (8) in Theorem 3.3.1, φ is surjective. From (9) in the same theorem $\ker \varphi \supset \mathbb{Q}^\times$. Hence $\ker \varphi \supset \mathbb{Q}^\times \tilde{N}$. Suppose $x \in \ker \varphi$. We must verify $x \in \mathbb{Q}^\times \tilde{N}$. Let $v_1, \cdots, v_{2n}$ be the places such that $(x_v, -\delta)_{\mathbb{Q}_v} = -1$. Let S be a finite set of primes such that for $p \notin S$, $p \neq 2$, v_i for $1 \leq i \leq 2n$ and $\delta, x_p \in \mathbb{Z}_p^\times$ hold. Put $e_p = \operatorname{ord}_p x_p$ and $h = \prod_{p \in S} p^{e_p}$. We take a prime q such that $q \equiv \eta h^{-1} x_p \bmod p^t$ for S, where t is a sufficiently large integer and $\eta = 1$ if all places v_i are finite, -1 otherwise. Let us show $q\eta h^{-1} x \in \tilde{N}$, i.e. $(q\eta h^{-1} x_v, -\delta)_{\mathbb{Q}_v} = 1$. If a prime p is not in S and $p \neq q$, then $p \neq 2$ and $q\eta h^{-1} x_p$ and δ are in $\mathbb{Z}_p^\times$, and so $(q\eta h^{-1} x_p, -\delta)_{\mathbb{Q}_p} = 1$. If $p \in S$, then $q\eta h^{-1} x_p$ is a square in $\mathbb{Q}_p^\times$ and so $(q\eta h^{-1} x_p, -\delta)_{\mathbb{Q}_p} = 1$. If all places v_i are finite, then $(q\eta h^{-1} x_\infty, -\delta)_{\mathbb{R}} = (x_\infty, -\delta)_{\mathbb{R}} = 1$. If some place v_i is infinite, then $\eta = -1$ and $(x_\infty, -\delta)_{\mathbb{R}} = -1$ and so $x_\infty < 0$. Thus $(q\eta h^{-1} x_\infty, -\delta)_{\mathbb{R}} = 1$. Finally we have

$$(q\eta h^{-1} x_q, -\delta)_{\mathbb{Q}_q} = \prod_{v \neq q} (q\eta h^{-1} x_v, -\delta)_{\mathbb{Q}_v} = 1.$$

Therefore we have shown $q\eta h^{-1} x \in \tilde{N}$ and completed the proof. □

Lemma 6.5.2. $[\mathbb{Q}_A^\times : \tilde{N}\theta(O^+(V))] = 2[\mathbb{Q}^\times : \theta(O^+(V))]$ *if* $-\delta \notin (\mathbb{Q}^\times)^2$.

Proof. The left-hand side is equal to

$$\begin{aligned}
&[\mathbb{Q}_A^\times : \tilde{N}\mathbb{Q}^\times][\tilde{N}\mathbb{Q}^\times : \tilde{N}\theta(O^+(V))] \\
&= 2[\tilde{N}\mathbb{Q}^\times : \tilde{N}\theta(O^+(V))] \\
&= 2[\mathbb{Q}^\times : \mathbb{Q}^\times \cap \tilde{N}\theta(O^+(V))].
\end{aligned}$$

If V_∞ is isotropic, then $\theta(O^+(V)) = \mathbb{Q}^\times$ by Proposition 6.3.1 and then $\mathbb{Q}^\times \cap \tilde{N}\theta(O^+(V)) = \mathbb{Q}^\times$. So in this case the assertion has been verified.

Suppose that V_∞ is anisotropic; we have only to show

$$\mathbb{Q}^\times \cap \tilde{N}\theta(O^+(V)) = \theta(O^+(V)) \ (= \{a \in \mathbb{Q}^\times \mid a > 0\}).$$

The left-hand side obviously contains the right-hand side. Suppose $b \in \mathbb{Q}^\times \cap \tilde{N}\theta(O^+(V))$ and write $b = cx$, $c \in \mathbb{Q}^\times, c > 0, x \in \tilde{N}$. We have only to prove $b \in \theta(O^+(V))$. Since V_∞ is anisotropic and $\dim U = 2$, we have $dU > 0$ and so $\delta > 0$ follows; then

$$x_\infty \in N_\infty = \{y \in \mathbb{R}^\times \mid (y, -\delta)_{\mathbb{R}} = 1\} = \mathbb{R}_+^\times.$$

Thus $x_\infty > 0$ and then $b > 0$, which means $b \in \theta(O^+(V))$ by Proposition 6.3.1. □

Lemma 6.5.3. *Suppose $-\delta \notin (\mathbb{Q}^\times)^2$. Then*

$$\theta(O_A^+(L))\mathbb{R}^\times \subset \tilde{N}\theta(O^+(V)) \Leftrightarrow \theta(O_A^+(L))\mathbb{R}^\times \subset \tilde{N}.$$

If we remove $\mathbb{R}^\times$ from both sides, it is still valid.

Proof. $\Leftarrow$ is clear. Let us prove $\Rightarrow$. Suppose that the right-hand side is false. Then there is a place v such that the v-component of $\theta(O_A^+(L))\mathbb{R}^\times$ is not in N_v. So we choose a v-component x_v of $\theta(O_A^+(L))\mathbb{R}^\times$ with $x_v \notin N_v$. We define $x = (x_u) \in \mathbb{Q}_A^\times$ by $x_u := 1$ if $u \neq v$. Since $x \in \theta(O_A^+(L))\mathbb{R}^\times$, we write $x = ay$, $a \in \theta(O^+(V))$, $y \in \tilde{N}$. For a place $u \neq v$, $x_u = 1$ implies $1 = ay_u$ and so $a = y_u^{-1} \in N_u$, i.e. $(a, -\delta)_{\mathbb{Q}_u} = 1$ for $u \neq v$. On the other hand, $\prod_v (a, -\delta)_{\mathbb{Q}_v} = 1$ yields

$$(a, -\delta)_{\mathbb{Q}_v} = \prod_{u \neq v} (a, -\delta)_{\mathbb{Q}_u} = 1.$$

Thus $a \in N_v$ and so $x_v = ay_v \in N_v$. This is a contradiction. The case without $\mathbb{R}^\times$ is quite similar. □

Lemma 6.5.4. *Suppose that $-\delta \notin (\mathbb{Q}^\times)^2$ and V_∞ is anisotropic. Then*

$$[\tilde{N}\theta(O^+(V))\theta(O_A^+(L))\mathbb{R}^\times : \tilde{N}\theta(O^+(V))] = 2$$

if and only if

$$[\tilde{N}\theta(O^+(V))\theta(O_A^+(L))\mathbb{R}^\times : \tilde{N}\theta(O^+(V))\theta(O_A^+(L))] = 2$$

and

$$\tilde{N}\theta(O^+(V))\theta(O_A^+(L)) = \tilde{N}\theta(O^+(V)).$$

Proof. The "if" part is clear. Let us verify the converse. If the conclusion is false, then we have

$$\tilde{N}\theta(O^+(V))\theta(O_A^+(L))\mathbb{R}^\times = \tilde{N}\theta(O^+(V))\theta(O_A^+(L))$$

and hence

$$\mathbb{R}^\times \subset N_\infty\theta(O^+(V))\theta(O^+(V_\infty)) = N_\infty\theta(O^+(V_\infty)).$$

Since V_∞ is anisotropic, $\theta(O^+(V_\infty)) = \mathbb{R}_+^\times$ and $\delta > 0$, and so

$$N_\infty = \{a \in \mathbb{R}^\times \mid (a, -\delta)_\mathbb{R} = 1\} = \mathbb{R}_+^\times.$$

This is a contradiction. □

Lemma 6.5.5. *Suppose* $-\delta \notin (\mathbb{Q}^\times)^2$. *Then*

$$[\mathbb{Q}_A^\times : \tilde{N}\theta(O^+(V))\theta(O_A^+(L))\mathbb{R}^\times] \leq 2,$$

and the equality holds if and only if

$$\theta(O_A^+(L)) \subset \tilde{N}.$$

Proof. Put

$$i := [\mathbb{Q}_A^\times : \tilde{N}\theta(O^+(V))\theta(O_A^+(L))\mathbb{R}^\times],$$
$$j := [\tilde{N}\theta(O^+(V))\theta(O_A^+(L))\mathbb{R}^\times : \tilde{N}\theta(O^+(V))].$$

Then by Lemma 6.5.2, we have

$$2[\mathbb{Q}^\times : \theta(O^+(V))] = ij.$$

If V_∞ is isotropic, then by Proposition 6.3.1 we have $\mathbb{Q}^\times = \theta(O^+(V))$ and so $ij = 2$. In particular we have $i \leq 2$. $i = 2$ if and only if $j = 1$, which is equivalent to $\theta(O_A^+(L))\mathbb{R}^\times \subset \tilde{N}\theta(O^+(V))$. Then Lemma 6.5.3 implies $i = 2 \Leftrightarrow \theta(O_A^+(L))\mathbb{R}^\times \subset \tilde{N}$. Noting $\theta(O^+(V_\infty)) = \mathbb{R}^\times$, we get the assertion.

Suppose that V_∞ is anisotropic; then Proposition 6.3.1 implies $ij = 4$. By Lemma 6.5.4 $i = 2$, i.e. $j = 2$, if and only if

$$\tilde{N}\theta(O^+(V))\theta(O_A^+(L)) = \tilde{N}\theta(O^+(V))$$

and

$$k := [\tilde{N}\theta(O^+(V))\theta(O_A^+(L))\mathbb{R}^\times : \tilde{N}\theta(O^+(V))\theta(O_A^+(L))] = 2.$$

k is clearly equal to

$$[\mathbb{R}^\times : \mathbb{R}^\times \cap \tilde{N}\theta(O^+(V))\theta(O_A^+(L))]$$

where $\mathbb{R}^\times$ is identified with elements in $\mathbb{Q}_A^\times$ with the component being 1 for every prime. Since V_∞ is anisotropic, $\delta > 0$ hence $N_\infty = \mathbb{R}_+^\times$, $\theta(O^+(V)) = \{a \in \mathbb{Q}^\times \mid a > 0\}$ by Proposition 6.3.1 and the ∞-part of $\theta(O_A^+(L))$ is equal to $\theta(O^+(V_\infty)) = \mathbb{R}_+^\times$. Thus

$$\mathbb{R}^\times \cap \tilde{N}\theta(O^+(V))\theta(O_A^+(L)) = \mathbb{R}_+^\times$$

and hence $k = 2$ is automatically satisfied if V_∞ is anisotropic. Thus $i = 2$ is equivalent to $\tilde{N}\theta(O^+(V))\theta(O_A^+(L)) = \tilde{N}\theta(O^+(V))$, which means

$$\theta(O_A^+(L)) \subset \tilde{N}\theta(O^+(V)).$$

By Lemma 6.5.3, this is equivalent to $\theta(O_A^+(L)) \subset \tilde{N}$. Lastly we must verify $i \leq 2$. Otherwise $i = 4$ and so $j = 1$, which implies $\mathbb{R}^\times \subset N_\infty\theta(O^+(V))$. Since V_∞ is anisotropic, we get $\delta > 0$, $N_\infty = \mathbb{R}_+^\times$ and

$$\theta(O^+(V)) = \{a \in \mathbb{Q}^\times \mid a > 0\}.$$

Thus the above inclusion does not hold. □

Lemma 6.5.6. *Put*

$$H_L := O_A^+(U)O^+(V)O'_A(V)O_A^+(L),$$

where $O_A^+(U)$ is canonically embedded in $O_A^+(V)$ by assuming that $O^+(U_v)$ acts trivially on W_v. If $-\delta \in (\mathbb{Q}^\times)^2$, then $O_A^+(V) = H_L$. If $-\delta \notin (\mathbb{Q}^\times)^2$, then

$$O_A^+(V)/H_L \cong \mathbb{Q}_A^\times/\tilde{N}\theta(O^+(V))\theta(O_A^+(L))\mathbb{R}^\times, \tag{1}$$

where the isomorphism is induced by θ.

Proof. First, we suppose $-\delta \in (\mathbb{Q}^\times)^2$; then U is a hyperbolic plane since $\delta \in \mathrm{d}U$. Take $g \in O_A^+(V)$. Since $\theta(O_v^+(U_v)) = \mathbb{Q}_v^\times$ and for a lattice L' on U $\theta(O^+(L'_p)) \supset \mathbb{Z}_p^\times$ holds for almost all p by Proposition 5.5.5, we have $\theta(O_A^+(U)) = \mathbb{Q}_A^\times$ and hence there is an isometry $h \in O_A^+(U)$ such that $\theta(gh) = (\mathbb{Q}_A^\times)^2$, that is $gh \in O'_A(V)$. Noting $O'_A(V)$ is a normal subgroup of $O_A^+(V)$, this case has been shown.

Next suppose $-\delta \notin (\mathbb{Q}^\times)^2$. Let us show first $\theta(O_A^+(U)) = \tilde{N}$. Write $U = \langle b\rangle \perp \langle b\delta\rangle$ $(b \in \mathbb{Q}^\times)$. Then $\theta(O^+(U_v)) = \theta(O^+(\langle 1\rangle \perp \langle \delta\rangle))$ holds. If U_v is isotropic, then $-\delta \in (\mathbb{Q}_v^\times)^2$ hence $\theta(O^+(U_v)) = \mathbb{Q}_v^\times$, and

$$N_v = \{a \in \mathbb{Q}_v^\times \mid (a, -\delta)_{\mathbb{Q}_v} = 1\} = \mathbb{Q}_v^\times.$$

If U_v is anisotropic, then $\theta(O^+(U_v)) = \{a^2 + b^2\delta \in \mathbb{Q}_v^\times \mid a, b \in \mathbb{Q}_v\}$ since

$$(a^2 + b^2\delta)(A^2 + B^2\delta) = (aA - bB\delta)^2 + (aB + Ab)^2\delta.$$

On the other hand, $x \in N_v$ if and only if $(x, -\delta)_{\mathbb{Q}_v} = 1$, which means $a^2x - b^2\delta = c^2$ for $a, b, c \in \mathbb{Q}_v$ which are not 0 at the same time. It is equivalent to $x = e^2 + d^2\delta$ for $e, d \in \mathbb{Q}_v$. Thus we have

$$\theta(O^+(U_v)) = N_v,$$

and then $\theta(O_A^+(U)) = \tilde{N}$ follows from Proposition 5.5.5 as above. So θ implies a homomorphism of the left-hand side and the right-hand side of (1). Since $\dim V \geq 3$, Proposition 5.5.1 and 5.5.5 yield the surjectivity.

Let us show the injectivity. Suppose that

$$\theta(f) \subset \tilde{N}\theta(O^+(V))\theta(O_A^+(L))\mathbb{R}^\times$$

for $f \in O_A^+(V)$. Multiplying by an element of $O_A^+(U)O^+(V)O_A^+(L)$, we may assume $\theta(f) \subset \mathbb{R}^\times(\mathbb{Q}_A^\times)^2$. Put $\theta(f) = r(\mathbb{Q}_A^\times)^2$ for $r \in \mathbb{R}^\times$. If V_∞ is anisotropic, then $\theta(O^+(V_\infty)) = \mathbb{R}_+^\times$ and so $r > 0$. Thus $\theta(f) = (\mathbb{Q}_A^\times)^2$ holds and hence $f \in O'_A(V)$. If V_∞ is isotropic, then $\theta(O^+(V_\infty)) = \mathbb{R}^\times$ holds, and therefore $O_A^+(L)$ contains $x = (x_v)$ such that $x_v = 1$ if $v \neq \infty$ and $\theta(x_v) = \{a \in \mathbb{R} \mid a < 0\}$ for $v = \infty$. If $r > 0$, then $\theta(f) = (\mathbb{Q}_A^\times)^2$. If $r < 0$, then $\theta(xf) = -r(\mathbb{Q}_A^\times)^2 = (\mathbb{Q}_A^\times)^2$. Thus θ is injective. □

Lemma 6.5.7. *Put*

$$i := [O_A^+(V) : H_L].$$

Then $i = 1$ *or* 2, *and* $i = 2$ *if and only if* $-\delta \notin (\mathbb{Q}^\times)^2$ *and* $\theta(O_A^+(L)) \subset \tilde{N}$,

Proof. By the last lemma, $-\delta \in (\mathbb{Q}^\times)^2$ implies $i = 1$. So we may assume $-\delta \notin (\mathbb{Q}^\times)^2$. Then, (1) in Lemma 6.5.6 and Lemma 6.5.5 imply the assertion. □

Proof of Theorem 6.5.1. First let us see that $O_A^+(V)/O^+(V)O'_A(V)O_A^+(L)$ acts transitively and faithfully on spn^+ in $gen(L)$. Take $L_1, L_2 \in gen(L)$ and $g \in O_A^+(V)$. If $L_1 \in spn^+(L_2)$, then $L_1 = \eta j(L_2)$ where $\eta \in O^+(V)$ and $j \in O'_A(V)$. Now

$$g(L_1) = \eta(\eta^{-1}g\eta g^{-1})(gjg^{-1})g(L_2)$$

means $g(L_1) \in spn^+ g(L_2)$ since $\eta^{-1}g\eta g^{-1}$, $gjg^{-1} \in O'_A(V)$. Thus $O_A^+(V)$ acts on spn^+ in $gen(L)$.

Next, $g(L_1) \in spn^+(L_1)$ means $g(L_1) = \eta j(L_1)$ for $\eta \in O^+(V)$ and $j \in O'_A(V)$. Writing $L_1 = k(L)$ for $k \in O_A^+(V)$, this means $(\eta jk)^{-1}gk \in O_A^+(L)$. Thus

$$g \in \eta jkO_A^+(L)k^{-1} \subset O^+(V)O'_A(V)O_A^+(L)$$

since $O'_A(V)$ contains the commutator subgroup of $O_A^+(V)$, i.e. the subgroup which fixes $spn^+(L_1)$ is $O^+(V)O'_A(V)O_A^+(L)$. The transitivity follows from the fact that $O_A^+(V)$ acts transitively on $gen(L)$. Thus the claim is verified.

For $j \in O_A^+(V)$, we have $O_A^+(jL) = jO_A^+(L)j^{-1} \subset O'_A(V)O_A^+(L)$ and hence $H_{jL} \subset H_L$. Similarly we have $H_L = H_{j^{-1}jL} \subset H_{jL}$ and so $H_L = H_{jL}$. Suppose that the lattice K on W in the theorem is represented by jL, that is there is an isometry $\sigma : K \hookrightarrow jL$. Extending it, we may assume $\sigma \in O(V)$. We will show that K is also represented by a lattice in $spn^+(j'jL)$ for $j' \in H_L$. Since

$$H_L = H_{jL} = \sigma^{-1}O_A^+(U)\sigma O^+(V)O'_A(V)O_A^+(jL),$$

we can write $j' = j_1j_2j_3j_4$ where

$$j_1 \in \sigma O_A^+(U)\sigma^{-1},\ j_2 \in O^+(V),\ j_3 \in O'_A(V),\ j_4 \in O_A^+(jL).$$

Then

$$j' = j_2(j_2^{-1}j_1j_2j_1^{-1})(j_1j_3j_1^{-1})j_1j_4 \in O^+(V)O'_A(V)j_1j_4$$

and so $spn^+(j'j(L)) \ni j_1j_4j(L) = j_1j(L)$. Here $j(L) \supset \sigma(K)$ implies $j(L) \supset \sigma(K) \perp N$ for some binary lattice. Then

$$j_1j(L) \supset j_1(\sigma(K) \perp N) = j_1\sigma(K \perp \sigma^{-1}(N))$$
$$= \sigma(\sigma^{-1}j_1\sigma)(K \perp \sigma^{-1}(N)) \supset \sigma(K),$$

since $\sigma^{-1}j_1\sigma$ is a trivial isometry on W. *Thus K is represented by $j_1j(L) \in spn^+(j'j(L))$ if $j' \in H_L$ and K is represented by jL.* By Lemma 6.5.7, we have $[O_A^+(V) : H_L] \leq 2$. Thus K is represented by either every spn^+ or exactly a half of the spn^+ in $gen(L)$.

The latter occurs if and only if $[O_A^+(V) : H_L] = 2$ and there exists some L' in $gen(L)$ such that K is not represented by $spn^+(L')$. $[O_A^+(V) : H_L] = 2$ is equivalent to (i) and (ii) by Lemma 6.5.7. So we have only to verify that the condition (iii) does not hold if and only if every spn^+ in $gen(L)$ represents K, assuming $[O_A^+(V) : H_L] = 2$. Before the proof of this, we note that $N_p = \theta(O^+(U_p)) \subset \theta(L_p; K_p)$ since an isometry $\varphi \in O^+(U_p)$ is extended by $\varphi|_{W_p} = \text{id}$ and then $\varphi \in O^+(V_p)$ and $\varphi(K_p) = K_p \subset L_p$. Therefore the condition (iii) is equivalent to $\theta(L_p; K_p) = N_p$ for every prime p.

From now on we assume $[O_A^+(V) : H_L] = 2$. First we suppose that K is represented by every spn^+ in $gen(L)$. We must show $\theta(L_p; K_p) \neq N_p$ for some prime p. By Lemma 6.5.6, there exists $\varphi \in O_A^+(V)$ such that

$$\theta(\varphi) \not\subset \tilde{N}\theta(O^+(V))\theta(O_A^+(L))\mathbb{R}^\times. \tag{2}$$

By the assumption, there is a lattice $L' \in spn^+(\varphi L)$ such that $K \hookrightarrow L'$. Multiplying φ by an element of $O^+(V)O'_A(V)$, we may assume $K \subset \varphi L$ without loss of generality. By the definition of $\theta(L_p; K_p)$, we have, for every prime p

$$\theta(\varphi_p) \subset \theta(L_p; K_p), \tag{3}$$

where φ_p is a p-component of φ. On the other hand, from (2) it follows that there is a prime q such that $\theta(\varphi_q) \not\subset N_q\theta(O^+(V))\theta(O^+(L_q))$, which yields $N_q \neq \theta(L_q; K_q)$ by (3).

Conversely we assume $\theta(L_q; K_q) \neq N_q$ for some prime q. Since $N_q \subset \theta(L_q; K_q)$ as above, $\theta(L_q; K_q) \not\subset N_q$ and hence there is $\varphi_q \in O^+(V_q)$ such that $\varphi_q(L_q) \supset K_q$ and $\theta(\varphi_q) \not\subset N_q$. We extend φ_q to $\varphi \in O_A^+(V)$ by putting $\varphi_v = \text{id}$ on V_v for every place $v \neq q$. Then $\varphi(L) \supset K$ is clear and $\theta(\varphi) \not\subset \tilde{N}$. Let us see

$$\theta(\varphi) \not\subset \tilde{N}\theta(O^+(V)). \tag{4}$$

If $\theta(\varphi) \subset \tilde{N}\theta(O^+(V))$, then there is a rational number b such that $b\theta(\varphi) \subset \tilde{N}$. For $v \neq q$, $b \in N_v$ is clear by the definition of φ_v. Thus $(b, -\delta)_{\mathbb{Q}_v} = 1$ holds for $v \neq q$, and then $(b, -\delta)_{\mathbb{Q}_q} = 1$ holds, which means $b \in N_q$. Thus $b \in \tilde{N}$ and $\theta(\varphi) \subset \tilde{N}$, which contradicts $\theta(\varphi_q) \not\subset N_q$. Next we show

$$\theta(\varphi) \not\subset \tilde{N}\theta(O^+(V))\theta(O_A^+(L))\mathbb{R}^\times. \tag{5}$$

If (5) is false, then $\theta(\varphi) \subset \tilde{N}\theta(O^+(V))\theta(O_A^+(L))\mathbb{R}^\times$, which implies $a\theta(\varphi) \subset \tilde{N}\theta(O_A^+(L))\mathbb{R}^\times$ for some $a \in \theta(O^+(V))$. Hence $a\theta(\varphi_p) \subset N_p\theta(O^+(L_p))$ for every prime p. Since $\varphi_\infty = \mathrm{id}$,

$$a\theta(\varphi_\infty) \subset \theta(O^+(V))(\mathbb{R}^\times)^2 \subset N_\infty\theta(O^+(V_\infty)).$$

Hence we have $a\theta(\varphi) \subset \tilde{N}\theta(O_A^+(L))$ and $\theta(\varphi) \subset \tilde{N}\theta(O^+(V))\theta(O_A^+(L))$. Then Lemma 6.5.7 implies $\theta(\varphi) \subset \tilde{N}\theta(O^+(V))$ which contradicts (4). Thus (5) is true and we have $\varphi \notin H_L$ by (1) in Lemma 6.5.6. Thus we have proved that $\varphi(L)$ in $spn^+(\varphi(L))$, which is not in the orbit of $spn^+(L)$ by H_L, represents K. Hence every spn^+ in $gen(L)$ represents K. □

Are there any diophantine-theoretic (not group-theoretic) results like [BöSP] in the codimension two case?

6.6 Representation of codimension ≥3

In the indefinite case, we have a satisfactory result.

Theorem 6.6.1. *Let V, W be regular quadratic spaces over $\mathbb{Q}$ with $\dim V + 3 \leq \dim W$ and L, M lattices on V, W respectively. Suppose that L_p is represented by M_p for every prime p, V_∞ is represented by W_∞ and W is indefinite. Then L is represented by M.*

Proof. By Corollary 4.4.2 V is represented by W. So we may assume $V \subset W$. By the assumption, there is an isometry σ_p from L_p to M_p, where we may assume $\sigma_p \in O(W_p)$. Multiplying by some isometry in $O(V_p^\perp)$ from the right if necessary, we may assume $\sigma_p \in O'(W_p)$ by Proposition 5.5.1. Applying Theorem 6.2.2 to W, M, there is an isometry $\sigma \in O'(V)$ such that

$$\begin{cases} \sigma(M_p) = M_p & \text{if } L_p \subset M_p \\ \sigma \text{ is sufficiently close to } \sigma_p & \text{if } L_p \not\subset M_p. \end{cases}$$

Hence we have $\sigma(L_p) \subset M_p$ for every prime p. Thus $\sigma(L) \subset M$. □

In the definite case, this Hasse principle does not hold in general. At present we can only show the following result.

Theorem 6.6.2. *Let W be a positive definite quadratic space over $\mathbb{Q}$ and N a lattice on it. Then there is a positive constant $c(N)$ such that a lattice M on a positive definite quadratic space V over $\mathbb{Q}$ is represented by N provided that*

$$\dim W \geq 2 \dim V + 3, \min(M) := \min\{Q(x) \mid x \in M, x \neq 0\} > c(N)$$

and M_p is represented by N_p for every prime p.

We need several preparatory lemmas.

Lemma 6.6.1. *Let $\mathbb{N}$ be the set of non-negative integers; we introduce a partial order in $\mathbb{N}^k$ defined by $(x_1, \cdots, x_k) \leq (y_1, \cdots, y_k)$ if $x_i \leq y_i$ for all i. Then every subset S of $\mathbb{N}^k$ contains only finitely many minimal elements.*

Proof. We use induction on k. The assertion is trivial for $k = 1$. Write

$$x = (p(x), x_k) \in \mathbb{N}^k$$

with $p(x) \in \mathbb{N}^{k-1}$, and put

$$S_n = \{x \in \mathbb{N}^{k-1} \mid (x, n) \in S\}.$$

Let M_n, M be the sets of the minimal vectors of $S_n, \cup_{n=0}^{\infty} S_n$ respectively. By the induction assumption, M_n and M are finite sets. For $y' \in M$ we fix any element $y \in S$ satisfying $y = (y', y_k)$ and denote by M' the finite set of such y. Put $m := \max\{y_k \mid y \in M'\}$. Suppose that $x \in S$ is minimal. Then from the definition $p(x) \in S_{x_k}$ there exists $y \in M'$ such that $p(y) \leq p(x)$. If $x_k \geq y_k$, then $x \geq y$ and so the minimality of x implies $x = y \in M'$. Suppose $x_k < y_k$ $(\leq m)$. Since x is minimal, $p(x)$ is so in S_{x_k}, i.e. $p(x) \in M_{x_k}$. Thus

$$x \in M' \cup (\cup_{n=0}^{m} (M_n, n)).$$

□

Lemma 6.6.2. *Let N_p be a regular quadratic lattice over $\mathbb{Z}_p$ of* rank $N_p =$ *$n \geq m$. Then there are only finitely many regular submodules $M_p(j)$ of* rank $M_p(j) = m$ *such that every regular submodule M_p of* rank $M_p = m$ *of N_p is represented by some $M_p(j)$.*

Proof. By scaling we may assume $\mathrm{s}(N_p) \subset \mathbb{Z}_p$. Let M_p be a regular quadratic lattice over $\mathbb{Z}_p$ of rank $M = m$ and $M_p = \perp_{i=1}^{t} L_i$ be a Jordan splitting. Since L_i is modular, there is an integer $b_i \in \mathbb{N}$ such that $p^{-b_i} L_i$

is unimodular or $p\mathbb{Z}_p$-modular. Since there are only finitely many non-isometric lattices over $\mathbb{Z}_p$ of unimodular or $p\mathbb{Z}_p$-modular lattices of fixed rank, there are only finitely many possibilities for the collection t, $\operatorname{rank} L_i$ and the isometry class K_i of $p^{-b_i}L_i$ $(i = 1, \cdots, t)$. Fix one of them and consider the corresponding $(b_1, \cdots, b_t)$. By the last lemma, there are only finitely many minimal ones. It is clear that if $M_p = \perp_{i=1}^t L_i$, $M'_p = \perp_{i=1}^t L'_i$,

$$p^{-b_i}L_i \cong p^{-b'_i}L'_i \cong K_i$$

and $b_i \leq b'_i$ for every i, then M'_p is represented by M_p. Hence M_p, running over all possible collections t, $\operatorname{rank} L_i$, K_i and minimal families $\{b_i\}$, constitutes a finite family with the required property. □

Lemma 6.6.3. *Let U be a positive definite quadratic space over $\mathbb{Q}$ and suppose that $\dim U \geq 3$ and U_q is isotropic for a prime q. Let L be a lattice on U such that L_p is maximal for every prime p. Then there is a natural number s such that L represents every positive definite quadratic lattice H for which H_p is represented by q^sL_p for every prime p.*

Proof. Let $\{L_i\}$ be a complete set of representatives of classes in $gen(L)$. We put $S = \{q, \infty\}$ and denote by L'_i the S-lattice

$$\cap_{p\neq q}(U \cap (L_i)_p).$$

Then Theorem 6.3.2 and Corollary 6.3.2 imply $cls^+(L'_i) = spn^+(L'_i) = gen_S(L'_i) = gen_S(L')$. Thus there is an isometry σ_i such that $(\sigma_i(L_i))_p = L_p$ for every prime $p \neq q$. Hence as a set of representatives of classes in $gen(L)$, we may assume $(L_i)_p = L_p$ for $p \neq q$. Then for a sufficiently large integer s, we have $q^sL_i \subset L$. For a lattice H in the theorem, H is represented by q^sL_i for some i by Theorem 6.1.3, noting that $\{q^sL_i\}$ is a complete set of representatives of the classes in $gen(q^sL)$. Thus H is represented by L. □

Lemma 6.6.4. *Let L, q, s be those in the last lemma, m an integer satisfying $\operatorname{rank} L \geq m+3$ and K a lattice on a positive definite quadratic space over $\mathbb{Q}$. Then there is a positive constant c such that $K \perp L$ represents a lattice $M = \mathbb{Z}[v_1, \cdots, v_m]$ if M_p is represented by $(K \perp q^sL)_p$ for every prime p and $(B(v_i, v_j)) > c1_m$.*

Proof. Let S be a finite set of primes such that $2, q \in S$ and for $p \notin S$, K_p and L_p are unimodular, and fix a natural number r such that

(1) $$p^r\,\mathrm{s}(K_p) \subset \mathrm{n}(q^sL_p) \quad \text{if} \quad p \in S.$$

Choose vectors $v_i^h \in K$ $(i = 1, \cdots, m,\ h = 1, \cdots, t)$ such that for any given $x_{1,p}, \cdots, x_{m,p} \in K_p$, there is some h $(1 \leq h \leq t)$ for which

$$(2) \qquad v_i^h \equiv x_{i,p} \bmod p^r K_p \quad \text{for } i = 1, \cdots, m \ \text{ and for all } \ p \in S.$$

We choose a positive number c such that

$$(3) \qquad c1_m > (B(v_i^h, v_j^h)) \quad \text{for } h = 1, \cdots, t.$$

Let M be a given lattice in the lemma. By the first condition on it, there exist $x_{i,p} \in K_p$, $y_{i,p} \in q^s L_p$ such that

$$(4) \qquad B(v_i, v_j) = B(x_{i,p}, x_{j,p}) + B(y_{i,p}, y_{j,p}) \quad \text{for all primes } p.$$

For some h satisfying (2) for these $x_{i,p}$, we put

$$(5) \qquad A := (B(v_i, v_j)) - (B(v_i^h, v_j^h)).$$

All entries of A are rational and

$$A = (B(v_i, v_j)) - c1_m + c1_m - (B(v_i^h, v_j^h))$$

implies that A is positive definite by virtue of the second condition on M and (3). By the choice of h, we can write $x_{i,p} = v_i^h + p^r z_{i,p}$ for some $z_{i,p} \in K_p$ for $i = 1, \cdots, m$, $p \in S$. Then (4) implies

$$\begin{aligned} A =& (B(x_{i,p}, x_{j,p})) + (B(y_{i,p}, y_{j,p})) - (B(v_i^h, v_j^h)) \\ =& (B(y_{i,p}, y_{j,p})) + (p^r B(v_i^h, z_{j,p}) + p^r B(z_{i,p}, v_j^h) + p^{2r} B(z_{i,p}, z_{j,p})). \end{aligned}$$

By (1) and v_i^h, $z_{i,p} \in K_p$, the (i, j) entry of A is congruent to $B(y_{i,p}, y_{j,p})$ $\bmod\, \mathrm{n}(q^s L_p)$ for $p \in S$. Then denoting by

$$H := \mathbb{Z}[u_1, \cdots, u_m]$$

a quadratic module defined by $(B(u_i, u_j)) := A$, $y_{i,p} \in q^s L_p$ yields $\mathrm{n}(H_p) \subset \mathrm{n}(q^s L_p)$ for $p \in S$.

Next, for $p \notin S$, $v_i \in M$, $v_i^h \in K$ and $\mathrm{n}(M_p) \subset \mathrm{n}(K_p \perp q^s L_p) = \mathbb{Z}_p$ imply

$$\mathrm{n}(H_p) \subset \mathrm{n}(M_p) + \mathrm{n}(K_p) \subset \mathbb{Z}_p = \mathrm{n}(q^s L_p)$$

by (5). Thus we have proved $\mathrm{n}(H_p) \subset \mathrm{n}(q^s L_p)$ for every prime p. Since

$$\dim \mathbb{Q}L - \dim \mathbb{Q}H = \operatorname{rank} L - m \geq 3$$

and $q^s L_p$ is maximal for every p, H_p is represented by $q^s L_p$ for every prime p by Theorem 5.2.3. Since A is positive definite, $(\mathbb{Q}H)_\infty$ is positive definite and then by the last lemma, H is represented by L. Thus for some $w_i \in L$, $(B(w_i, w_j)) = A$ and hence

$$(B(v_i, v_j)) = (B(v_i^h, v_j^h)) + (B(w_i, w_j)).$$

This means that M is represented by $K \perp L$. □

Proof of Theorem 6.6.2.

Let N be the lattice on W in the theorem. Let S be a finite set of primes such that $2 \in S$, N_p is unimodular for $p \notin S$ and N_q is unimodular for some prime q ($\neq 2$) in S. We will construct a set of submodules $K(J), L(J)$ of N as in Lemma 6.6.4 and show that a given M in the theorem satisfies the condition in Lemma 6.6.4 for some J. Put $m = \dim V$. For each $p \in S$, we choose finitely many submodules $M_p(j_p)$ of rank $= m$ in N_p according to Lemma 6.6.2. To each collection $J = (j_p)_{p \in S}$, we take a submodule $K(J)$ of rank $= m$ in N satisfying the conditions $K(J)_p \cong M_p(j_p)$ for $p \in S$ and $\mathrm{d}\,K(J) \in \mathbb{Z}_p^\times$ or $p\mathbb{Z}_p^\times$ for $p \notin S$ by Theorem 6.2.1 and Corollary 5.4.5. We construct a submodule $L(J)$ of

$$\operatorname{rank} L(J) = \operatorname{rank} N - m \, (\geq m + 3)$$

in $\{x \in N \mid B(x, K(J)) = 0\}$ by local conditions as follows: We put

$$L(J)_p = K(J)_p^\perp \text{ in } N_p \quad \text{for } p \notin S.$$

Since $\mathrm{s}(K(J)_p) \subset \mathbb{Z}_p$ and $\mathrm{d}\,K(J)_p \in \mathbb{Z}_p^\times \cup p\mathbb{Z}_p^\times$, Proposition 5.3.3 implies $\mathrm{d}\,L(J)_p \subset \mathbb{Z}_p^\times \cup p\mathbb{Z}_p^\times$. Thus $L(J)_p$ is $\mathbb{Z}_p$-maximal if $p \notin S$. For $p \in S$, we take any maximal submodule of $K(J)_p^\perp$ as $L(J)_p$. From Proposition 5.5.3 and Corollary 6.3.2, it follows that $gen(L) = spn(L)$. We show that $L(J)_q$ is isotropic. If $\operatorname{rank} L(J)_q \geq 5$, then $L(J)_q$ is isotropic. Otherwise, we have $\operatorname{rank} L(J)_q = 4, m = 1, \operatorname{rank} N_q = 5$. By the assumption on q ($\neq 2$), N_q is unimodular. Hence we have

$$N_q \cong \langle \begin{pmatrix} 0 & 1 \\ 1 & 0 \end{pmatrix} \rangle \perp \langle \begin{pmatrix} 0 & 1 \\ 1 & 0 \end{pmatrix} \rangle \perp \text{(some lattice)}. \tag{7}$$

If $L(J)_q$ is anisotropic, $\mathbb{Q}_q N \supset \mathbb{Q}_q L(J)_q$ contains a 4-dimensional anisotropic subspace and hence $\operatorname{ind} \mathbb{Q}_q N = 1$, which contradicts (7). Thus $L(J)_q$ is isotropic. So, we have constructed submodules $K(J), L(J)$ of N such that $\operatorname{rank} K(J) = \dim V$, $L(J)$ is locally maximal, especially $\mathbb{Z}_p$-maximal if $p \notin S$, $L(J)_q$ is isotropic, and finally every submodule $M(p)$ in N_p with $\operatorname{rank} M(p) = \dim V$ is represented by $K(J)_p$ for $p \in S$ for some J.

Let M be a lattice of rank m of a positive definite quadratic space over $\mathbb{Q}$ such that M_p is represented by N_p for every p. Suppose $p \notin S$; then N_p is unimodular. Hence $\mathrm{n}(M_p) \subset \mathbb{Z}_p$ holds. Since $L(J)_p$ is $\mathbb{Z}_p$-maximal and $\operatorname{rank} L(J)_p \geq m+3$, M_p is represented by $L(J)_p = (q^s L(J))_p$ by Theorem 5.2.3 for every J where s is the maximal integer among those integers s given in Lemma 6.6.3 for $L(J)$. For some J, M_p is represented by $K(J)_p$ for all $p \in S$. Thus M_p is represented by $(K(J) \perp q^s L(J))_p$ for every prime p. By Lemma 6.6.4, there is a constant $c(J)$ such that M is represented by $K(J) \perp L(J) \subset N$ if $(B(v_i, v_j)) > c(J)1_m$ for some basis $\{v_i\}$ of M. Put $c' := \max_J c(J)$. By the reduction theory in Chapter 2, there is a basis $\{v_i\}$ of M such that $(B(v_i, v_j)) \in S_{4/3,1/2}$ (the Siegel domain; see Chapter 2), and then

$$(B(v_i, v_j)) > \tilde{c}\,\mathrm{diag}(Q(v_1), \cdots, Q(v_m))$$

for some positive constant $\tilde{c}$ by Theorem 2.1.2. If $\min_{0 \neq v \in M} Q(v)$ is sufficiently large, then we have $(B(v_i, v_j)) > c'1_m$ and hence M is represented by N. This completes the proof. □

Exercise.

Let K, L be regular quadratic lattices over $\mathbb{Z}$ with $\operatorname{rank} L \geq \operatorname{rank} K + 3$. If K is represented by $gen(L)$, show K is represented by $spn^+(L)$.

6.7 Orthogonal decomposition

If a quadratic module L has a non-trivial orthogonal decomposition

$$L = L_1 \perp L_2 \ (L_1 \neq 0, L_2 \neq 0),$$

then L is called *decomposable.* Otherwise it is called *indecomposable.* The aim of this section is to show the following two facts.

Theorem 6.7.1. *Let V be a positive definite quadratic space over $\mathbb{Q}$. If $L = L_1 \perp \cdots \perp L_t$ is an orthogonal splitting into indecomposable submodules of a lattice L on V, then the set of $\{L_i\}$ is uniquely determined by L.*

Theorem 6.7.2. *Let V be an indefinite quadratic space over $\mathbb{Q}$. Then there is no indecomposable lattice on V if $\dim V \geq 12$.*

The first is due to Eichler and the second is due to Watson.

Proof of Theorem 6.7.1.

Let L be a lattice on V. By scaling of V if necessary, we may assume $\mathrm{s}(L) \subset \mathbb{Z}$. An element x in L is called *reducible* if there are non-zero elements $y, z \in L$ such that $x = y + z$ and $B(y, z) = 0$. Otherwise x is called *irreducible.* Let S be the set of all irreducible elements of L. Since V is positive definite and $\mathrm{s}(L) \subset \mathbb{Z}$, $x = y + z$ $(y \neq 0, z \neq 0)$ and $B(y,z) = 0$ imply $1 \leq Q(y), Q(z) < Q(x)$ for $x, y, z \in L$. Thus every non-zero element in L is a sum of irreducible elements, and S spans L. We introduce an equivalence relation $\sim$ in S. For $x, y \in S$, we write $x \sim y$ if there is a chain of elements $z_1, \cdots, z_r$ in S such that $z_1 = x, z_r = y$ and $B(z_i, z_{i+1}) \neq 0$ for $i = 1, \cdots, r-1$. Let $\{S_i\}_{i=1}^t$ be the equivalence classes of this equivalence relation. Note that $B(S_i, S_j) = 0$ if $i \neq j$. Let L_i be the submodule spanned by S_i. Then $B(L_i, L_j) = 0$ if $i \neq j$. Since S spans L, we have $L = \perp_{i=1}^t L_i$. Thus t is a finite number. Let $L = \perp_{i=1}^s K_i$ be another orthogonal decomposition with K_i indecomposable. Take an element x in S_k. Since x is irreducible, x is in some K_j. For an element y in S_k, there is a chain $z_1, \cdots, z_r \in S_k$ such that $z_1 = x$, $z_r = y$, $B(z_i, z_{i+1}) \neq 0$ $(i = 1, \cdots, r-1)$. Hence $x = z_1, z_2, \cdots, z_r = y \in K_j$ follows. Thus S_k is contained in some K_j. Therefore $L_k \subset K_j$ holds. Put $C_j = \{k \mid L_k \subset K_j\}$; then

$$L = \perp_{j=1}^s (\perp_{k \in C_j} L_k) \subset \perp_{j=1}^s K_j = L,$$

which yields $K_j = \perp_{k \in C_j} L_k$. Since K_j is indecomposable, $\sharp C_j = 1$ follows for all j. This is the assertion of the theorem. □

To prove Theorem 6.7.2, we divide the proof into several steps.

Lemma 6.7.1. *Let K be a regular quadratic lattice over $\mathbb{Z}_p$ with $n :=$* rank K. *Then there is a splitting $K = K_1 \perp K_2$ such that* $\mathrm{d}\, K_1 \in (\mathbb{Q}_p^\times)^2$ *and moreover* rank $K_1 = 2$ *if $p \neq 2$ and $n \geq 5$,* rank $K_1 = 2$ *or* 4 *if $p = 2$ and $n \geq 8$.*

Proof. First suppose $p \neq 2$. Let us recall $\sharp(\mathbb{Q}_p^\times/(\mathbb{Q}_p^\times)^2) = 4$ and K has an orthogonal splitting $\perp_{i=1}^n \langle \eta_i \rangle$ $(\eta_i \in \mathbb{Q}_p^\times)$. For $n \geq 5$, there is a pair (i,j) such that $\eta_i \eta_j \in (\mathbb{Q}_p^\times)^2$. We have only to take $\langle \eta_i \rangle \perp \langle \eta_j \rangle$ as K_1.

Next suppose $p = 2$ and write

$$K = \perp_{i=1}^{m_1} \langle \beta_i \rangle \perp (\perp_{i=1}^{m_2} \langle 2^{a_i} \begin{pmatrix} 0 & 1 \\ 1 & 0 \end{pmatrix} \rangle) \perp (\perp_{i=1}^{m_3} \langle 2^{b_i} \begin{pmatrix} 2 & 1 \\ 1 & 2 \end{pmatrix} \rangle),$$

where $m_i \geq 0$ and rank $K = n = m_1 + 2(m_2 + m_3) \geq 8$. For any given five different classes in $\mathbb{Q}_2^\times/(\mathbb{Q}_2^\times)^2$, we can choose four classes whose product is in $(\mathbb{Q}_2^\times)^2$ by Corollary 3.1.4. Thus we may assume $m_1 \leq 4$ and the product $\beta_i \beta_j$ is not in $(\mathbb{Q}_2^\times)^2$ for any $i \neq j$. If $m_2 \geq 2$, then we can take $(\langle 2^{a_1} \begin{pmatrix} 0 & 1 \\ 1 & 0 \end{pmatrix} \rangle \perp \langle 2^{a_2} \begin{pmatrix} 0 & 1 \\ 1 & 0 \end{pmatrix} \rangle)$ as K_1. Thus we may assume m_2 and similarly $m_3 \leq 1$. Now $n \geq 8$ implies $m_1 = 4, m_2 = m_3 = 1$. If the product of two elements of β_i $(1 \leq i \leq 4)$ is in $-(\mathbb{Q}_2^\times)^2$ or $3(\mathbb{Q}_2^\times)^2$, then taking an orthogonal sum with $\langle 2^{a_1}\begin{pmatrix}0&1\\1&0\end{pmatrix}\rangle$ or $\langle 2^{b_1}\begin{pmatrix}2&1\\1&2\end{pmatrix}\rangle$, we get a required submodule K_1 with $\operatorname{rank} K_1 = 4$. Thus it remains to show the non-existence of the subset S (represented by β_i) in $\mathbb{Q}_2^\times/(\mathbb{Q}_2^\times)^2$ with $\sharp S = 4$ such that the product of any two distinct elements of S is different from $\pm 1, 3$ in $\mathbb{Q}_2^\times/(\mathbb{Q}_2^\times)^2$ and the product of all elements is not a square.

Write $S = \{s_i\}_{i=1}^4$. If $\operatorname{ord}_2 s_1 \equiv \operatorname{ord}_2 s_2 \equiv \operatorname{ord}_2 s_3 \bmod 2$, then

$$\operatorname{ord}_2 s_1 s_2 \equiv \operatorname{ord}_2 s_1 s_3 \equiv 0 \bmod 2$$

and s_1s_2 and s_1s_3 are different classes in $\mathbb{Q}_2^\times/(\mathbb{Q}_2^\times)^2$. One of them must be equal to ± 1 or $3 \bmod (\mathbb{Q}_p^\times)^2$. This is a contradiction. Thus we may assume $\operatorname{ord}_2 s_i \equiv 0 \bmod 2$ $(i = 1, 2)$, $\operatorname{ord}_2 s_i \equiv 1 \bmod 2$ $(i = 3, 4)$ without loss of generality. Since s_1s_2 and $s_3s_4 \neq \pm 1, 3 \bmod (\mathbb{Q}_2^\times)^2$, $\{s_1, s_2\} = \{1, 5\}$ or $\{3, 7\}$ and $\{s_3, s_4\} = \{2, 2\cdot 5\}$ or $\{2\cdot 3, 2\cdot 7\}$. However this yields

$$\prod_{i=1}^{4} s_i \in (\mathbb{Q}_2^\times)^2.$$

This contradicts the assumption on S and then completes the proof of the lemma. □

Lemma 6.7.2. *Let W be an isotropic regular quadratic space over $\mathbb{R}$ with* $\dim W \geq 11$, *and r be* 2, 4, 6 *or* 8. *Suppose*

$$W \cong (\perp_s \langle 1\rangle) \perp (\perp_t \langle -1\rangle)$$

with $s \geq t$. Then there is a subspace $W_0 \subset W$ such that $\operatorname{d} W_0 = 1$, *the Hasse invariant of* $W_0 = 1$ *and* $\dim W_0 = r$.

Proof. Since $s \geq t$ and $s + t \geq 11$, we have $s \geq 6$. If $r \leq 6$, then put $W_0 = \perp_r \langle 1\rangle$. Suppose $r = 8$. If $s \geq 8$, then put $W_0 = \perp_r \langle 1\rangle$. If $s = 6$ or 7, then $t \geq 4$ and so put $W_0 = (\perp_4 \langle 1\rangle) \perp (\perp_4 \langle -1\rangle)$. □

Lemma 6.7.3. *Let L be a lattice on the quadratic space given in Theorem 6.7.2. Then there is a decomposable lattice $M \in gen(L)$.*

Proof. By assumption, $\dim V \geq 12$ and V_∞ is isotropic. Scaling of V by -1 if necessary, we may assume $V_\infty \cong (\perp_s \langle 1\rangle) \perp (\perp_t \langle -1\rangle)$ with $s \geq t$.

Suppose that there is an integer $r = 2, 4, 6$ or 8 such that there is a splitting for every prime p

(C1) $\quad L_p = L_{p,1} \perp L_{p,2}$ with $\operatorname{rank} L_{p,1} = r$, $\operatorname{d} L_{p,1} \in (\mathbb{Q}_p^\times)^2$.

If L_p is unimodular, then so is $L_{p,1}$, hence the Hasse invariant $S_p(\mathbb{Q}_p L_{p,1}) = 1$ under the additional assumption $p \neq 2$. Thus $S_p(\mathbb{Q}_p L_{p,1}) = 1$ for almost all primes, and

$$s_\infty := \prod_p S_p(\mathbb{Q}_p L_{p,1})$$

is well-defined. Next, we suppose

$$\text{(C2)} \qquad s_\infty = 1.$$

First we show that the assumptions (C1), (C2) give a required lattice M. By the last lemma, there is a subspace W_0 of V_∞ such that $\dim W_0 = r, \mathrm{d}\, W_0 = 1$ and $S_\infty(W_0) = 1$. Thus by Theorem 4.4.2 there is a quadratic space U over $\mathbb{Q}$ such that $U_p \cong \mathbb{Q}_p L_{p,1}, U_\infty \cong W_0$. We may assume $U \subset V$. Taking any lattice K on U and N on $U^\perp$, we construct a new lattice M by local conditions as follows; we put

$$M_p := K_p \perp N_p \quad \text{if} \quad K_p \perp N_p = L_p.$$

This is true for almost all primes.

Suppose $K_p \perp N_p \neq L_p$. By the definition of U, there is a lattice K'_p on U_p isometric to $L_{p,1}$. Therefore there is an isometry $\eta_p \in O(V_p)$ such that $\eta_p(L_{p,1}) \subset U_p$. By using it, we put

$$M_p := \eta_p(L_p) = \eta_p(L_{p,1}) \perp \eta_p(L_{p,2}) \quad \text{if} \quad K_p \perp N_p \neq L_p.$$

Then $K_p, \eta_p(L_{p,1})$ and $N_p, \eta_p(L_{p,2})$ define the lattices M_1, M_2 on $U, U^\perp$ respectively, $M = M_1 \perp M_2 \in gen(L)$ is clear.

It remains to verify the conditions (C1) and (C2) for $r = 2, 4, 6$ or 8. Applying Lemma 6.7.1 to L_p, we get a splitting

$$L_p = L'_{p,1} \perp L'_{p,2} \perp \text{(some lattice) with rank } L'_{p,i} = 2 \text{ or } 4,\ \mathrm{d}\, L'_{p,i} \in (\mathbb{Q}_p^\times)^2$$

and moreover rank $L_{p,1}$, rank $L_{p,2}$ are independent of p when p runs over all primes. Put $s_i = \prod_p S_p(\mathbb{Q}_p L'_{p,i})$ $(i = 1, 2)$; then

$$s_1 s_2 = \prod S_p(\mathbb{Q}_p(L'_{p,1} \perp L'_{p,2}))$$

holds. Hence at least one of s_1, s_2 and $s_1 s_2$ is equal to 1, and so taking $L'_{p,1}, L'_{p,2}$ or $L'_{p,1} \perp L'_{p,2}$ as $L_{p,1}$, the conditions (C1) and (C2) are satisfied. □

Proof of Theorem 6.7.2.

Let L be a lattice on V and $M = M_1 \perp M_2$ a decomposable lattice in $gen(L)$ as given in Lemma 6.7.3. We may assume $\operatorname{rank} M_2 \geq 3$. Let σ_p be an isometry from L_p to M_p. Since $\dim \mathbb{Q}_p M_2 \geq 3$, there is an isometry $\eta_p \in O(\mathbb{Q}_p M_2)$ such that considering $\eta_p = \text{id}$ on $\mathbb{Q}_p M_1$, $\eta_p \sigma_p \in O'(V_p)$ holds. Putting $u_p = \eta_p \sigma_p$, we have $u_p \in O'(V_p)$ and

$$u_p(L_p) = (M_1)_p \perp \text{ (some lattice)}.$$

We define a lattice N on V by

$$N_p = \begin{cases} L_p & \text{if } L_p = (M_1)_p \perp (M_2)_p, \\ u_p(L_p) & \text{otherwise.} \end{cases}$$

Then $N \in spn^+ L$ and $N_p = (M_1)_p \perp$ (some lattice) for every prime p. Thus $N = M_1 \perp$ (some lattice) follows. Since V_∞ is isotropic, Theorem 6.3.2 implies $spn^+ = cls^+$ and hence $N \cong L$. □

6.8 The Minkowski-Siegel formula

In this section we prove the definite case of the so-called Minkowski-Siegel formula by Siegel's original proof without its analytic part. But we will refer to some analytic ideas, one of which will convince readers who are familiar with the theory of automorphic forms.

For a lattice K on a positive definite quadratic space over $\mathbb{Q}$, we put

$$E(K) := \sharp O(K) \quad \text{and} \quad w(K) := \sum E(K_i)^{-1}$$

where K_i runs over a complete set of representatives of isometry classes in $gen(K)$. $w(K)$ is called the *weight* of K.

Theorem 6.8.1. *Let M, N be lattices on positive definite quadratic spaces U, V and put $m := \operatorname{rank} M$, $n := \operatorname{rank} N$ and*

$$\epsilon_{m.n} := \begin{cases} 1/2 & \text{if either} \quad n = m+1 \quad \text{or} \quad n = m > 1, \\ 1 & \text{otherwise} \end{cases}$$

Then we have

$$w(M)^{-1} = \epsilon_{m,n} \pi^{m(m+1)/4} \prod_{i=0}^{m-1} \Gamma((m-i)/2))^{-1} (\mathrm{d}\, M)^{-(m+1)/2}$$

$$\text{(i)} \qquad \times \prod_p \alpha_p(M_p, M_p),$$

(ii)

$$w(N)^{-1}\sum_i \frac{r(M,N_i)}{E(N_i)}$$
$$= \epsilon_{m,n}\pi^{m(2n-m+1)/4}\prod_{i=0}^{m-1}\Gamma((n-i)/2))^{-1}(\mathrm{d}\,N)^{-m/2}(\mathrm{d}\,M)^{(n-m-1)/2}$$
$$\times\prod_p \alpha_p(M_p,N_p),$$

(iii)

$$w(N)^{-1}\sum_i \frac{r_{pr}(M,N_i)}{E(N_i)}$$
$$= \epsilon_{m,n}\pi^{m(2n-m+1)/4}\prod_{i=0}^{m-1}\Gamma((n-i)/2))^{-1}(\mathrm{d}\,N)^{-m/2}(\mathrm{d}\,M)^{(n-m-1)/2}$$
$$\times\prod_p D_p(M_p,N_p),$$

where N_i runs over a complete set of representatives of the isometry classes in gen(N), $r(M,N_i)$ is the number of the isometries from M to N_i, Γ is the ordinary gamma function, $r_{pr}(M,N_i)$ is the number of isometries σ from M to N_i such that $\sigma(M)$ is primitive in N_i and finally

$$D_p(M_p,N_p) = 2^{m\delta_{2,p}-\delta_{m,n}}d_p(M_p,N_p).$$

We note that (i) is a special case of (ii), by putting $N = M$. We call the left-hand side of (ii) (resp.(iii)) the mean value $m(M,N)$ (resp. $m_{pr}(M,N)$) of the (resp. primitive) representation of M by N. We need several lemmas. Consider the following situation (*).

(*): *Let the ring R and the field F be either $\mathbb{Z}$ and $\mathbb{Q}$, or $\mathbb{Z}_p$ and $\mathbb{Q}_p$, and N, M, K be quadratic lattices over R such that $FN = FM \perp FK$ is regular, M is a primitive submodule of N, and $K = proj(N)$ where proj is the projection mapping from FN to FK along $FN = FM \oplus FK$. We put $m :=$ rank $M, n :=$ rank N and $k := n - m =$ rank K.*

Lemma 6.8.1. *Under the condition (*), there is a unique linear mapping φ from $K \to FM/M$ such that $N = M + R[x+\varphi(x) \mid x \in K]$. Conversely, let ϕ be a linear mapping from K to FM/M. Then M is primitive in $N_\phi = N_\phi(M,K) := M + R[x+\phi(x) \mid x \in K]$ and $projN_\phi = K$, and $N_\phi = N$ if and only if $\phi = \varphi$.*

Proof. Let $\{v_i\}$ be a basis of K. By the assumption $K = proj(N)$, there are elements $\varphi(v_i) \in FM$ such that $\varphi(v_i)+v_i \in N$. That is φ gives a linear mapping from K to FM. Let x be an element of N, $proj(x)$ is in K and hence $proj(x) = \sum a_i v_i$ for some $a_i \in R$. Then

$$x - (\sum a_i v_i + \varphi(\sum a_i v_i)) = (x - proj(x)) - \varphi(\sum a_i v_i) \in FM \cap N = M.$$

Thus

$$x \in M + R[y + \varphi(y) \mid y \in K],$$

and so

$$N \subset M + R[y + \varphi(y) \mid y \in K] \subset N.$$

So the existence of φ has been shown. Suppose that two linear mappings $\varphi_i : K \to FM/M$ satisfy $N = M + R[y + \varphi_i(y) \mid y \in K]$. Then for $x \in K$, $x + \varphi_i(x) \in N$ $(i = 1, 2)$ implies $\varphi_1(x) - \varphi_2(x) \in N \cap FM = M$. Thus φ_1 and φ_2 determine the same mapping from K to FM/M. For bases $\{v_i\}, \{u_i\}$ of M, K, $\{v_i\} \cup \{u_i + \phi(u_i)\}$ is a basis of N_ϕ. So M is primitive in N_ϕ and $projN_\phi = K$. □

Lemma 6.8.2. *Under the condition (*),*

$$\mathrm{d}\,N = \mathrm{d}\,M\,\mathrm{d}\,K,\ \mathrm{s}(K) \subset \mathrm{s}(N)^2\,\mathrm{s}(M^\sharp) + \mathrm{s}(N)$$

and $[K^\sharp : K^\sharp \cap K]$ *is bounded from above by a constant dependent only on* M, N. *Moreover* $[\varphi(K) + M : M]$ *is bounded by some constant depending only on* M, N *where* φ *is the linear mapping in the last lemma. If* M *and* N *are unimodular, then* $N = M \perp K$.

Proof. We take a basis $\{v_i\}_{i=1}^k$ and $\{u_i\}_{i=1}^m$ of K and M such that $(\varphi(v_1), \cdots, \varphi(v_k)) = (q_1^{-1}u_1, \cdots, q_r^{-1}u_r, 0, \cdots, 0)$ for some r and for positive integers $q_1, \cdots, q_r$ or a power q_i of the prime p according to $R = \mathbb{Z}$ or $\mathbb{Z}_p$ respectively. Put

(1) $$\tilde{N} = R[q_1^{-1}u_1, \cdots, q_r^{-1}u_r] + N.$$

Then $N = M + R[y + \varphi(y) \mid y \in K]$ implies

(2) $$\tilde{N} = R[q_1^{-1}u_1, \cdots, q_r^{-1}u_r, u_{r+1}, \cdots, u_m] \perp K.$$

Hence $\mathrm{d}\,M\,\mathrm{d}\,K = (q_1 \cdots q_r)^2\,\mathrm{d}\,\tilde{N}$. Since M is primitive in N, the basis $\{u_i\}_{i=1}^m$ of M can be extended to a basis of N, and hence (1) implies $[\tilde{N} : N] = q_1 \cdots q_r$. Thus we have $\mathrm{d}\,M\,\mathrm{d}\,K = \mathrm{d}\,N$. Next $v_i + \varphi(v_i) \in N$ for $v_i \in K$ implies $B(M, \varphi(v_i)) = B(M, v_i + \varphi(v_i)) \subset B(M, N)$ and hence $q_i^{-1}B(M, u_i) = B(M, \varphi(v_i)) \subset B(M, N)$ for $i \le r$. Thus

(3) $$u_i \in q_i B(M, N) M^\sharp \quad \text{for} \quad i \le r.$$

Hence q_i is bounded from above and so the boundedness of $[\varphi(K)+M : M]$ is clear. On the other hand

$$B(v_i, v_j) + B(\varphi(v_i), \varphi(v_j)) = B(v_i + \varphi(v_i), v_j + \varphi(v_j)) \in \mathrm{s}(N)$$

yields $B(v_i, v_j) \in \mathrm{s}(N)$ if i or $j > r$, otherwise

$$\begin{aligned} B(v_i, v_j) &\in -B(q_i^{-1}u_i, q_j^{-1}u_j) + \mathrm{s}(N) \\ &\subset B(M,N)^2\,\mathrm{s}(M^\sharp) + \mathrm{s}(N) \subset \mathrm{s}(N)^2\,\mathrm{s}(M^\sharp) + \mathrm{s}(N). \end{aligned}$$

Thus $\mathrm{s}(K) \subset \mathrm{s}(N)^2\,\mathrm{s}(M^\sharp) + \mathrm{s}(N)$ holds. Let a be a natural number such that $a^2(\mathrm{s}(N)^2\,\mathrm{s}(M^\sharp) + \mathrm{s}(N)) \subset R$. Then we have $\mathrm{s}(aK) \subset R$ and

$$\begin{aligned} [K^\sharp : K^\sharp \cap K] &= [a^2(aK)^\sharp : a^2(aK)^\sharp \cap aK] \\ &\le [a^2(aK)^\sharp : a^2((aK)^\sharp \cap aK)] = [(aK)^\sharp : aK] \\ &= (\mathrm{d}(aK)/\,\mathrm{d}((aK)^\sharp))^{1/2}, \end{aligned}$$

which is dependent only on M and N. Suppose that M and N are unimodular; then (3) implies $u_i \in q_i M$ for $i \le r$ and hence $q_i = 1$. Thus $N = \tilde{N} = M \perp K$ follows from (1) and (2). □

For a linear mapping φ from K to FM/M, we put $N_\varphi = N_\varphi(M, K) := M + R[x + \varphi(x) \mid x \in K]$.

Lemma 6.8.3. *We keep the condition (*). Denote by Φ the set of the linear mappings φ from K to FM/M such that there is an isometry $\sigma : N_\varphi \cong N$ and by I the set of the isometries $\sigma : M \hookrightarrow N$ such that $\sigma(M)$ is primitive in N and $proj_{\sigma(M)^\perp} N \cong K$, where $proj_{\sigma(M)^\perp}$ denotes the projection to $F\sigma(M)^\perp$. Then the mapping $\xi : (\varphi, \sigma) \mapsto \sigma|_M$ from $\sqcup_{\varphi\in\Phi}\{\sigma : N_\varphi \cong N\}$ to I is surjective, and $\xi((\varphi_1, \sigma_1)) = \xi((\varphi_2, \sigma_2))$ if and only if $\varphi_1 = \varphi_2\alpha$ and $\sigma_1 = \sigma_2(\mathrm{id}|_M \perp \alpha)$ as isometries on FN for $\alpha \in O(K)$.*

Proof. For $\varphi \in \Phi$, let σ be an isometry from N_φ on N. Since M is primitive in N_φ, $\sigma(M)$ is so in N. Moreover we have

$$\sigma(K) = \sigma(proj_{M^\perp} N_\varphi) = proj_{\sigma(M)^\perp} N.$$

Thus ξ is well-defined. Let us show the surjectivity. Let $\eta \in I$ and put $H = proj_{\eta(M)^\perp} N$; then by the definition of I, there is an isometry

$$\alpha : K \cong H.$$

Put $\sigma := \eta \perp \alpha$. $M = \sigma^{-1}(\eta(M))$ is primitive in $\sigma^{-1}(N)$, since $\eta(M)$ is so in N. Now $K = \sigma^{-1}(H)$ is the projection of $\sigma^{-1}(N)$ to $FM^\perp = FK$, since

H is the projection of N to $F\eta(M)^\perp$. The condition (*) is satisfied for $M, K, \sigma^{-1}(N)$. Applying Lemma 6.8.1 to $\sigma^{-1}(N)$ instead of N, we have $\sigma^{-1}(N) = N_\varphi$ for some $\varphi : K \to FM/M$. Since $\xi((\varphi, \sigma)) = \sigma|_M = \eta$, the mapping ξ is surjective.

Next suppose $\sigma_i : N_{\varphi_i} \cong N$ $(i = 1, 2)$ and $\xi((\varphi_1, \sigma_1)) = \xi((\varphi_2, \sigma_2))$; then $\sigma := \sigma_2^{-1}\sigma_1 : N_{\varphi_1} \cong N_{\varphi_2}$ and $\sigma|_M = \mathrm{id}$. Thus $\sigma(M^\perp) = M^\perp$ and so $\sigma(FK) = FK$. Considering the projection of the both sides to FK, we have $\sigma(K) = K$. Put $\sigma|_K = \alpha$; then $\alpha \in O(K)$ and $\sigma_2^{-1}\sigma_1 = \sigma = \mathrm{id}\,|_M \perp \alpha$. Thus $\sigma_1 = \sigma_2(\mathrm{id}\,|_M \perp \alpha)$ and for $x \in K$,

$$\sigma(x + \varphi_1(x)) = \alpha(x) + \varphi_1(x) = \alpha(x) + \varphi_2(\alpha(x)) + (\varphi_1 - \varphi_2\alpha)(x)$$

implies

$$(\varphi_1 - \varphi_2\alpha)(x) = \sigma(x + \varphi_1(x)) - (\alpha(x) + \varphi_2(\alpha(x))) \in N_{\varphi_2} \cap FM = M.$$

Thus $\varphi_1 = \varphi_2\alpha \bmod M$.

Conversely $\varphi_1 = \varphi_2\alpha$ and $\sigma_1 = \sigma_2(\mathrm{id}\,|_M \perp \alpha)$ for $\alpha \in O(K)$ imply obviously $\xi((\varphi_1, \sigma_1)) = \sigma_1|_M = \sigma_2|_M = \xi((\varphi_2, \sigma_2))$. □

Lemma 6.8.4. *Keep the condition (*) with $R = \mathbb{Z}$ and assume that $\mathbb{Q}N$ is positive definite. Then*

$$E(N)^{-1}\sharp\{\sigma : M \hookrightarrow N \mid \sigma(M) \textit{ is primitive in } N \textit{ and } proj_{\sigma(M)^\perp} N \cong K\}$$
$$= E(K)^{-1}\sharp\{\varphi : K \to \mathbb{Q}M/M \mid N_\varphi \cong N\}.$$

Proof. Using the notation of Lemma 6.8.3, we have

$$\sharp I = E(K)^{-1} \times \sharp\Phi \times E(N),$$

which is our assertion. □

Lemma 6.8.5. *Keep the condition (*) with $R = \mathbb{Z}_p$ and let q be a sufficiently large power of p. Then in Lemma 6.8.3,*

$$(4) \qquad \xi((\varphi_1, \sigma_1))(x) \equiv \xi((\varphi_2, \sigma_2))(x) \bmod qN^\sharp \quad \textit{for } x \in M$$

if and only if

$$\sigma_1 = \beta\sigma_2(\mathrm{id}\,|_M \perp \alpha) \textit{ on } FN, \quad \varphi_1 = \varphi_2\alpha$$

for $\alpha \in O(K)$ and $\beta \in O(N)$ satisfying $\beta(x) \equiv x \bmod qN^\sharp$ for $x \in N$.

Proof. Suppose (4) for $\varphi_i \in \Phi$ and $\sigma_i : N_{\varphi_i} \cong N$. Then

$$\sigma_1(x) \equiv \sigma_2(x) \bmod qN^\sharp$$

for $x \in M$ and hence $\sigma_2^{-1}\sigma_1(x) \equiv x \bmod qN_{\varphi_2}^\sharp$ for $x \in M$. $\sigma_2^{-1}\sigma_1(M)$ is primitive in $N_{\varphi_2} = \sigma_2^{-1}\sigma_1(N_{\varphi_1})$ since M is so in N_{φ_1}. Then applying Theorem 5.4.1 to $\sigma_2^{-1}\sigma_1 : M \cong \sigma_2^{-1}\sigma_1(M)$ in N_{φ_2}, there is an isometry $\eta \in O(N_{\varphi_2})$ such that $\eta|_M = \sigma_2^{-1}\sigma_1|_M$ and

$$(5) \qquad \eta(x) \equiv x \bmod qN_{\varphi_2}^\sharp \quad \text{for} \quad x \in N_{\varphi_2}.$$

Then $\eta^{-1}\sigma_2^{-1}\sigma_1 : N_{\varphi_1} \cong N_{\varphi_2}$ and $\eta^{-1}\sigma_2^{-1}\sigma_1|_M = \mathrm{id}$. Taking the projection on $FK = FM^\perp$ of N, we have $\eta^{-1}\sigma_2^{-1}\sigma_1(K) = K$. Define $\alpha \in O(K)$ by $\alpha := \eta^{-1}\sigma_2^{-1}\sigma_1|_K$; then $\eta^{-1}\sigma_2^{-1}\sigma_1 = \mathrm{id}_M \perp \alpha$. Put $\beta := \sigma_2\eta\sigma_2^{-1} \in O(N)$; then $\sigma_1 = \beta\sigma_2(\mathrm{id}|_M \perp \alpha)$ and (5) implies $\beta(y) \equiv y \bmod qN^\sharp$ for $y \in N$. To complete the proof of the " only if " part, it remains to verify $\varphi_1 = \varphi_2\alpha$. For $x \in K$, we have

$$\eta^{-1}\sigma_2^{-1}\sigma_1(x + \varphi_1(x)) = \alpha(x) + \varphi_1(x) = \alpha(x) + \varphi_2(\alpha(x)) + (\varphi_1 - \varphi_2\alpha)(x),$$
$$\eta^{-1}\sigma_2^{-1}\sigma_1(x + \varphi_1(x)) \in \eta^{-1}\sigma_2^{-1}\sigma_1(N_{\varphi_1}) = N_{\varphi_2},$$
$$\alpha(x) + \varphi_2(\alpha(x)) \in N_{\varphi_2}.$$

Thus we have $(\varphi_1 - \varphi_2\alpha)(x) \in N_{\varphi_2} \cap FM = M$, and so $\varphi_1 = \varphi_2\alpha$ as mappings from K to FM/M. Thus the " only if " part has been proved. The " if " part is clear because of $\xi((\varphi_i, \sigma_i)) = \sigma_i|_M$. □

Lemma 6.8.6. *We keep the notation of Lemma 6.8.3 and assume $R = \mathbb{Z}_p$. Then we have, for a sufficiently large power q of p,*

$$\sharp\Phi \times \sharp(O(N)/\sim_q) = \sharp(I/\sim_q)\sharp(O(K)/\sim_q);$$

where for $\sigma_1, \sigma_2 \in O(H)$ for $H = N$ or K; $\sigma_1 \sim_q \sigma_2$ is defined by

$$\sigma_1\sigma_2^{-1}(x) \equiv x \bmod qH^\sharp \quad \textit{for} \quad x \in H;$$

for $\eta_1, \eta_2 \in I$, $\eta_1 \sim_q \eta_2$ is defined by

$$\eta_1(x) \equiv \eta_2(x) \bmod qN^\sharp \quad \textit{for} \quad x \in M.$$

Proof. Put

$$A := \sqcup_{\varphi\in\Phi}(\{\sigma : N_\varphi \cong N\}/\sim_q),$$

where for $\sigma_i : N_\varphi \cong N$, $\sigma_1 \sim_q \sigma_2$ is defined by $\sigma_1\sigma_2^{-1} \sim_q \mathrm{id}|_N$. Then

$$\sharp A = \sharp\Phi \cdot \sharp(O(N)/\sim_q)$$

is clear. Consider the mapping ξ' from A to $I/\sim_q$ by $(\varphi, \sigma) \mapsto \sigma|_M$. This is well-defined. Since ξ' is surjective by Lemma 6.8.3, we have only to show that the number of inverse images is $\sharp(O(K)/\sim_q)$. By the last lemma, $\xi'((\varphi_1, \sigma_1)) = \xi'((\varphi_2, \sigma_2))$ if and only if

$$\sigma_1 = \beta\sigma_2(\mathrm{id}|_M \perp \alpha), \quad \varphi_1 = \varphi_2\alpha$$

for $\alpha \in O(K), \beta \in O(N)$ with $\beta \sim_q \mathrm{id}|_N$. So $\xi'((\varphi_1, \sigma_1)) = \xi'((\varphi_2, \sigma_2))$ means $\sigma_1 \sim_q \sigma_2(\mathrm{id}|_M \perp \alpha), \varphi_1 = \varphi_2\alpha$. On the other hand, $\alpha \in O(K)$ acts on A by $(\varphi, \sigma) \mapsto (\varphi\alpha, \sigma(\mathrm{id}|_M \perp \alpha))$, and so putting $G := \{\alpha \in O(K) \mid \varphi\alpha = \varphi, \sigma(\mathrm{id}|_M \perp \alpha) \sim_q \sigma\}$; the number of inverse images is the number of elements in an orbit by $O(K)$ and so equal to $[O(K) : G]$. Thus we have only to verify $G = \{\alpha \in O(K) \mid \alpha \sim_q \mathrm{id}|_K\}$. First let us show

$$qN_\varphi^\sharp \cap K = qK^\sharp. \tag{6}$$

For $y \in FK$,

$$\begin{aligned} y \text{ is in the left-hand side } &\Leftrightarrow y \in K \text{ and } B(q^{-1}y, N_\varphi) = B(q^{-1}y, K) \subset \mathbb{Z}_p \\ &\Leftrightarrow y \in K \cap qK^\sharp = qK^\sharp \end{aligned}$$

since q is sufficiently large.

Suppose $\alpha \sim_q \mathrm{id}|_K$. Let $\sigma : N_\varphi \cong N$ for $\varphi \in \Phi$. For $\alpha \in O(K)$, put $\eta = \sigma(\mathrm{id}|_M \perp \alpha)\sigma^{-1}$; then

$$\begin{aligned} \sigma(\mathrm{id}|_M \perp \alpha) \sim_q \sigma &\Leftrightarrow \eta(x) \equiv x \bmod qN^\sharp \quad \text{for} \quad x \in N \\ &\Leftrightarrow (\mathrm{id}|_M \perp \alpha)(y) \equiv y \bmod qN_\varphi^\sharp \quad \text{for} \quad y \in N_\varphi \\ &\Leftrightarrow (\mathrm{id}|_M \perp \alpha)(y + z + \varphi(z)) \\ &\qquad \equiv y + z + \varphi(z) \bmod qN_\varphi^\sharp \quad \text{for} \quad y \in M, z \in K \\ &\Leftrightarrow \alpha(z) \equiv z \bmod qN_\varphi^\sharp \quad \text{for} \quad z \in K \\ &\Leftrightarrow \alpha(z) - z \in qK^\sharp \quad \text{for} \quad z \in K \quad \text{by (6)} \\ &\Leftrightarrow \alpha \sim_q \mathrm{id}|_K. \end{aligned}$$

To complete the proof, we have only to show that $\alpha \sim_q \mathrm{id}|_K$ implies $\varphi = \varphi\alpha$. By Lemma 6.8.2 $[K^\sharp : K^\sharp \cap K]$ is bounded by a number dependent only on M, N. Thus

$$\varphi(qK^\sharp) \subset q[K^\sharp : K^\sharp \cap K]^{-1}\varphi(K^\sharp \cap K) \subset q[K^\sharp : K^\sharp \cap K]^{-1}\varphi(K) \subset M,$$

and so $\alpha \sim_q \mathrm{id}|_K$ implies $\varphi((1 - \alpha)(K)) \subset \varphi(qK^\sharp) \subset M$, i.e. $\varphi = \varphi\alpha$.

Conversely suppose $\alpha \in G$; then

$$\sigma(\mathrm{id}|_M \perp \alpha)(x + y + \varphi(y)) \equiv \sigma(x + y + \varphi(y)) \bmod qN^\sharp$$

for $x \in M, y \in K$, and $\sigma\alpha(y) \equiv \sigma(y) \bmod qN^\sharp$ for $y \in K$, which yields $\alpha(y) - y \in q\sigma^{-1}(N)^\sharp \cap K = qK^\sharp$ by (6). Thus $\alpha \sim_q \mathrm{id}|_K$. □

Lemma 6.8.7. *Continuing the last lemma, we have*

$$\begin{aligned}\sharp\Phi =& q^{m(m+1)/2-mn} p^{n \operatorname{ord}_p \mathrm{d}\, N+(m-n)\operatorname{ord}_p \mathrm{d}\, K}\beta_p(N,N)^{-1}\beta_p(K,K)\\ &\times\sharp\{\sigma : M \to N/qN^\sharp \mid Q(\sigma(x)) \equiv Q(x) \bmod 2q, \sigma(M) \text{ is primitive in } N,\\ &\qquad \text{and } proj_{\sigma(M)^\perp} N \cong K\}.\end{aligned}$$

Proof. Since for a linear mapping $\sigma : N \to N$ such that

$$Q(\sigma(x)) \equiv Q(x) \bmod 2q,$$

there exists an isometry $\sigma' \in O(N)$ such that $\sigma'(x) \equiv \sigma(x) \bmod qN^\sharp$, by applying Theorem 5.4.2 to $L = N, G = N^\sharp, k = \operatorname{ord}_p q$, we see the canonical mapping from $O(N)/\sim_q$ to

$$\{\sigma : N \to N/qN^\sharp \mid Q(\sigma(x)) \equiv Q(x) \bmod 2q\}$$

is bijective. Thus

$$\sharp(O(N)/\sim_q) = p^{-n \operatorname{ord}_p \mathrm{d}\, N} q^{n^2-n(n+1)/2}\beta_p(N,N).$$

Similarly

$$\sharp(O(K)/\sim_q) = p^{-(n-m)\operatorname{ord}_p \mathrm{d}\, K} q^{(n-m)(n-m-1)/2}\beta_p(K,K).$$

By virtue of the last lemma, we have only to verify $\sharp(I/\sim_q)$ is the number of linear mappings $\sigma : M \to N/qN^\sharp$ such that

$$Q(\sigma(x)) \equiv Q(x) \bmod 2q,$$

$\sigma(M)$ is primitive in N and $proj_{\sigma(M)^\perp} N \cong K$. Here we note that the conditions are independent of taking representatives of the images in N (keeping the linearity of σ), because the first and second are clear and the third follows from Corollary 5.4.5. So to complete the proof, we have only to show that for a linear mapping $\sigma : M \to N/qN^\sharp$ such that $Q(\sigma(x)) \equiv Q(x) \bmod 2q$, $\sigma(M)$ is primitive and $proj_{\sigma(M)^\perp} N \cong K$, there is an element $\eta \in I$ such that $\eta(x) \equiv \sigma(x) \bmod qN^\sharp$ for $x \in M$. Applying Theorem 5.4.2 to $\sigma, G = N^\sharp$, there is an isometry $\eta : M \hookrightarrow N$ such that $\eta(x) \equiv \sigma(x) \bmod qN^\sharp$ for $x \in M$. It is clear that $\eta(M)$ is primitive in N since $\sigma(M)$ is so. By Corollary 5.4.5, there is an isometry $\gamma \in O(N)$ such that $\gamma\eta(M) = \sigma(M)$. So

$$proj_{\eta(M)^\perp} N \cong proj_{\gamma\eta(M)^\perp} N \cong K,$$

and $\eta \in I$. □

Lemma 6.8.8. *Keep the condition (*) with $F = \mathbb{Q}$, and let S be a finite set of primes such that if $p \notin S$, then M_p, N_p are unimodular. Then*

$$\{\varphi : K \to \mathbb{Q}M/M \mid N_\varphi \in gen(N)\}$$

and

$$\prod_{p \in S} \{\varphi_p : K_p \to \mathbb{Q}_p M_p/M_p \mid N_{\varphi_p}(M_p, K_p) \cong N_p\}$$

are bijective.

Proof. First, suppose that φ_p is given for $p \in S$. Let $\{v_i\}_{i=1}^k$ and $\{u_i\}_{i=1}^m$ be bases of K and M respectively; making use of them we define $A(\varphi_p) \in M_{m,k}(\mathbb{Q}_p)$ by

$$(\varphi_p(v_1), \cdots, \varphi_p(v_k)) = (u_1, \cdots, u_m)A(\varphi_p),$$

and take a matrix $\tilde{A} \in M_{m,k}(\mathbb{Q})$ such that

$$\begin{cases} \tilde{A} \equiv A(\varphi_p) \bmod M_{m,k}(\mathbb{Z}_p) & \text{for} \quad p \in S, \\ \tilde{A} \in M_{m,k}(\mathbb{Z}_p) & \text{otherwise.} \end{cases}$$

Define the mapping φ by

$$(\varphi(v_1), \cdots, \varphi(v_k)) := (u_1, \cdots, u_m)\tilde{A}.$$

We show $N_\varphi \in gen(N)$. For $p \in S$

$$\begin{aligned} &((\varphi - \varphi_p)(v_1), \cdots, (\varphi - \varphi_p)(v_k)) \\ &\qquad = (u_1, \cdots, u_m)(\tilde{A} - A(\varphi_p)) \in (M_p, \cdots, M_p) \end{aligned}$$

implies $(\varphi - \varphi_p)(K) \subset M_p$ and

$$(N_\varphi)_p = M_p + \mathbb{Z}_p[x + \varphi(x) \mid x \in K_p] = M_p + \mathbb{Z}_p[x + \varphi_p(x) \mid x \in K_p] \cong N_p.$$

If $p \notin S$, then M_p and N_p are unimodular and so Lemma 6.8.2 yields that $N_p = M_p \perp K_p$. By the definition of $\tilde{A}$,

$$(N_\varphi)_p = M_p + \mathbb{Z}_p[x + \varphi(x) \mid x \in K_p] = M_p \perp K_p,$$

and so we have $(N_\varphi)_p = N_p$. Thus we have proved $N_\varphi \in gen(N)$.

Conversely a linear mapping

$$\varphi : K \to \mathbb{Q}M/M$$

such that $N_\varphi \in gen(N)$ induces canonically $\varphi_p : K_p \to \mathbb{Q}_p M_p/M_p$ such that $N_{\varphi_p}(M_p, K_p) \cong N_p$. If $p \notin S$, then N_{φ_p} and N_p equal $M_p \perp K_p$ by Lemma 6.8.2 and in particular $\varphi_p(K_p) \subset M_p$. The correspondence $\varphi \to \{\varphi_p\}$ gives the inverse of the above correspondence. Thus we have proved the lemma. □

Lemma 6.8.9. *Under the condition (*) with $p \neq 2$ and $R = \mathbb{Z}_p$, assume that M, N are unimodular. Then we have $\sharp\Phi = 1$ and in Lemma 6.8.7 we can remove the condition on σ that $\sigma(M)$ is primitive in N and $proj_{\sigma(M)^\perp} N \cong K$.*

Proof. Take $\varphi \in \Phi$; then there is an isometry $\eta : N_\varphi \cong N$. Since M and N are unimodular, $N_\varphi = M \perp M^\perp$. For $x \in K$, $x + \varphi(x) \in N_\varphi$ implies $\varphi(x) \in M$. It means $\varphi(K) \subset M$. Thus Φ contains only one linear mapping from K to FM/M. Moreover $\varphi(K) \subset M$ implies $N_\varphi = M \perp K$. Suppose $\sigma : M \to N/qN^\sharp$ such that $Q(\sigma(x)) \equiv Q(x) \bmod 2q$. We assume that σ is a linear mapping to N. Let us show that the other two conditions on σ in the assertion of Lemma 6.8.7 are automatically satisfied. By Corollary 5.4.2, there is an isometry $\beta : M \hookrightarrow N$ such that $\sigma(M) = \beta(M)$. Since M is unimodular, $\beta(M)$ is so and hence $N = \beta(M) \perp *$. Thus $\sigma(M) = \beta(M)$ is primitive in N. Since $p \neq 2$, $\beta(M)$ is transformed to M by $O(N)$ by Corollary 5.4.1. Thus

$$proj_{\sigma(M)^\perp} N \cong proj_{M^\perp} N = K.$$

□

Proof of Theorem 6.8.1.

We denote by $\rho(M)$ the quotient of the left-hand side by the right-hand side of (i). First let us show $\rho(M) = 1$ if $m = \operatorname{rank} M = 1$. Since both sides of (i) do not change under scaling by Proposition 5.6.1, we may assume $M \cong \langle 1 \rangle$. Then $gen(M) = cls(M)$ and $w(M) = 2^{-1}$ are clear. For an odd prime p, by Proposition 5.6.1

$$\begin{aligned}\alpha_p(M, M) &= 2^{-1}\beta_p(M, M) = 2^{-1}\sharp A_p(1, 1) \\ &= 2^{-1}\sharp\{x \bmod p \mid x^2 \equiv 1 \bmod p\} \\ &= 1.\end{aligned}$$

If $p = 2$, then

$$\begin{aligned}\alpha_2(M, M) &= \beta_2(M, M) = \sharp A_4(1, 1) \\ &= \sharp\{x \bmod 4 \mid x^2 \equiv 1 \bmod 8\} \\ &= 2.\end{aligned}$$

Since $\Gamma(1/2) = \sqrt{\pi}$, we get $\rho(M) = 1$.

Now we assume that the equation in (i) holds for any $m < n$ and suppose $m < n$. When N_i runs over a complete set of representatives of isometry classes in $gen(N)$ as in the theorem, we have

$$\begin{aligned} l &:= \sum_i \frac{r_{pr}(M, N_i)}{E(N_i)} \\ &= \sum_i \sum_K E(N_i)^{-1} \sharp\{\sigma : M \hookrightarrow N_i \mid \sigma(M) \quad \text{is primitive in } N_i \text{ and} \\ &\qquad\qquad proj_{\sigma(M)^\perp} N_i \cong K\}, \end{aligned}$$

where K runs over all possible isometry classes (that is K is a lattice on a positive definite quadratic space such that $\mathbb{Q}N \cong \mathbb{Q}M \perp \mathbb{Q}K$) $\mathrm{d}\, N = \mathrm{d}\, M \,\mathrm{d}\, K$, $\mathrm{s}(K) \subset \mathrm{s}(N)^2\, \mathrm{s}(M^\sharp) + \mathrm{s}(N)$ and $\operatorname{rank} K = n - m$ by Lemma 6.8.2; then by Lemma 6.8.4

(R1)

$$\begin{aligned} l &= \sum_i \sum_K E(K)^{-1} \sharp\{\varphi : K \to \mathbb{Q}M/M \mid N_\varphi(M, K) \cong N_i\} \\ &= \sum_K E(K)^{-1} \sharp\{\varphi : K \to \mathbb{Q}M/M \mid N_\varphi(M, K) \in gen(N)\} \\ &= \sum_K E(K)^{-1} \prod_{p \in S} \sharp\{\varphi_p : K_p \to \mathbb{Q}_p M_p/M_p \mid N_{\varphi_p}(M_p, K_p) \cong N_p\}, \end{aligned}$$

by Lemma 6.8.8 where S is a finite set of primes such that if $p \notin S$, then M_p, N_p are unimodular. Now Lemma 6.8.7 implies

$$\begin{aligned} l = \sum_K E(K)^{-1} \prod_{p \in S} &\{q^{m(m+1)/2 - mn} p^{n \operatorname{ord}_p \mathrm{d}\, N + (m-n) \operatorname{ord}_p \mathrm{d}\, K} \\ &\times \beta_p(N_p, N_p)^{-1} \beta_p(K_p, K_p) \sharp\{\sigma : M_p \to N_p/qN_p^\sharp \mid Q(\sigma(x)) \equiv \\ &Q(x) \bmod 2q, \sigma(M_p) \text{ is primitive in } N_p, \text{ and } proj_{\sigma(M_p)^\perp} N_p \cong K_p\}\} \end{aligned}$$

where q is a sufficiently large power of p,

$$\begin{aligned} = \sum_K w(K) \prod_{p \in S} &\{q^{m(m+1)/2 - mn} p^{n \operatorname{ord}_p \mathrm{d}\, N + (m-n) \operatorname{ord}_p \mathrm{d}\, K} \\ &\times \beta_p(N_p, N_p)^{-1} \beta_p(K_p, K_p) \sharp\{\sigma : M_p \to N_p/qN_p^\sharp \mid Q(\sigma(x)) \equiv \\ &Q(x) \bmod 2q, \sigma(M_p) \text{ is primitive in } N_p, \text{ and } proj_{\sigma(M_p)^\perp} N_p \cong K_p\}\} \end{aligned}$$

where K runs over a complete set of representatives of possible genera. Since

$$\beta_p(M_p, N_p) = 2^{-m\delta_{2,p} + \delta_{m,n}} \alpha_p(M_p, N_p),$$

we have

$$\beta_p(N_p,N_p)^{-1}\beta_p(K_p,K_p)=2^{m\delta_{2,p}}\alpha_p(N_p,N_p)^{-1}\alpha_p(K_p,K_p).$$

By Proposition 5.6.2 and the remark in the proof, the infinite products $\prod_p \alpha_p(N_p,N_p)$, $\prod_p \alpha_p(K_p,K_p)$ converge. Using Lemma 6.8.9, we have

$$\begin{aligned} l=&\sum_K w(K)(\prod_p \alpha_p(K_p,K_p))(\mathrm{d}\,K)^{m-n}(\mathrm{d}\,N)^n 2^m \prod_p \alpha_p(N_p,N_p)^{-1}\\ &\times \prod_p \{q^{m(m+1)/2-mn}\sharp\{\sigma : M_p \to N_p/qN_p^\sharp \mid Q(\sigma(x))\equiv Q(x) \bmod 2q,\\ &\qquad \sigma(M_p) \text{ is primitive in } N_p \text{ and } proj_{\sigma(M_p)^\perp} N_p \cong K_p\}\}\end{aligned}$$

Using the induction hypothesis for K, we have

$$\begin{aligned} &w(K)\prod_p \alpha_p(K_p,K_p)(\mathrm{d}\,K)^{m-n}\\ &\quad =\epsilon^{-1}\pi^{-(n-m)(n-m+1)/4}\prod_{i=0}^{n-m-1}\Gamma((n-m-i)/2)(\mathrm{d}\,K)^{(m+1-n)/2},\end{aligned}$$

where $\epsilon=1/2$ if $n-m>1$, 1 otherwise. Thus l is equal to

$$\begin{aligned} &\epsilon^{-1}\pi^{-(n-m)(n-m+1)/4}\prod_{i=0}^{n-m-1}\Gamma((n-m-i)/2)2^m(\mathrm{d}\,N)^n\prod_p \alpha_p(N_p,N_p)^{-1}\\ &\quad\times \sum_K \prod_p p^{(\mathrm{ord}_p\,\mathrm{d}\,K)(m+1-n)/2} q^{m(m+1)/2-mn}\\ &\quad\times \sharp\{\sigma : M_p \to N_p/qN_p^\sharp \mid Q(\sigma(x))\equiv Q(x) \bmod 2q,\\ &\qquad \sigma(M_p) \text{ is primitive in } N_p, \text{ and } proj_{\sigma(M_p)^\perp} N_p \cong K_p\}.\end{aligned}$$

Since $\mathbb{Q}N\cong\mathbb{Q}M\perp\mathbb{Q}K$, the isometry class of $\mathbb{Q}K$ is unique. By Theorem 6.1.1, K runs over a complete set of representatives of possible genera if and only if K_p runs over a complete set of representatives of possible isometry classes for $p\in S$, noting $\mathrm{d}\,N_p=\mathrm{d}\,M_p\,\mathrm{d}\,K_p$ and

$$\mathrm{s}(K_p)\subset \mathrm{s}(N_p)^2\,\mathrm{s}(M_p^\sharp)+\mathrm{s}(N_p).$$

Finally we have

$$
\begin{aligned}
l = & \epsilon^{-1}\pi^{-(n-m)(n-m+1)/4}\prod_{i=0}^{n-m-1}\Gamma((n-m-i)/2)2^m(\mathrm{d}N)^n\prod_p \alpha_p(N_p,N_p)^{-1}\\
& \times \prod_p p^{(\mathrm{ord}_p\,\mathrm{d}\,N-\mathrm{ord}_p\,\mathrm{d}\,M)(m+1-n)/2} q^{m(m+1)/2-mn}\\
& \times \sharp\{\sigma : M_p \to N_p/qN_p^\sharp \mid Q(\sigma(x)) \equiv Q(x) \bmod 2q,\\
& \qquad \sigma(M_p) \text{ is primitive in } N_p\}\\
= & \epsilon^{-1}\pi^{-(n-m)(n-m+1)/4}\prod_{i=0}^{n-m-1}\Gamma((n-m-i)/2)2^m(\mathrm{d}\,N)^{(m+1+n)/2}\\
& \times (\mathrm{d}\,M)^{(n-m-1)/2}\prod_p \alpha_p(N_p,N_p)^{-1}\prod_p p^{-m\,\mathrm{ord}_p\,\mathrm{d}\,N} d_p(M,N)\\
= & \epsilon^{-1}\pi^{-(n-m)(n-m+1)/4}\prod_{i=0}^{n-m-1}\Gamma((n-m-i)/2)(\mathrm{d}\,M)^{(n-m-1)/2}\\
& \times (\mathrm{d}\,N)^{(n-m+1)/2}\prod_p \alpha_p(N_p,N_p)^{-1}\prod_p D_p(M_p,N_p).
\end{aligned}
$$

By the definition of ρ, we have

$$\prod_p \alpha_p(N_p,N_p) = \rho(N)^{-1}2\pi^{-n(n+1)/4}\prod_{i=0}^{n-1}\Gamma((n-i)/2)(\mathrm{d}\,N)^{(n+1)/2}w(N)^{-1}$$

and hence

$$
\begin{aligned}
l/w(N) = & \rho(N)2^{-1}\epsilon^{-1}\pi^{m(2n-m+1)/4}\prod_{i=0}^{m-1}\Gamma((n-i)/2)^{-1}\\
& \times (\mathrm{d}\,M)^{(n-m-1)/2}(\mathrm{d}\,N)^{-m/2}\prod_p D_p(M_p,N_p).
\end{aligned}
$$

Thus we have $\rho(N) = 1$ if and only if (iii) is true when $m < n$. Here we claim that $\rho(N)$ is also the quotient of the left-hand side of (ii) by the right-hand side of (ii). To see it, we note that

$$r(M,N_i) = \sum_{\mathbb{Q}M \supset M' \supset M} r_{pr}(M',N_i)$$

since, considering $\sigma : M \hookrightarrow N_i$ as an isometry from $\mathbb{Q}M$ to $\mathbb{Q}N_i$, it induces

$$\sigma : M' := \mathbb{Q}M \cap \sigma^{-1}(N_i) \hookrightarrow N_i$$

with $\sigma(M')$ being primitive in N_i, and that locally

$$\alpha_p(M_p,(N_i)_p) = \sum_{\mathbb{Q}_p M_p \supset M'_p \supset M_p} [M'_p : M_p]^{m+1-n} D_p(M'_p,(N_i)_p)$$

and

$$(\mathrm{d}\,M_p)^{(n-m-1)/2}[M'_p : M_p]^{m+1-n} = (\mathrm{d}\,M'_p)^{(n-m-1)/2}.$$

Since the correspondence between global and local lattices are given by Theorem 6.1.1, we get the claim. In particular, (ii) follows from (iii).

Thus denoting by $\rho(M)$ the quotient of the left-hand side by the right-hand side of (i), we have proved that, if $\operatorname{rank} K = 1$, then $\rho(K) = 1$, and if $\rho(K) = 1$ with $\operatorname{rank} K < n$, then $\rho(N)$ is the quotient of the left-hand side by the right-hand side of (ii) and (iii) for $m < n$; so the ratio is independent of M. It remains to verify $\rho(N) = 1$ but we give only the basic ideas of the proof.

By scaling, we may assume $\mathrm{s}(N) \subset 2\mathbb{Z}$. Denoting the left-hand side and right-hand side of (ii) by $L(M,N)$ and $R(M,N)$ respectively, we have shown $L(M,N) = \rho(N)R(M,N)$. For $\tau \in \mathbb{C}$ with imaginary part of $\tau > 0$, we put

$$\begin{aligned}\theta(\tau,N) : &= \sum_{x\in N} e(Q(x)\tau)\\ &= \sum_{k=0}^{\infty} r(\langle k\rangle,N)e(k\tau),\end{aligned}$$

where $e(z)$ denotes $e^{2\pi i z}$, and

$$\begin{aligned}\Theta(\tau,N) : &= w(N)^{-1}\sum_i E(N_i)^{-1}\theta(\tau,N_i)\\ &= 1 + \sum_{k=1}^{\infty} m(\langle k\rangle,N)e(k\tau).\end{aligned}$$

On the other hand, we put

$$E(\tau,N) := 1 + \epsilon_{1,n}\pi^{n/2}\Gamma(n/2)^{-1}(\mathrm{d}\,N)^{-1/2}\sum_{k=1}^{\infty} k^{n/2-1}\prod_p \alpha_p(\langle k\rangle,N_p)e(k\tau).$$

Then we have proved that $\Theta(\tau,N) - \rho(N)E(\tau,N)$ is a constant number. Siegel showed that for $m > 4$

$$E(\tau,N) = 1 + \sum_{a,b} H(N,a/b)(b\tau - a)^{-m/2} \tag{7}$$

where a, b run over relatively prime integers with $b > 0$ and

$$H(N, a/b) := (i/b)^{m/2}(\mathrm{d}\,N)^{-1/2} \sum_{x \in N/bN} e(\frac{a}{2b}Q(x)).$$

Then $\Theta(\tau, N)$ and $E(\tau, N)$ are modular forms (see the definition in the Notes) and if a constant number is a modular form, then it is identically zero. Thus $\Theta(\tau, N) = \rho(N)E(\tau, N)$ holds. Take $\lim_{y\to\infty}$ on both sides with $\tau = iy$. Comparing both sides, we have $\rho(N) = 1$. For $m \leq 4$, (7) is not absolutely convergent. But, when we multiply each term by $|b\tau - a|^{-s}$, it is absolutely convergent for real part of s sufficiently large, and it has an analytic continuation on the whole s-plane. We can then define $E(\tau, N)$ to be the value at $s = 0$. This is an idea of Hecke. We need a modification in the case $m = 3$.

Siegel's original idea is as follows.

The Dirichlet series $D(s)$ corresponding to $\Theta(\tau, N)$ is (through the Mellin transform) defined by

$$\sum_{k=1}^{\infty} m(\langle k\rangle, N)k^{-s};$$

it is holomorphic for $\mathrm{Re}\, s > n/2$ and has a simple pole at $a := n/2$ with the residue

$$b := (\mathrm{d}\,N)^{-1/2}\pi^{n/2}\Gamma(n/2)^{-1}.$$

By the Tauberian theorem,

$$\lim_{x\to\infty} x^{-a} \sum_{k<x} m(\langle k\rangle, N) = b.$$

But, for technical reasons, Siegel considered instead the Dirichlet series

$$\sum_{k=1}^{\infty} \frac{m(\langle k\rangle, N)}{k^{a-1}} k^{-s}.$$

Then it has a simple pole at $s = 1$ and

(8) $$\lim_{x\to\infty} x^{-1} \sum_{k<x} m(\langle k\rangle, N)k^{-(a-1)} = b.$$

These follow from classical analytic number theory. Replacing $m(\langle k\rangle, N)$ by the right-hand side of (ii), Siegel proved the formula similar to (8) and got $\rho(N) = 1$.

Remark.

If we accept the existence of a unimodular Haar measure on $O_A^+(V)$, it is not hard to see that the partial sum $w_s(K) = \sum_{K_i} E(K_i)^{-1}$ where K_i runs over a complete set of representatives of isometry classes in $spn(K)$ and is dependent only on $gen(K)$. Note that in the equation (R1),

$$\sharp\{\varphi : K \to \mathbb{Q}M/M \mid N_\varphi \in spn(N)\}$$

depends only on $spn(K)$. To see this, let $K' = \sigma\eta(K)$ for $\sigma \in O(\mathbb{Q}K)$, $\eta \in O'_A(\mathbb{Q}K)$. Putting $\sigma = \eta = \mathrm{id}$ on $\mathbb{Q}M$,

$$N_{\varphi\eta^{-1}\sigma^{-1}}(M, K') = \sigma\eta N_\varphi \in spn(N).$$

Thus the correspondence $\varphi \mapsto \varphi\eta^{-1}\sigma^{-1}$ gives the invariance of the number in question. Then restricting N_i in $spn(N)$ and grouping the isometry classes, the second line of (R1) becomes

$$\text{(9)} \qquad \sum_K w_s(K) \sum_{K'} \sharp\{\varphi : K' \to \mathbb{Q}M/M \mid N_\varphi(M, K') \in spn(N)\}$$

where K runs over the different genera and K' runs over the different spinor genera in $gen(K)$. If either $n - m \geq 3$ or $n - m = 2$ and

$$O_A^+(V) = O_A^+(W)O^+(V)O'_A(V)O_A^+(N)$$

where $V \cong U \perp W$ (cf. Lemma 6.5.7), then the inner sum on K' is dependent only on $gen(N)$, since, for $\sigma \in O_A^+(V)$, we can take $\eta \in O_A^+(\mathbb{Q}K)$, $\alpha \in O^+(V)$, $\gamma \in O'_A(V)$, $\beta \in O_A^+(N)$ such that $\sigma = \gamma\eta(\alpha\beta)^{-1}$. Then

$$N_{\varphi\eta^{-1}}(M, \eta K') = \eta N_\varphi(M, K') \in spn(\eta N) = spn(\sigma N).$$

Therefore (9) depends only on $gen(N)$, considering the correspondence $K', \phi \mapsto \eta K', \phi\eta^{-1}$. Thus the left-hand sides of (ii), (iii), restricting $N_i \in spn(N)$, are dependent only on $gen(N)$ if $n - m \geq 3$ or the above condition is satisfied for when $n - m = 2$.

7

Some functorial properties of positive definite quadratic forms

In this chapter, we investigate functorial properties of positive definite quadratic forms with respect to the tensor product and scalar extension.

Let L, M and N be quadratic modules over $\mathbb{Z}$ and R_K the maximal order of an algebraic number field K. Some of the fundamental problems are:

Does $L \otimes M \cong L \otimes N$ imply $M \cong N$?
What can we say about M, N when $R_K M \cong R_K N$, for example, is $M \cong N$?

When we consider these in the category containing indefinite quadratic forms, it is not interesting. For example, let M, N be unimodular positive definite quadratic modules over $\mathbb{Z}$ with the same rank and $\mathrm{n}(M) = \mathrm{n}(N) = 2\mathbb{Z}$, then $L \otimes M \cong L \otimes N$ holds for $L = \langle 1 \rangle \perp \langle -1 \rangle$. So the condition $L \otimes M \cong L \otimes N$ says nothing. If the field K is not totally real, $R_K M$ is isotropic at an infinite imaginary place of K and hence the isometry class of $R_K M$ is nothing but its spinor genus (by the generalization of Theorem 6.3.2). Therefore $R_K M \cong R_K N$ holds. So the above problems are meaningless.

However if we confine ourselves to the category of positive definite quadratic forms and totally real algebraic number fields, the answer seems affirmative. At least there is no counter-example so far. Let us give some results in this chapter.

In this chapter, by a positive lattice we mean a lattice on a positive definite quadratic space over $\mathbb{Q}$. *Hence, if* L *is a positive lattice, then for a basis* $\{e_i\}$ *of* L *over* $\mathbb{Z}$, $B(e_i, e_j) \in \mathbb{Q}$ *and the matrix* $(B(e_i, e_j))$ *is positive definite.*

R_K *denotes the maximal order of an algebraic number field* K.

For quadratic spaces U, V over $\mathbb{Q}$, we defined the tensor product $U \otimes V$ after Proposition 1.5.4. It is easy to see, taking orthogonal bases of U, V, that if U, V are positive definite quadratic spaces, then $U \otimes V$ is also positive definite. We introduce a quadratic form on the tensor product $L \otimes M$ of positive lattices L, M by its restriction from $\mathbb{Q}L \otimes \mathbb{Q}M$. Since for bases $\{e_i\}, \{f_i\}$ of L, M, respectively, $\{e_i \otimes f_j\}$ is a basis of $L \otimes M$, we see that $L \otimes M$ is a positive lattice if L, M are positive lattices, and the corresponding matrix is

$$B((e_i \otimes f_j, e_k \otimes f_h)) = (B(e_i, e_j)) \otimes (B(f_i, f_j))$$

(tensor product of matrices). We recall that

$$B(x_1 \otimes y_1, x_2 \otimes y_2) = B(x_1, x_2)B(y_1, y_2) \text{ for } x_i \in L, y_i \in M.$$

In 7.1, we introduce the notion of a positive lattice of E-type, which at the present time is the most convenient one for our problem. Sections 7.2 and 7.3 are preparatory. In 7.4 and 7.5, the main results in this chapter are given.

7.1 Positive lattices of E-type

For a positive lattice L, we put

$$\begin{aligned} \min(L) &= \min\{Q(x) \mid x \in L, x \neq 0\}, \\ \mathcal{M}(L) &= \{x \in L \mid Q(x) = \min(L)\}. \end{aligned}$$

$\mathcal{M}(L)$ is obviously a finite set. An element of $\mathcal{M}(L)$ is called a *minimal vector* of L. For positive lattices L, M, we have $Q(x \otimes y) = Q(x)Q(y)$ and so if $x \in \mathcal{M}(L), y \in \mathcal{M}(M)$, then

(1) $$\min(L \otimes M) \leq Q(x \otimes y) = Q(x)Q(y) = \min(L)\min(M).$$

Definition. *A positive lattice* L *is of E-type if and only if*

$$\mathcal{M}(L \otimes M) \subset \{x \otimes y \mid x \in L, y \in M\} \quad \textit{for any positive lattice } M$$

holds.

Note that elements of $L \otimes M$ cannot necessarily be written as $x \otimes y$. Elements $x \otimes y$ in $L \otimes M$ are called *split*.

Positive lattices of E-type play an essential role in this chapter. The first basic result is

Lemma 7.1.1. *Let L, M be positive lattices. Then we have*

$$(2) \qquad \min(L \otimes M) \quad \leq \quad \min(L)\min(M)$$

and if L is of E-type, then we have

$$(3) \quad \min(L \otimes M) = \min(L)\min(M), \quad \mathcal{M}(L \otimes M) = \mathcal{M}(L) \otimes \mathcal{M}(M),$$

where $\mathcal{M}(L) \otimes \mathcal{M}(M)$ denotes $\{x \otimes y \mid x \in \mathcal{M}(L), y \in \mathcal{M}(M)\}$ for abbreviation.

Moreover we have

(i) *If $\operatorname{rank}(L) = 1$, then L is of E-type.*
(ii) *If L, M are of E-type, then $L \perp M$ and $L \otimes M$ are also of E-type.*
(iii) *If L is of E-type and M is a submodule of L with $\min(M) = \min(L)$, then M is of E-type.*

Proof. The first equality in (2) is nothing but (1). Let L be a positive lattice of E-type and M be a positive lattice. For a minimal vector $x \otimes y$ of $L \otimes M$ $(x \in L, y \in M)$, we have

$$\min(L \otimes M) = Q(x \otimes y) = Q(x)Q(y) \geq \min(L)\min(M),$$

and then by (2), $\min(L \otimes M) = \min(L)\min(M)$. Thus $x \in \mathcal{M}(L), y \in \mathcal{M}(M)$ follow and so

$$\begin{aligned} \min(L \otimes M) &= \min(L)\min(M), \\ \mathcal{M}(L \otimes M) &= \{x \otimes y \,|\, x \in \mathcal{M}(L), y \in \mathcal{M}(M)\}. \end{aligned}$$

Thus (3) has been proved.
(i) is obvious. We prove (ii); let N be a positive lattice. Let $v = x + y$ $(x \in L \otimes N, y \in M \otimes N)$ be a minimal vector of $(L \perp M) \otimes N$. Since $Q(v) = Q(x) + Q(y)$ and $Q(v)$ is minimal, $x = 0$ or $y = 0$, i.e. $v = y \in M \otimes N$ or $= x \in L \otimes N$. Suppose that v is in $L \otimes N$ for example; then

$$\min((L \perp M) \otimes N) = Q(v) \geq \min(L \otimes N) \geq \min((L \perp M) \otimes N),$$

and

$$\min((L \perp M) \otimes N) = Q(v) = \min(L \otimes N)$$

and

$$v \in \mathcal{M}(L \otimes N) = \mathcal{M}(L) \otimes \mathcal{M}(N)$$

follow. Hence, v is split in $L \otimes N \subset (L \perp M) \otimes N$. Thus $L \perp M$ is of E-type.

Next let u be a minimal vector of $L \otimes M \otimes N$; then

$$u = z \otimes w \ \ (z \in \mathcal{M}(L), w \in \mathcal{M}(M \otimes N))$$

follows from (3) and w is split since M is of E-type. Hence u is split in $(L \otimes M) \otimes N$. This means that $L \otimes M$ is of E-type.

Let us prove the third assertion. For a positive lattice N, we have

$$\begin{aligned} \min(L)\min(N) &= \min(L \otimes N) \\ &\leq \min(M \otimes N) \\ &\leq \min(M)\min(N) \quad \text{by (2)} \\ &= \min(L)\min(N) \quad \text{by the assumption.} \end{aligned}$$

Thus we have $\min(M \otimes N) = \min(L \otimes N)$. Therefore a minimal vector v of $M \otimes N$ is also a minimal one of $L \otimes N$ and so it is split in $L \otimes N$. Write $v = x \otimes y, x \in \mathcal{M}(L), y \in \mathcal{M}(N)$; then y is primitive and so $\{y\}$ is extended a basis of N and hence $x \otimes y \in M \otimes N$ implies $x \in M$. Thus v is split in $M \otimes N$. □

We note that if L is a positive lattice of E-type, then a scaling of L is also of E-type by (ii).

Lemma 7.1.2. *$\mu_r < \sqrt{r}$ holds for $2 \leq r \leq 43$, where μ_r is the Hermite constant in Chapter 2.*

Proof. By virtue of Theorem 2.2.1, we have only to show

$$(\mu_r \leq\ )\ 2\pi^{-1}\Gamma(2 + r/2)^{2/r} < \sqrt{r} \quad \text{for } 2 \leq r \leq 43.$$

This is verified by a direct calculation for $2 \leq r \leq 8$. Since

$$\Gamma(x) < \sqrt{2\pi}x^{x-1/2}e^{-x+1/12x}$$

is known, we put $f(x) := \log\sqrt{x} - \log\{2\pi^{-1}g(2 + x/2)^{2/x}\}$, denoting the right-hand side of the above inequality for $\Gamma(x)$ by $g(x)$, and we show $f(x) > 0$ for $8 \leq x \leq 43$. It is easy to see that

$$\begin{aligned} f(x) =& 2^{-1}\log x - \log(2/\pi) - 2x^{-1}\log\sqrt{2\pi} - (3/x + 1)\log(2 + x/2) \\ &+ 4/x + 1 - 1/(12x) + 1/(12(4 + x)), \\ x^2 f'(x) =& 2\log\sqrt{2\pi} - 3\log 2 - 3 - x/2 + 3\log(x + 4) - 10/(3(x + 4)) \\ &- 4/(3(4 + x)^2) \\ =& 2\log\sqrt{2\pi} - 3\log 2 - 1 + 3\log z - z/2 - 10/(3z) - 4/(3z^2) \\ & \text{where we put } z := x + 4, \\ (x^2 f'(x))' =& (3z^3)^{-1}(-3/2 \cdot z^3 + 9z^2 + 10z + 8), \\ (-3/2 \cdot z^3 + & 9z^2 + 10z + 8)' = -9/2(z - 2 + 2\sqrt{14}/3)(z - 2 - 2\sqrt{14}/3) \\ -3/2 \cdot z^3 + & 9z^2 + 10z + 8 = -1168 < 0 \quad \text{for} \quad z = 8 + 4 > 2 + 2\sqrt{14}/3. \end{aligned}$$

Hence drawing the graph of $3z^3(x^2f'(x))'$, it is negative for $x \geq 8$ and hence $x^2f'(x)$ is monotonically decreasing there and $f'(8) < 0$ implies that $f'(x) < 0$ for $x \geq 8$. Then, $f(43) > 0$ implies $f(x) > 0$ for $8 \leq x \leq 43$. □

Lemma 7.1.3. *Let A, B be positive definite symmetric matrices of degree n; then* $\operatorname{tr} AB \geq n(|A||B|)^{1/n}$.

Proof. Write $B = D[T]$ for a diagonal matrix $D = \operatorname{diag}(d_1, \cdots, d_n)$ and an orthogonal matrix T. Denoting by $a_1, \cdots, a_n$ the diagonal entries of TA^tT, we have

$$\begin{aligned}\operatorname{tr} AB =& \operatorname{tr} TA^tTD = \sum a_i d_i \geq n(\prod(a_i d_i))^{1/n} \\ =& n|B|^{1/n}(\prod a_i)^{1/n} \geq n|B|^{1/n}|TA^tT|^{1/n} \\ =& n|B|^{1/n}|A|^{1/n},\end{aligned}$$

where we used the fact that the arithmetic mean is greater than or equal to the harmonic mean and that for a positive definite matrix $c = (c_{ij})$, $\prod c_{ii} \geq \det c$.

Theorem 7.1.1. *If L is a positive lattice with* $\operatorname{rank} L \leq 43$, *then L is of E-type.*

Proof. Let M be a positive lattice and $v = \sum_{i=1}^r x_i \otimes y_i$ be a minimal vector of $L \otimes M$ and assume that r is minimal among these expressions; then $\{x_1, \cdots, x_r\}, \{y_1, \cdots, y_r\}$ are linearly independent and so

$$L_1 := \mathbb{Z}[x_1, \cdots, x_r], M_1 := \mathbb{Z}[y_1, \cdots, y_r]$$

are positive lattices of rank $= r$. Then we have

$$\begin{aligned}\min(L_1)\min(M_1) &\geq \min(L)\min(M) \geq \min(L \otimes M) = Q(v) \\ &= B(v,v) = \sum_{i,j} B(x_i, x_j)B(y_i, y_j) \\ &= \operatorname{tr}(B(x_i,x_j))(B(y_i,y_j)) \\ &\geq r(\mathrm{d}\, L_1\, \mathrm{d}\, M_1)^{1/r} \geq r\mu_r^{-2}\min(L_1)\min(M_1), \qquad (4)\end{aligned}$$

which yields $r \leq \mu_r^2$ and so by Lemma 7.1.2, $r = 1$, that is, v is split. □

Corollary 7.1.1. *Put $c_n := \min_{1 \leq r \leq n} r\mu_r^{-2}$; then if L, M are positive lattices with* $\operatorname{rank} L = n$, *then*

$$\min(L \otimes M) \geq c_n \min(L)\min(M).$$

Proof. In the proof of the previous lemma, (4) immediately implies the assertion. □

Remark.

$\mu_n > cn$ for some constant c is known. Hence c_n tends to zero.

Lemma 7.1.4. *If $n \geq 40$, then $\mu_n < n/6$.*

Proof. As in the proof of Lemma 7.1.2, we have only to show

$$\begin{aligned} h(x) :&= \log x/6 - \log\{2\pi^{-1}\Gamma(2+x/2)^{2/x}\} \\ &= f(x) - 2^{-1}\log x + \log(x/6) > 0 \end{aligned}$$

for $x \geq 40$, where f is the function defined in the proof of Lemma 7.1.2. It is easy to see that $x^2h'(x) > 0$ for $x \geq 40$ and $h(40) > 0$, which complete the proof. □

Theorem 7.1.2. *Let L be a positive lattice such that* $\mathrm{s}\, L \subset \mathbb{Z}$ *and* $\min(L) \leq 6$; *then L is of E-type.*

Proof. Let M be a positive lattice and $v = \sum_{i=1}^r x_i \otimes y_i$ be a minimal vector of $L \otimes M$ with r being minimal; then as in the proof of Theorem 7.1.1, we have, since $B(x_i, x_j) \in \mathbb{Z}$,

$$\begin{aligned} \min(L \otimes M) = Q(v) &\geq r(\det(B(x_i, x_j))\det(B(y_i, y_j)))^{1/r} \\ &\geq r(\det(B(y_i, y_j)))^{1/r}. \end{aligned}$$

On the other hand, $\min(L\otimes M) \leq \min(L)\min(M) \leq 6\min(M_1)$, where we put $M_1 := \mathbb{Z}[y_1, \cdots, y_r]$. Thus we get $r(\mathrm{d}\, M_1)^{1/r} \leq 6\min(M_1)$ and hence $r/6 \leq \min(M_1)(\mathrm{d}\, M_1)^{-1/r} \leq \mu_r$. From the previous lemma follows $r < 40$ and then Theorem 7.1.1 yields that M_1 is of E-type and hence $v \in L \otimes M_1$ is split, since v is a minimal vector of $L \otimes M$ and hence $L \otimes M_1$. □

Example 1.

Let $\{\mathbf{e}_1, \cdots, \mathbf{e}_{n+1}\}$ be an orthogonal basis of $\mathbb{R}^{n+1}$.
Let

$$A_n := \{\sum_{i=1}^{n+1} x_i\mathbf{e}_i \mid x_i \in \mathbb{Z}, \sum_{i=1}^{n+1} x_i = 0\}$$

be the lattice spanned by $\mathbf{e}_i - \mathbf{e}_j$ $(1 \leq i, j \leq n+1)$. Then $\operatorname{rank} A_n = n$ and $\mathrm{d}\, A_n = n+1$.
Let

$$D_n := \{\sum_{i=1}^{n} x_i\mathbf{e}_i | x_i \in \mathbb{Z}, \sum_{i=1}^{n} x_i \text{ is even}\} \ (n \geq 4)$$

be the lattice spanned by $\pm\mathbf{e}_i \pm \mathbf{e}_j (1 \leq i \neq j \leq n)$ in $\mathbb{R}^n$. Note $\operatorname{rank} D_n = n$ and $\mathrm{d}\, D_n = 4$.

Denote by E_8:= the lattice spanned by D_8 and $\frac{1}{2}\sum_{i=1}^{8} \mathbf{e}_i$. Then rank $E_8 = 8$ and $\mathrm{d}\, E_8 = 1$.

Denote by E_7:= the orthogonal complement of $\frac{1}{2}\sum_{i=1}^{8} \mathbf{e}_i$ in E_8. Then

$$E_7 = \{\sum_{i=1}^{8} x_i \mathbf{e}_i \in E_8 \mid \sum_{i=1}^{8} x_i = 0\}$$

is isometric to the orthogonal complement in E_8 of any minimal vector of E_8. Note rank $E_7 = 7$ and $\mathrm{d}\, E_7 = 2$.

Denote by E_6:=the orthogonal complement of $\mathbb{Z}[\mathbf{e}_1 + \mathbf{e}_8, \frac{1}{2}\sum_{i=1}^{8} \mathbf{e}_i]$ in E_8. Then

$$E_6 = \{\sum_{i=1}^{8} x_i \mathbf{e}_i \in E_8 \mid x_1 + x_8 = x_2 + \cdots + x_7 = 0\}$$

is isometric to the orthogonal complement in E_8 of any submodule of E_8 isometric to A_2. We have rank $E_6 = 6$ and $\mathrm{d}\, E_6 = 3$.

For these lattices L, it is easy to see that L is spanned by $\mathcal{M}(L)$ and is an indecomposable positive lattice with $\mathrm{s}(L) = \mathbb{Z}, \min L = 2$. Conversely, such lattices are isometric to one of the above lattices, by using the theory of classification of root systems of Lie algebras. Anyway, by Theorem 7.1.2 they are of E-type. Next we show

Theorem 7.1.3. *Let F be a finite abelian extension of $\mathbb{Q}$, and R_F the ring of integers in F. We introduce the symmetric bilinear form B_F by $B_F(x, y) := \mathrm{tr}_{F/\mathbb{Q}}\, x\bar{y}$ where $\bar{y}$ denotes the complex conjugate of y. Then (R_F, B_F) is a positive lattice of E-type.*

To prove it, we need several lemmas. Since F is abelian, $Q(x) := B_F(x, x) = \mathrm{tr}\, |x|^2 = \sum |\sigma(x)|^2$ where σ runs over the Galois group $\mathrm{Gal}(F/\mathbb{Q})$; so R_F is clearly a positive lattice.

Lemma 7.1.5. $\min R_F = [F : \mathbb{Q}]$ *and $\mathcal{M}(R_F)$ are roots of unity in F.*

Proof. Put $n := [F : \mathbb{Q}]$; then for $0 \neq x \in R_F$, we have

$$Q(x) = \sum |\sigma(x)|^2 \geq n(\prod |\sigma(x)|^2)^{1/n} = n(\mathrm{N}_{F/\mathbb{Q}}(x))^{2/n} \geq n,$$

where $\mathrm{N}_{F/\mathbb{Q}}$ is the norm from F to $\mathbb{Q}$. Now $Q(1) = n$ is clear and hence $\min R_F = n$. If $Q(x) = n$ for $x \in R_F$, then in the above inequality $|\sigma(x)|$ is independent of σ so $|\sigma(x)| = 1$ for all $\sigma \in \mathrm{Gal}(F/\mathbb{Q})$. Thus x is a root of unity since $x \in R_F$. □

Lemma 7.1.6. *Let p be a prime and $L := \mathbb{Z}[u_1, \cdots, u_{p-1}]$ a quadratic lattice defined by $B(u_i, u_j) = -1$ if $i \neq j$, $B(u_i, u_i) = p - 1$ for $1 \leq i, j \leq p - 1$. Then L is a positive lattice of E-type.*

Proof. Let M be a positive lattice, and for $x = \sum u_i \otimes w_i \in L \otimes M$ ($u_i \in L, w_i \in M$), we have

$$\begin{aligned} Q(x) &= \sum_{i,j} B(u_i, u_j) B(w_i, w_j) \\ &= (p-1) \sum Q(w_i) - \sum_{i \neq j} B(w_i, w_j) \\ &= (p-1) \sum Q(w_i) + \sum_{i<j} Q(w_i - w_j) - (p-2) \sum Q(w_i) \\ &= \sum Q(w_i) + \sum_{i<j} Q(w_i - w_j). \qquad (5) \end{aligned}$$

Hence $Q(x) \geq 0$ and moreover $Q(x) = 0$ yields $w_i = 0$, so $x = 0$. Thus $L \otimes M$ is a positive lattice and hence L is also, taking $M = \langle 1 \rangle$. It is clear that $\min(L \otimes M) \leq (p-1) \min M$. To show the converse inequality, we may assume $w_1, \cdots, w_k \neq 0, w_{k+1} = \cdots = w_{p-1} = 0$ $(1 \leq k \leq p-1)$ for the above $x \neq 0$, since all permutations of $\{u_i\}$ are isometries of L. Then the $p-1$ elements $w_1, \cdots, w_k, w_1 - w_{k+1}, \cdots, w_1 - w_{p-1}$ are not zero, and (5) implies $Q(x) \geq (p-1) \min M$. Thus we have $\min(L \otimes M) = (p-1) \min M$. Suppose $x \in \mathcal{M}(L \otimes M)$; then all elements $w_i, w_i - w_j$ except for the above $p-1$ elements should be zero. Thus $w_1 - w_k = 0$ and $w_i - w_j = 0$ for $i = 2 < j$, so $w_1 = w_k, w_2 = \cdots = w_{p-1}$. If $k = 1$, then by definition of k, $w_2 = \cdots = w_{p-1} = 0$ and so $x = u_1 \otimes w_1$ is split. If $k > 1$, then $w_1 = w_k = w_2$ and so $w_1 = \cdots = w_{p-1}$ follows. This means that $x = \sum u_i \otimes w_1$ is split. Thus L is of E-type. □

Lemma 7.1.7. *Let p be a prime and $n \geq 2$ and put $F = \mathbb{Q}(\zeta)$, where ζ is a primitive p^n-th root of unity. Then R_F is a positive lattice of E-type.*

Proof. It is well known that R_F has a basis $v_i := \zeta^{i-1}, 1 \leq i \leq p^{n-1}(p-1)$. Then $B(v_i, v_j) = \mathrm{tr}_{F/\mathbb{Q}}\, \zeta^{i-j}$ and

$$\mathrm{tr}_{F/\mathbb{Q}}\, \zeta^m = \sum_{i \bmod p^n} \zeta^{im} - \sum_{i \bmod p^{n-1}} \zeta^{ipm} = \begin{cases} p^{n-1}(p-1) & \text{if } p^n | m, \\ -p^{n-1} & \text{if } p^{n-1} | m, p^n \nmid m, \\ 0 & \text{if } p^{n-1} \nmid m. \end{cases}$$

Let L be a positive lattice in the previous lemma. Let

$$M := \mathbb{Z}[w_1, \cdots, w_{p^{n-1}}]$$

be a positive lattice defined by $(B(w_i, w_j)) = p^{n-1}1_{p^{n-1}}$. We define a basis $\{z_i\}$ of $L \otimes M$ by

$$z_i := u_{b+1} \otimes w_a \ (i = a + bp^{n-1}, 1 \leq a \leq p^{n-1}, 0 \leq b \leq p-2)$$

for $1 \leq i \leq (p-1)p^{n-1}$. If

$$i = a + bp^{n-1}, j = c + dp^{n-1}(1 \leq a, c \leq p^{n-1}, 0 \leq b, d \leq p-2),$$

then we have

$$\begin{aligned} B(z_i, z_j) =& B(u_{b+1} \otimes w_a, u_{d+1} \otimes w_c) \\ =& B(u_{b+1}, u_{d+1}) B(w_a, w_c) \\ =& \begin{Bmatrix} -1 & \text{if } b \neq d \\ p-1 & \text{if } b = d \end{Bmatrix} \times p^{n-1}\delta_{ac}, \end{aligned}$$

where δ is Kronecker's delta. It is easy to see that $i = j$ if and only if $a = c$ and $b = d$, and that $i \equiv j \bmod p^{n-1}$ if and only if $a = c$. Hence $B(z_i, z_j) = B(v_i, v_j)$ which means $R_F \cong L \otimes M$. By the previous lemma, L is of E-type, and by Lemma 7.1.1, M and so $L \otimes M$ are of E-type. □

Proof of Theorem 7.1.3.

Let F be an abelian field, then F is contained in $K := \mathbb{Q}(\zeta_m)$ for a certain primitive m-th root ζ_m of unity, which is known as Kronecker's theorem. Let $m = p_1^{a_1} \cdots p_n^{a_n}$ be the prime decomposition of m; then R_K is the product of the ring of integers R_i of $K_i := \mathbb{Q}(\zeta_{p_i^{a_i}})$ and

$$\mathrm{tr}_{K/\mathbb{Q}} \prod_i x_i = \prod_i \mathrm{tr}_{K_i/\mathbb{Q}}\, x_i \text{ for } x_i \in R_{K_i}$$

implies that the quadratic lattice R_K is the tensor product of the positive lattices R_i. Thus R_K is of E-type and $\min R_K = \prod p_i^{a_i - 1}(p_i - 1)$ by Lemma 7.1.5. The same lemma shows $1 \in \mathcal{M}(R_K) \cap R_F$ and then from Lemma 7.1.1 it follows that a submodule R_F of R_K is of E-type. Since $B_F = [K : F]^{-1} B_K$ on R_F, (R_F, B_F) is also of E-type. □

Exercise 1.

Let $L := \mathbb{Z}[e_1, \cdots, e_n]$ be a quadratic lattice defined by (i) $B(e_i, e_j) \in \mathbb{Q}$, (ii) $B(e_i, e_i) = 1$ and $|B(e_i, e_j)| < 1/n$ for $i \neq j$. Show L is a positive lattice of E-type and $\mathcal{M}(L) = \{\pm e_i\}$.
(Hint: For a positive lattice M and $x = \sum e_i \otimes u_i \in L \otimes M$ $(x \neq 0)$, put $S = \{i \,|\, u_i \neq 0\} \neq \emptyset$ and $s = \sharp S$, and use

$$\begin{aligned} Q(x) \geq& \sum_{i \in S} Q(u_i) - \sum_{i \in S} (\sum_{\substack{j \in S \\ j \neq i}} |a_{ij}|) Q(u_i) \\ \geq& (\min M)(s - (s-1)s/n) \geq \min M). \end{aligned}$$

Exercise 2.

Let L be a positive lattice such that $L \cong \langle a[n] \rangle$, where

$$a = \mathrm{diag}(a_1, \cdots, a_m) > 0$$

and $n = (n_{ij})$ is an upper triangular matrix with all diagonals being 1. If $a_i \geq a_1$ for all i, show L is of E-type.
(Hint: For a positive lattice M and $x = \sum e_i \otimes u_i \in \mathcal{M}(L \otimes M)$, define k by $u_k \neq 0$ and $u_{k+1} = \cdots = 0$. Then show

$$\begin{aligned}\min L \min M &\geq \min(L \otimes M) = Q(x)\\ &= \mathrm{tr}(a \cdot (B(u_i, u_j))[{}^t n])\\ &= \sum_{i<k} a_i b_{ii} + a_k Q(u_k) \geq a_1 Q(u_k) = Q(e_1 \otimes u_k),\end{aligned}$$

where $(B(u_i, u_j))[{}^t n] = (b_{ij})$, and $Q(u_k) = b_{kk}$. The minimality of $Q(x)$ yields $b_{ii} = 0$ for all $i < k$ and then $b_{ij} = 0$ unless $i = j = k$. Hence $\mathrm{rank}(B(u_i, u_j)) = 1$, so $Q(u_i)Q(u_k) - B(u_i, u_k)^2 = 0$ and then $u_i = B(u_i, u_k)Q(u_k)^{-1}u_k$.)

Exercise 3.

Let m be a natural number. Prove there exists a positive constant c_m such that if L is a positive lattice of rank $L = m$, then there is a submodule L_0 of L satisfying (i) $[L : L_0] < c_m$, (ii) $\min L = \min L_0$ and (iii) L_0 is of E-type.
(Hint: Choose a basis $\{e_i\}$ of L such that $(B(e_i, e_j))$ is in Siegel's domain $S_{4/3,1/2}$ and $Q(e_1) = \min L$ and put $L_0 := \mathbb{Z}[e_1, \delta_2 e_2, \cdots, \delta_m e_m]$ for integers $\delta_i \geq (4/3)^{(i-1)/2}$ and use Exercise 2.)

Exercise 4.

Let L be a positive lattice such that $L \cong \langle a[n] \rangle$ ($a = \mathrm{diag}(a_1, \cdots, a_m)$, $a[n] \in S_{4/3,1/2}$). If $a_2 \geq (4/3)^{m-2} a_1$, prove L is of E-type.
(Hint: Use Exercise 2.)

The following is due to Hsia.

Exercise 5.

For any given natural number m show there always exists a unimodular positive lattice L with $\min(L) = m$.
(Hint: For m even, use induction and the fact that

$$\min(E_8 \otimes K) = 2 \min(K).$$

For m odd, choose unimodular lattices L_1 with $\min L_1 = 2(m-1)$ and L_2 with $\min(L_2) = 2(m+1)$. Put $L' := L_1 \perp L_2$. For $v_i \in \mathcal{M}(L_i)$ let

$$v = (v_1 + v_2)/2.$$

Then

$$J := \mathbb{Z}v + \{x_1 + x_2 \mid x_i \in L_i,\ B(x_1, v_1) + B(x_2, v_2) \equiv 0 \bmod 2\}$$

is a required one.)

7.2 A fundamental lemma

For positive lattices L, M, we say that M is *divisible* by L if $M \cong L \otimes K$ for some positive lattice K.

Let L be an indecomposable positive lattice. We consider two assertions (A), (B):

(A) For any indecomposable positive lattice X, the following assertions (A1), (A2), and (A3) hold.

(A1) $L \otimes X$ is indecomposable.

(A2) $L \otimes X \cong L \otimes Y$ yields $X \cong Y$ for a positive lattice Y.

(A3) If $X = \otimes^t L \otimes X'$ and X' is not divisible by L, then $O(L \otimes X)$ is generated by $O(L), O(X')$ and interchanges of L.

(B) Let M, N be positive lattices, and suppose that $\sigma : L \otimes M \cong L \otimes N$ and that $\sigma(L \otimes m) = L \otimes n$ $(m \in M, n \in N)$ yields $m = 0, n = 0$. Then there exists a subset $\{v_1, \cdots, v_t\}$ of L dependent on M, N, σ which satisfies the following three assertions:

(B1) v_i's are primitive in L and the submodule spanned by $\{v_1, \cdots, v_t\}$ is of finite index in L.

(B2) Putting

$$M_i = M_{v_i} := \{m \in M \mid \sigma(L \otimes m) \subset v_i \otimes N\},$$
$$N_i = N_{v_i} := \{n \in N \mid \sigma^{-1}(L \otimes n) \subset v_i \otimes M\},$$

we have $\operatorname{rank} M_i = \operatorname{rank} N_i = \operatorname{rank} M / \operatorname{rank} L = \operatorname{rank} N / \operatorname{rank} L$ for all i.

(B3) $\sigma(\mathbb{Q}(v_i \otimes M_i)) \subset \mathbb{Q}(v_i \otimes N_i)$.

The aim of this section is to prove

Fundamental Lemma. *For an indecomposable positive lattice L, the assertions* (A) *and* (B) *are equivalent.*

Using this lemma, we will verify in 7.4 that the lattices in Example 1 of 7.1, for example satisfy the above assertion (A).

First we prove (A) $\Rightarrow$ (B). Let $\{v_i\}$ be a basis of L. The condition (B1) is clearly satisfied. Let M, N be positive lattices; suppose that $\sigma : L \otimes M \cong L \otimes N$ and that

$$\text{(1)} \qquad \sigma(L \otimes m) = L \otimes n \ \text{ yields } m = 0, n = 0.$$

First we assume that M, N are indecomposable. Moreover, we put $M = \otimes^p L \otimes M', N = \otimes^q L \otimes N'$ and assume that M', N' are not divisible by L. Since M, N are indecomposable, putting

$$X_r := \otimes^{p-r} L \otimes M', Y_r := \otimes^{q-r} L \otimes N' \ (r = 0, 1, \cdots),$$

X_r and Y_r are also indecomposable, and then (A2) implies inductively $X_r \cong Y_r$ starting from $L \otimes M \cong L \otimes N$. Hence, we have $p = q$ and $M' \cong N'$. Therefore we may suppose $M' = N', M = N$. Then (A3) yields, for $x_i \in L, m \in M'$

$$\text{(2)} \quad \sigma(x_1 \otimes \cdots \otimes x_{p+1} \otimes m) = \sigma_1(x_{s(1)}) \otimes \cdots \otimes \sigma_{p+1}(x_{s(p+1)}) \otimes \beta(m),$$

for some permutation s of $\{1, \cdots, p+1\}$, $\sigma_i \in O(L)$ and $\beta \in O(M')$. If $s(1) = 1$, then (2) implies

$$\sigma(L \otimes x_2 \otimes \cdots \otimes x_{p+1} \otimes m) = L \otimes \sigma_2(x_{s(2)}) \otimes \cdots \otimes \sigma_{p+1}(x_{s(p+1)}) \otimes \beta(m)$$

for $x_i \in L, m \in M'$. This contradicts (1); thus $s(1) > 1$. Then (2) implies

$$\begin{aligned} \sigma(v_i \otimes M) &= \sigma(v_i \otimes L \otimes \cdots \otimes L \otimes M') \\ \text{(3)} \qquad &= L \otimes \cdots L \otimes \sigma_{s^{-1}(1)}(v_i) \otimes L \otimes \cdots \otimes L \otimes N' \\ \sigma^{-1}(v_i \otimes N) &= \sigma^{-1}(v_i \otimes L \otimes \cdots \otimes L \otimes N') \\ \text{(4)} \qquad &= L \otimes \cdots L \otimes \sigma_1^{-1}(v_i) \otimes L \otimes \cdots \otimes L \otimes M' \end{aligned}$$

where $\sigma_{s^{-1}(1)}(v_i)$ (resp. $\sigma_1^{-1}(v_i)$) is on the $s^{-1}(1)$-th (resp. $s(1)$-th) coordinate. Here, denoting by N_i, M_i positive lattices such that $L \otimes N_i, L \otimes M_i$ are the right-hand side modules of (3), (4), respectively, we have

$$\text{(3')} \qquad \sigma^{-1}(L \otimes N_i) = v_i \otimes M$$
$$\text{(4')} \qquad \sigma(L \otimes M_i) = v_i \otimes N,$$

and then $\operatorname{rank} N_i = \operatorname{rank} M / \operatorname{rank} L = \operatorname{rank} N / \operatorname{rank} L = \operatorname{rank} M_i$, and

$$\begin{aligned} M_i &= \{m \in M \mid \sigma(L \otimes m) \subset v_i \otimes N\}, \\ N_i &= \{n \in N \mid \sigma^{-1}(L \otimes n) \subset v_i \otimes M\}. \end{aligned}$$

Thus the condition (B2) is satisfied. From (3′), (4′) it follows that

$$\sigma(v_i \otimes M_i) = \sigma(v_i \otimes M \cap L \otimes M_i) \subset L \otimes N_i \cap v_i \otimes N = v_i \otimes N_i$$

and hence (B3).

Now let M, N be (not necessarily indecomposable) positive lattices and suppose $\sigma : L \otimes M \cong L \otimes N$ with the condition (1). Let $M = \perp_{j=1}^{k} X_j, N = \perp_{j=1}^{h} Y_j$ be decompositions into indecomposable lattices. By the condition (A1), $L \otimes X_j, L \otimes Y_j$ are indecomposable. Hence by virtue of Theorem 6.7.1, we have $k = h$ and may assume $\sigma(L \otimes X_j) = L \otimes Y_j$, changing suffices if necessary. We can apply the above to X_j, Y_j, σ. Putting

$$\begin{aligned} X_{ji} =& \{x \in X_j \mid \sigma(L \otimes x) \subset v_i \otimes Y_j\}, \\ Y_{ji} =& \{y \in Y_j \mid \sigma^{-1}(L \otimes y) \subset v_i \otimes X_j\}, \\ M_i =& \{m \in M \mid \sigma(L \otimes m) \subset v_i \otimes N\}, \\ N_i =& \{n \in N \mid \sigma^{-1}(L \otimes n) \subset v_i \otimes M\}, \end{aligned}$$

we show $M_i = \perp_j X_{ji}, N_i = \perp_j Y_{ji}$. The left-hand sides obviously contain the right-hand sides. Suppose $m = \sum x_j \in M_i$ $(x_j \in X_j)$. For $z \in L$, $\sigma(z \otimes m) = \sum \sigma(z \otimes x_j) \in v_i \otimes N$ yields $\sigma(z \otimes x_j) \subset v_i \otimes Y_j$ for all $z \in L$. Hence $x_j \in X_{ji}$ holds and so $M_i \subset \perp_j X_{ji}$. Similarly $N_i \subset \perp_j Y_{ji}$, which is what we wanted. Since $\operatorname{rank} M_i = \sum_j \operatorname{rank} X_{ji} = \sum_j \operatorname{rank} Y_{ji} = \operatorname{rank} N_i$ and $\operatorname{rank} X_{ji} = \operatorname{rank} X_j / \operatorname{rank} L$, we have $\operatorname{rank} M_i = \operatorname{rank} M / \operatorname{rank} L$ and similarly $\operatorname{rank} N_i = \operatorname{rank} N / \operatorname{rank} L$, which implies the condition (B2). (B3) follows from $\sigma(\mathbb{Q}(v_i \otimes X_{ji})) \subset \mathbb{Q}(v_i \otimes Y_{ji})$ and $M_i = \perp_j X_{ji}, N_i = \perp_j Y_{ji}$. □

Next we show (B) ⇒ (A). We need several lemmas. *In the rest of this section, L denotes an indecomposable positive lattice such that the assertion (B) is true.* If $\operatorname{rank} L = 1$, then we have nothing to do. So we assume $\operatorname{rank} L > 1$.

Lemma 7.2.1. *Letting $M, N, M_i, N_i, \sigma, v_i$ be those in (B), we have $M \cong N$, $\sigma(L \otimes M_i) = v_i \otimes N$ and $\sigma^{-1}(L \otimes N_i) = v_i \otimes M$. In particular,* $\operatorname{rank} M = \operatorname{rank} N \geq \operatorname{rank} L$.

Proof. By virtue of (B2), the primitiveness of v_i implies that of M_i, N_i in M, N, respectively, that is we have

$$\begin{aligned} L \otimes M_i =& \mathbb{Q}(L \otimes M_i) \cap L \otimes M, \\ L \otimes N_i =& \mathbb{Q}(L \otimes N_i) \cap L \otimes N. \end{aligned} \tag{5}$$

(B2) yields $\sigma(L \otimes M_i) \subset v_i \otimes N$ and $\operatorname{rank} M_i = \operatorname{rank} N / \operatorname{rank} L$ and hence $\sigma(\mathbb{Q}(L \otimes M_i)) = \mathbb{Q}(v_i \otimes N)$. From (5) follows

$$\begin{aligned} \sigma(L \otimes M_i) =& \sigma(\mathbb{Q}(L \otimes M_i)) \cap \sigma(L \otimes M) \\ =& \mathbb{Q}(v_i \otimes N) \cap L \otimes N = v_i \otimes N, \end{aligned} \tag{6}$$

and similarly

$$(6') \qquad \sigma^{-1}(L \otimes N_i) = v_i \otimes M.$$

Now $\dim \mathbb{Q}(v_i \otimes M_i) = \dim \mathbb{Q}(v_i \otimes N_i)$ follows from (B2) and then (B3) implies $\sigma(\mathbb{Q}(v_i \otimes M_i)) = \mathbb{Q}(v_i \otimes N_i)$, which, since $v_i \otimes M_i, v_i \otimes N_i$ are primitive in $L \otimes M, L \otimes N$, respectively, implies

$$\begin{aligned}\sigma(v_i \otimes M_i) &= \sigma(\mathbb{Q}(v_i \otimes M_i) \cap L \otimes M) = \mathbb{Q}(v_i \otimes N_i) \cap L \otimes N \\ &= v_i \otimes N_i.\end{aligned}$$

Hence we can define an isometry $\mu_i : M_i \cong N_i$ by

$$(7) \qquad \sigma(v_i \otimes m) = v_i \otimes \mu_i(m).$$

We will construct an isometry $\mu : M \cong N$ by making use of μ_i. We first show

$$(8) \qquad B(a,b) = B(\mu_i(a), \mu_j(b)) \text{ for } a \in M_i, b \in M_j.$$

Since

$$\begin{aligned}B(v_i, v_j)B(a,b) =& B(v_i \otimes a, v_j \otimes b) \\ =& B(v_i \otimes \mu_i(a), v_j \otimes \mu_j(b)) \quad \text{by (7)} \\ =& B(v_i, v_j)B(\mu_i(a), \mu_j(b)),\end{aligned}$$

(8) holds if $B(v_i, v_j) \neq 0$.

Suppose $B(v_i, v_j) = 0$; then

$$\begin{aligned}B(L,L)B(M_i,M_j) =& B(L \otimes M_i, L \otimes M_j) = B(v_i \otimes N, v_j \otimes N) \quad \text{by (6)} \\ =& B(v_i, v_j)B(N,N) = 0,\end{aligned}$$

and then $B(M_i, M_j) = 0$ follows.

Similarly,

$$\begin{aligned}&B(L,L)B(\mu_i(M_i), \mu_j(M_j)) = B(L,L)B(N_i,N_j) \\ =& B(L \otimes N_i, L \otimes N_j) = B(v_i \otimes M, v_j \otimes M) \quad \text{by (6')} \\ =& B(v_i, v_j)B(M,M) = 0.\end{aligned}$$

Thus we have $B(\mu_i(M_i), \mu_j(M_j)) = 0$ and hence $B(\mu_i(M_i), \mu_j(M_j)) = B(M_i, M_j) = 0$ if $B(v_i, v_j) = 0$ which completes the proof of (8).

By (B1), we may assume that a subset $\{v_1, \cdots, v_n\}$ is a basis of $\mathbb{Q}L$, changing suffices if necessary; then by (6), $\oplus_{i=1}^n M_i$ is a submodule of M of finite index and

$$\mathbb{Q}M = \bigoplus_{i=1}^{n} \mathbb{Q}M_i.$$

Hence by virtue of (8), we can define an isometry $\mu : \mathbb{Q}M \cong \mathbb{Q}N$ by $\mu(\sum m_i) = \sum \mu_i(m_i)$ for $m_i \in \mathbb{Q}M_i$. We have only to show $\mu(M) = N$. Let $\{e_i\}$ be a basis of L and $e_i = \sum_{j=1}^n a_{ij}v_j, v_i = \sum_{h=1}^n b_{ih}e_h,\ a_{ij} \in \mathbb{Q}, b_{ih} \in \mathbb{Z}$; then

$$\sum_{k=1}^{n} b_{ik}a_{kj} = \delta_{ij}. \tag{9}$$

For $m \in M$, we write $m = \sum_{i=1}^n m_i,\ m_i \in \mathbb{Q}M_i$, and $\sigma(v_j \otimes m_i) = v_i \otimes n_{ji},\ n_{ji} \in \mathbb{Q}N$ by (6); then (7) yields $n_{ii} = \mu_i(m_i)$. Now we have

$$\begin{aligned}\sigma(e_k \otimes m) &= \sigma(\sum_j a_{kj}v_j \otimes \sum_i m_i) = \sum_{i,j} a_{kj}v_i \otimes n_{ji} \\ &= \sum_h e_h \otimes (\sum_{i,j} a_{kj}b_{ih}n_{ji}).\end{aligned}$$

From $\sigma(e_k \otimes m) \in L \otimes N$ follows $\sum_{i,j} a_{kj}b_{ih}n_{ji} \in N$. Summing them up with $k = h$, (9) implies $\sum_i n_{ii} \in N$ and so $\mu(m) = \sum \mu_i(m_i) = \sum n_{ii} \in N$. Thus we have shown $\mu(M) \subset N$. The assumption $L \otimes M \cong L \otimes N$ implies $\mathrm{d}\,M = \mathrm{d}\,N$ and so $\mu(M) = N$. □

The proof shows the following, putting $u_i = v_i$ for $1 \le i \le n$.

Corollary 7.2.1. *If a subset $\{u_j\}_{j=1}^n$ of $\{v_i\}_{i=1}^t$ is a basis of $\mathbb{Q}L$, then $\sigma(u_i \otimes m) = u_i \otimes \mu(m)$ for $m \in M_{u_i}$ for some isometry $\mu : M \cong N$ and $\mu(M_{u_i}) = N_{u_i}$.*

Lemma 7.2.2. *Let p be a prime number and let A be a matrix in $M_n(\mathbb{Z})$ satisfying $A \equiv 1_n \bmod p$. If $A^k = 1_n$ for $k \ge 1$, then $A = 1_n$ or $A = W \operatorname{diag}(\pm 1, \cdots, \pm 1)W^{-1}$ for $W \in GL_n(\mathbb{Z})$. The latter case occurs only when $p = 2$.*

Proof. Let the order of A be $p^r h$ $(h,p) = 1$, and put $B = A^{p^r}$; then $B^h = 1_n$ and $B \equiv 1_n \bmod p$.

Suppose $h \ne 1$, i.e. $B \ne 1_n$ and define an integer m (≥ 1) by $B \equiv 1_n \bmod p^m$ but $B \not\equiv 1_n \bmod p^{m+1}$. Since $(h,p) = 1$,

$$1_n = B^h = (1_n + (B - 1_n))^h \equiv 1_n + h(B - 1_n) \bmod p^{m+1}$$

gives $B \equiv 1_n \bmod p^{m+1}$, contradicting the defining property of m. Therefore we get $B = 1_n$ and $h = 1$.

Suppose $p \neq 2$. Putting $C := A^{p^{r-1}}$, we have $C \equiv 1_n \bmod p$, $C^p = 1, C \neq 1_n$.

Writing $C = 1_n + p^t D$ $(t \geq 1, D \in M_n(\mathbb{Z}), D \not\equiv 0 \bmod p)$,

$$\begin{aligned} 1_n =& C^p = 1_n + p^{t+1}D + \sum_{s=2}^{p-1} \binom{p}{s} (p^t D)^s + p^{tp}D^p \\ \equiv& 1_n + p^{t+1}D \bmod p^{t+2} \end{aligned}$$

yields $D \equiv 0 \bmod p$ which is a contradiction. Thus we have completed the case of $p \neq 2$.

Suppose $p = 2$ and $r \geq 2$; then putting $F := A^{2^{r-2}}$, $F \equiv 1_n \bmod 2$ and the order of F is 4. Writing $F = 1_n + 2G$, we have $F^2 \equiv 1_n \bmod 4$. Defining D by the same equation as in the case of $p \neq 2$ for $C := F^2$, we have $t \geq 2$ and then $1_n = C^2 = 1_n + 2^{t+1}D + 2^{2t}D^2 \equiv 1_n + 2^{t+1}D \bmod 2^{t+2}$ yields $D \equiv 0 \bmod 2$, again a contradiction. Thus we have $r \leq 1$, i.e. $A^2 = 1_n$. Put $L = \mathbb{Z}^n$(column vectors) and let A act on L canonically. Set $L_\pm = \{x \in L \mid Ax = \pm x\}$. Then the identity $x = (x + Ax)/2 + (x - Ax)/2$ gives $L = L_+ \oplus L_-$ since $Ax \equiv x \bmod 2L$. Hence we have only to take W as a base change matrix. □

Lemma 7.2.3. *Let K, M, N be positive lattices and suppose that K is indecomposable and there exist submodules M_i', N_i' in M, N such that*

$$[M : M'] < \infty \text{ for } M' := M_1' \perp \cdots \perp M_n',$$
$$[N : N'] < \infty \text{ for } N' := N_1' \perp \cdots \perp N_n',$$

and that there is an isometry $\sigma : K \otimes M \cong K \otimes N$ such that

$$\sigma = \sigma_i \otimes \mu_i : K \otimes M_i' \cong K \otimes N_i',$$

where $\sigma_i \in O(K), \mu_i : M_i' \cong N_i'$. Then there exist decompositions

$$M = M_1 \perp \cdots \perp M_m, N = N_1 \perp \cdots \perp N_m$$

such that $\sigma = \alpha_i \otimes \beta_i : K \otimes M_i \cong K \otimes N_i$, where $\alpha_i \in O(K), \beta_i : M_i \cong N_i$. In particular, $M \cong N$ follows.

Proof. We may assume $\mathrm{s}\,K = \mathbb{Z}$, $\mathrm{s}\,M, \mathrm{s}\,N \subset \mathbb{Z}$ without loss of generality. We use induction on $\operatorname{rank} M = \operatorname{rank} N$. If $\operatorname{rank} M = 1$, then the assertion is clear since $K \otimes M \cong K \otimes N$ implies $M \cong \langle \mathrm{d}\,M \rangle = \langle \mathrm{d}\,N \rangle \cong N$. Suppose rank $M > 1$. Put $\tilde{M}_1 = \mathbb{Q}M'_1 \cap M, \tilde{N}_1 = \mathbb{Q}N'_1 \cap N$; then $\sigma(K \otimes M'_1) = K \otimes N'_1$ and $\sigma(K \otimes M) = K \otimes N$ implies $\sigma(K \otimes \tilde{M}_1) = K \otimes \tilde{N}_1$ and so $\mu_1(\tilde{M}_1) = \tilde{N}_1$, regarded as $\mu_1 : \mathbb{Q}M'_1 \cong \mathbb{Q}N'_1$. Replacing M'_1, N'_1 by $\tilde{M}_1, \tilde{N}_1$, respectively, we may assume that M'_1, N'_1 are primitive in M, N, respectively. Let u be a primitive element in M'_1 and put $v = \mu_1(u)$. From the assumption follows $\sigma(K \otimes u) = K \otimes v$ and so $\sigma(K \otimes u^\perp) = K \otimes v^\perp$ by virtue of $(K \otimes u)^\perp = K \otimes u^\perp$. Applying the induction hypothesis to $\sigma : K \otimes u^\perp \cong K \otimes v^\perp$, we may assume moreover that

$$M'_1 = \mathbb{Z}u_1, u_1^\perp = M'_2 \perp \cdots \perp M'_n, M' = \mathbb{Z}u_1 \perp u_1^\perp,$$
$$N'_1 = \mathbb{Z}v_1, v_1^\perp = N'_2 \perp \cdots \perp N'_n, N' = \mathbb{Z}v_1 \perp v_1^\perp,$$

where we put $u_1 := u, v_1 := v$, and they are primitive in M, N respectively. Here we note that $\sigma(L \otimes M') = L \otimes N'$ implies $[M : M'] = [N : N']$. Hence if $[M : M'] = 1$, then we have nothing to do, and so we may assume $[M : M'] > 1$.

If $\sigma_i = \pm \operatorname{id}$ for every i, then we may assume $\sigma_i = \operatorname{id}$ by virtue of $(-\operatorname{id}) \otimes \mu_i = \operatorname{id} \otimes (-\mu_i)$, and then $\sigma = \operatorname{id} \otimes \mu$ for some isometry $\mu : M' \cong N'$. Then $\sigma(K \otimes M) = K \otimes N$ implies $\mu(M) = N$. Hence we have

$$\sigma = \operatorname{id} \otimes \mu, \mu : M \cong N$$

and we have completed the proof in this case. Thus, to prove the lemma, by multiplying $\pm(\sigma_1 \otimes \operatorname{id})^{-1}$ we may assume further that

$$(10) \qquad \begin{cases} \sigma(z \otimes u_1) = z \otimes v_1 \text{ for } z \in K, \\ \sigma_k \neq \pm \operatorname{id} \quad \text{for some } k > 1, \\ [M : M'] > 1. \end{cases}$$

We will show that $M = M'_k \perp (M'_k)^\perp$ under these assumptions. Once this has been proved, we can then apply the induction assumption to

$$\sigma : K \otimes (M'_k)^\perp \cong K \otimes (N'_k)^\perp,$$

which completes the proof.

Noting that $u = u_1$ is primitive, choose a basis $\{u_i\}$ of M such that $B(u_1, u_i) = 0$ for $i \geq 3$. If $B(u_1, u_2) = 0$, then $M = M'_1 \perp (M'_1)^\perp$ and this contradicts the assumption $[M : M'] > 1$. Thus we have $B(u_1, u_2) \neq 0$. Considering σ^{-1}, we can choose a similar basis $\{v_i\}$ of N and hence for a basis of $\{u_i\}$ of M and $\{v_i\}$ of N, we have

$$(11) \qquad \begin{cases} B(u_1, u_2) \neq 0, B(u_1, u_i) = 0 \text{ for } i \geq 3, \\ B(v_1, v_2) \neq 0, B(v_1, v_i) = 0 \text{ for } i \geq 3. \end{cases}$$

Take elements $x_i, y_i \in K$ such that

$$\sum B(x_i, y_i) = 1, \tag{12}$$

by virtue of the assumption $\mathrm{s}\,K = \mathbb{Z}$. For $m \in M$, by (12)

$$B(u_1, m) = \sum B(x_i \otimes u_1, y_i \otimes m) \in B(K \otimes u_1, K \otimes M)$$

is clear and hence $B(u_1, M) \subset B(K \otimes u_1, K \otimes M)$ follows. The converse inclusion is clear, and so $B(u_1, M) = B(K \otimes u_1, K \otimes M)$ and similarly

$$B(v_1, N) = B(K \otimes v_1, K \otimes N).$$

Hence we have

$$\begin{aligned} B(u_1, M) &= B(K \otimes u_1, K \otimes M) = B(\sigma(K \otimes u_1), \sigma(K \otimes M)) \\ &= B(K \otimes v_1, K \otimes N) = B(v_1, N), \end{aligned}$$

which implies

$$(Q(u_1), B(u_1, u_2)) = (Q(v_1), B(v_1, v_2)) \quad \text{by (11).}$$

Since $Q(u_1) = Q(v_1)$, the reduced denominator of $B(u_1, u_2)/Q(u_1)$ and $B(v_1, v_2)/Q(v_1)$ are the same; we denote it by r. If $r = 1$, then $B(u_1, u_2 - B(u_1, u_2)Q(u_1)^{-1}u_1) = 0$ and $u_2 - B(u_1, u_2)Q(u_1)^{-1}u_1 \in M$ implies $M = \mathbb{Z}u_1 \perp u_1^{\perp}$. This contradicts $[M : M'] > 1$ in (10) and so $r > 1$ holds. Let

$$u_1^{\perp} = \mathbb{Z}[x, u_3, \cdots], v_1^{\perp} = \mathbb{Z}[y, v_3, \cdots]$$

where

$$x := r(u_2 - au_1), y := r(v_2 - a'v_1)$$

for $a := B(u_1, u_2)/Q(u_1), a' := B(v_1, v_2)/Q(v_1)$, and put

$$x = \sum_{i \geq 2} m_i \ (m_i \in M_i'), \quad y = \sum_{i \geq 2} n_i \ (n_i \in N_i').$$

Fix any prime p dividing r. Then a/a' is a p-adic unit, and for $z \in K$, we have

$$\begin{aligned} \sigma(z \otimes u_2) &= \sigma(z \otimes (r^{-1}x + au_1)) = r^{-1}\sigma(z \otimes x) + a\sigma(z \otimes u_1) \\ &= r^{-1}\sigma(z \otimes x) + az \otimes v_1 = r^{-1}\sigma(z \otimes x) + az \otimes a'^{-1}(v_2 - r^{-1}y) \\ &= (a/a')z \otimes v_2 + r^{-1}\Big(\sum_{i \geq 2} \sigma_i(z) \otimes \mu_i(m_i) - a/a' \sum_{i \geq 2} z \otimes n_i\Big) \end{aligned}$$

and then

$$r^{-1}\Big(\sum_{i\geq 2}\sigma_i(z)\otimes\mu_i(m_i)-a/a'\sum_{i\geq 2}z\otimes n_i\Big)=\sigma(z\otimes u_2)-(a/a')z\otimes v_2$$
$$\in\mathbb{Z}_p(K\otimes N)\cap\mathbb{Q}(K\otimes v_1^{\perp})\subset\mathbb{Z}_p(K\otimes v_1^{\perp}).$$

Thus we get, taking the k-th component,

(13)
$$\sigma_k(z)\otimes\mu_k(m_k)-(a/a')\,z\otimes n_k\in r\mathbb{Z}_p(K\otimes N_k')\text{ for }z\in K.$$

Taking a basis $\{w_i\}$ of N_k' such that

$$n_k=a_k w_1,\mu_k(m_k)=b_k w_1+c_k w_2$$

for $a_k,b_k,c_k\in\mathbb{Z}$, (13) means

$$(b_k\sigma_k(z)-(a/a')\,a_k z)\otimes w_1+c_k\sigma_k(z)\otimes w_2\in r\mathbb{Z}_p(K\otimes N_k')$$

for $z\in K$, and hence

(14) $\quad b_k\sigma_k(z)\equiv(a/a')\,a_k z\bmod r\mathbb{Z}_pK,\ c_k\sigma_k(z)\in r\mathbb{Z}_pK$ for $z\in K$.

We show that $\sigma_k\neq\pm\mathrm{id}$ yields $a_k\equiv 0\bmod r\mathbb{Z}_p$. Suppose $a_k\not\equiv 0\bmod r\mathbb{Z}_p$; then it implies $\sigma_k(z)\equiv cz\bmod p\mathbb{Z}_pK$ for $c=(a/a')\,a_k/b_k\in\mathbb{Z}_p^{\times}$ by (14). Making use of (12), $1=\sum B(\sigma_k(x_i),\sigma_k(y_i))\equiv c^2\bmod p$ and hence $c\equiv\pm 1\bmod p$. Thus we get $\sigma_k(z)\equiv\pm z\bmod p\mathbb{Z}_pK$ where the sign is determined by c and hence it is independent of each $z\in K$. Since the order of σ_k is finite, the previous lemma implies $\sigma_k=\pm\mathrm{id}$ or $K=K_+\perp K_-$ where $K_\pm=\{z\in K\mid\sigma_k z=\pm z\}$. These are contradictions. Thus we have shown $a_k\equiv 0\bmod r\mathbb{Z}_p$. Then (14) implies $b_k\equiv c_k\equiv 0\bmod r\mathbb{Z}_p$ and so $n_k,\mu_k(m_k)\in r\mathbb{Z}_pN_k'$. Since p is any prime number dividing r, we have $n_k,\mu_k(m_k)\in rN_k'$, which implies

$$m_k\in rM_k'\subset rM.$$

Then

$$u_2-r^{-1}m_k=r^{-1}x+au_1-r^{-1}m_k$$
$$=r^{-1}\sum_{\substack{i\geq 2\\ i\neq k}}m_i+au_1\in M\cap\mathbb{Q}(M_k')^{\perp}=(M_k')^{\perp}.$$

With it,

$$M_k'\perp(M_k')^{\perp}\supset M_1'\perp\cdots\perp M_n'\supset M_2'\perp\cdots\perp M_n'=u_1^{\perp}$$

implies

$$M_k'\perp(M_k')^{\perp}\ni u_1,u_2-r^{-1}m_k,u_3,\cdots$$

and hence $M_k'\perp(M_k')^{\perp}\ni u_i$ for all i since $r^{-1}m_k\in M_k'$. This means $M=M_k'\perp(M_k')^{\perp}$, since $\{u_i\}$ is a basis of M. This completes the proof of the lemma as we noted before. □

Lemma 7.2.4. *Let K, X, Y be positive lattices and suppose that K is indecomposable and $\sigma : K \otimes X \cong K \otimes Y$. Then there exist submodules M_0, M of X and N_0, N of Y such that:*

(i) *M_0, M (resp. N_0, N) are primitive submodules in X (resp. Y) (unless they are $\{0\}$) such that*

$$[X : M_0 \perp M] < \infty, [Y : N_0 \perp N] < \infty,$$
$$\sigma(K \otimes M_0) = K \otimes N_0, \ \sigma(K \otimes M) = K \otimes N.$$

(ii) *There exist decompositions $M_0 = \perp_i M_{0,i}, N_0 = \perp_i N_{0,i}$ such that for every i*

$$\sigma = \alpha_i \otimes \beta_i : K \otimes M_{0,i} \cong K \otimes N_{0,i}$$
$$\text{for } \alpha_i \in O(K) \text{ and } \beta_i : M_{0,i} \cong N_{0,i}.$$

In particular, $M_0 \cong N_0$.

(iii) *$\sigma(K \otimes m) = K \otimes n$ for $m \in M, n \in N$ implies $m = 0, n = 0$.*

Proof. Let $m_1, \cdots, m_r$ be mutually orthogonal elements of X such that $\sigma(K \otimes m_i) = K \otimes n_i$ for some $n_i \in Y$, and we assume that r is the maximal number. Then putting

$$M := \{m_1, \cdots, m_r\}^\perp, N := \{n_1, \cdots, n_r\}^\perp, M_0 := M^\perp \text{ and } N_0 := N^\perp,$$

(i), (iii) are clear, since

$$K \otimes M_0 = K \otimes \mathbb{Q}[m_1, \cdots, m_r] \cap K \otimes X$$

and

$$\sigma(K \otimes M_0) = \sigma(K \otimes \mathbb{Q}[m_1, \cdots, m_r] \cap K \otimes X)$$
$$= K \otimes \mathbb{Q}[n_1, \cdots, n_r] \cap K \otimes Y = K \otimes N_0.$$

(ii) follows from Lemma 7.2.3. □

Lemma 7.2.5. *Let M, N be indecomposable lattices and suppose $M \otimes N = K_1 \perp K_2$ ($K_1 \neq 0, K_2 \neq 0$). Then the isometry α of $M \otimes N$ defined by $\alpha = \mathrm{id}$ on K_1, $= -\mathrm{id}$ on K_2 is not contained in $O(M) \otimes O(N)$.*

Proof. Suppose $\alpha = \sigma \otimes \mu$ ($\sigma \in O(M), \mu \in O(N)$).

$$\alpha \equiv \mathrm{id} \bmod 2(M \otimes N)$$

implies

$$\sigma(x) \otimes \mu(y) \equiv x \otimes y \bmod 2(M \otimes N) \text{ for } x \in M, y \in N. \tag{15}$$

We take a primitive vector v_1 of N, and if $\mu(v_1) \notin \mathbb{Q}v_1$, then we choose $v_2 \in N$ such that $\{v_1, v_2\}$ is a basis of $\mathbb{Q}[v_1, \mu(v_1)] \cap N$. Hence there are integers a, b such that $\mu(v_1) = av_1 + bv_2$, which is also valid even in the case of $\mu(v_1) \in \mathbb{Q}v_1$. Then (15) implies

$$\sigma(x) \otimes \mu(v_1) - x \otimes v_1 = (a\sigma(x) - x) \otimes v_1 + b\sigma(x) \otimes v_2 \in 2(M \otimes N).$$

Since $\{v_1, v_2\}$ can be extended to a basis of N, it yields $a\sigma(x) \equiv x \bmod 2M$ for $x \in M$. Thus $a \equiv 1 \bmod 2$ and $\sigma(x) \equiv x \bmod 2M$. By virtue of Lemma 7.2.2, there is a decomposition $M = M_1 \oplus M_2$ such that

$$\sigma = \mathrm{id} \text{ on } M_1, \sigma = -\mathrm{id} \text{ on } M_2.$$

For $x_i \in M_i$ $(i = 1, 2)$, $B(x_1, x_2) = B(\sigma(x_1), \sigma(x_2)) = -B(x_1, x_2)$ is clear and this yields that M_1 and M_2 are orthogonal with each other. The indecomposability of M yields M_1 or $M_2 = 0$, i.e. $\sigma = \pm\,\mathrm{id}$ on M. Similarly we have $\mu = \pm\,\mathrm{id}$ on N and then $\alpha = \pm\,\mathrm{id}$ on $M \otimes N$ which contradicts the definition of α. □

Lemma 7.2.6. *$L \otimes L$ is indecomposable and $O(L \otimes L)$ is generated by $O(L)$ and the interchanges of L.*

Proof. Let $\sigma \in O(L \otimes L)$. In Lemma 7.2.4 we put $K = X = Y = L$, and let M, M_0, N, N_0 be those there. We show $L = M_0$ or M. Suppose $M = 0$; then (i) in Lemma 7.2.4 implies $L = M_0$. Suppose $M \neq 0$. By (i), we have $\sigma : L \otimes M \cong L \otimes N$, and by (iii) we can apply Lemma 7.2.1 to $\sigma : L \otimes M \cong L \otimes N$, we have $\operatorname{rank} M \geq \operatorname{rank} L$. Since M is a primitive submodule of L, $M = N = L$ holds. Thus we have shown $L = M_0$ or M.

If $L = M_0$, then (ii) yields $\sigma = \alpha \otimes \beta$ for $\alpha, \beta \in O(L)$ since L is indecomposable.

If $L = M$, then $M_0 = 0$ and applying Lemma 7.2.1 by (iii) there exists a submodule $M_i \neq 0$ and $v_i \in L$ such that $\sigma(L \otimes M_i) = v_i \otimes L$. Comparing ranks, we see that $\operatorname{rank} M_i = 1$. Thus we have shown $\sigma(L \otimes u) = v \otimes L$ for some $u, v \in L, u \neq 0, v \neq 0$. Let $\mu \in O(L \otimes L)$ be the isometry defined by $\mu(x \otimes y) = y \otimes x$ $(x, y \in L)$. Then

$$\mu\sigma(L \otimes u) = L \otimes v$$

follows. Applying the above to $\mu\sigma : L \otimes L \cong L \otimes L$, and denoting by M', M'_0, N', N'_0 the corresponding submodules in Lemma 7.2.4 for $\mu\sigma$, we have $M'_0 \neq 0$ so $M'_0 = L$, and $\mu\sigma \in O(L) \otimes O(L)$ as above. Thus $O(L \otimes L)$ is generated by $O(L)$ and the interchange isometry μ.

Next suppose that $L \otimes L$ is decomposable, i.e.

$$L \otimes L = K_1 \perp K_2, K_1 \neq 0, K_2 \neq 0,$$

and define an isometry $\alpha \in O(L \otimes L)$ by $\alpha = \mathrm{id}$ on $K_1, = -\mathrm{id}$ on K_2. By the previous lemma and the above, we have

$$(16) \qquad \alpha = \mu(\beta \otimes \sigma)$$

for $\beta, \sigma \in O(L)$ and the above interchange isometry μ. $\alpha^2 = \mathrm{id}$ implies, for $x, y \in L$, $x \otimes y = \alpha^2(x \otimes y) = \mu(\beta \otimes \sigma)(\sigma(y) \otimes \beta(x)) = \sigma\beta(x) \otimes \beta\sigma(y)$. Hence $x \otimes L = \sigma\beta(x) \otimes L$ follows thus $\sigma\beta(x) = \pm x$ for $x \in L$ and so $\sigma\beta(x) \equiv x \bmod 2L$ for $x \in L$. Since L is indecomposable, taking $-\alpha$ if necessary, we may assume

$$(17) \qquad \sigma\beta = \mathrm{id} \text{ on } L.$$

Let $\{v_i\}$ be a basis of L and put $x = \sigma(v_i) \otimes v_j$ for $i \neq j$ (we are assuming $\operatorname{rank} L > 1$); then $x = (x + \alpha(x))/2 + (x - \alpha(x))/2$, and

$$(x + \alpha(x))/2 \in K_1, (x - \alpha(x))/2 \in K_2$$

follows from $\alpha^2 = 1$ and $\alpha \equiv \mathrm{id} \bmod 2(L \otimes L)$ and then (16), (17) imply

$$(\sigma(v_i) \otimes v_j + \sigma(v_j) \otimes v_i)/2 = (x + \alpha(x))/2 \in K_1.$$

This contradicts that $\{\sigma(v_h) \otimes v_k \mid h, k\}$ is a basis of $L \otimes L$. □

Lemma 7.2.7. *Let m be a natural number. Then if $O(\otimes^m L)$ is generated by $O(L)$ and interchanges of L, and if $\otimes^{m-1} L$ is indecomposable, then $\otimes^m L$ is indecomposable.*

Proof. By the previous lemma, we may assume $m \geq 3$. We assume $\otimes^m L = K_1 \perp K_2$, $K_i \neq 0$ $(i = 1, 2)$ and define an isometry $\alpha \in O(\otimes^m L)$ by

$$(18) \qquad \alpha = \mathrm{id} \text{ on } K_1, = -\mathrm{id} \text{ on } K_2.$$

From the assumption, we can write $\alpha = (\otimes \sigma_i)\mu$, where $\sigma_i \in O(L)$ and $\mu(x_1 \otimes \cdots \otimes x_m) = x_{\mu(1)} \otimes \cdots \otimes x_{\mu(m)}$ $(x_i \in L)$. Here we identified the isometry μ and the permutation of the set $\{1, \cdots, m\}$. Now $\alpha^2 = 1$ implies

$$\begin{aligned} x_1 \otimes \cdots \otimes x_m &= \alpha^2(x_1 \otimes \cdots \otimes x_m) \\ &= \alpha(\sigma_1(x_{\mu(1)}) \otimes \cdots \otimes \sigma_m(x_{\mu(m)})) \\ &= \sigma_1(\sigma_{\mu(1)}(x_{\mu^2(1)})) \otimes \cdots \otimes \sigma_m(\sigma_{\mu(m)}(x_{\mu^2(m)})), \end{aligned}$$

which implies $\mu^2 = \mathrm{id}$. If $\mu(1) = 1$, then $\alpha \in O(L) \otimes O(\otimes^{m-1} L)$ which contradicts Lemma 7.2.5 by the assumption that $\otimes^{m-1} L$ is indecomposable. If $\mu(1) = 2$, then $\mu(2) = 1$ so $\alpha \in O(\otimes^2 L) \otimes O(\otimes^{m-2} L)$ follows. It also contradicts Lemma 7.2.5, since $\otimes^2 L$ is indecomposable by Lemma 7.2.6. Put $j = \mu(1) \geq 3$ and define an isometry η by

$$\eta(x_1 \otimes x_2 \otimes \cdots \otimes x_j \otimes \cdots) = x_2 \otimes x_j \otimes \cdots \otimes x_1 \otimes \cdots,$$

where all elements except for the first, the second and the j-th coordinates are fixed. Then we have

$$\begin{aligned}\eta^{-1}\alpha\eta(x_1 \otimes \cdots \otimes x_m) =& \eta^{-1}\alpha(x_2 \otimes x_j \otimes \cdots \otimes x_1 \otimes \cdots) \\ =& \eta^{-1}(\sigma_1(x_1) \otimes \cdots \otimes \sigma_j(x_2) \otimes \cdots) \\ & (\sigma_j(x_2) \text{ is on the } j\text{-th coordinate}) \\ =& \sigma_j(x_2) \otimes \sigma_1(x_1) \otimes \cdots,\end{aligned}$$

and hence $\eta^{-1}\alpha\eta \in O(\otimes^2 L) \otimes O(\otimes^{m-2} L)$. Since $\eta^{-1}\alpha\eta = \mathrm{id}$ on $\eta^{-1}(K_1)$ and is $-\mathrm{id}$ on $\eta^{-1}(K_2)$ by (18), we have a contradiction to Lemma 7.2.5 as above. □

Lemma 7.2.8. *Let K, X, Y be positive lattices. Suppose that K is indecomposable with* rank $K > 1$ *and $\sigma : K \otimes X \cong K \otimes Y$, and that there exist submodules M_0, M of X and N_0, N of Y satisfying the conditions (i), (ii), (iii) below. Then $X = M_0 \perp M$, $Y = N_0 \perp N$ and $X \cong Y$.*

(i) *M_0, M (resp. N_0, N) are primitive submodules of X (resp. Y) (unless they are $\{0\}$) such that*

$$[X : M_0 \perp M] < \infty, \quad [Y : N_0 \perp N] < \infty,$$

$$\sigma(K \otimes M_0) = K \otimes N_0, \quad \sigma(K \otimes M) = K \otimes N. \tag{19}$$

(ii) *There exist decompositions $M_0 = \perp_i M_{0,i}, N_0 = \perp_i N_{0,i}$ such that for every i*

$$\sigma = \alpha_i \otimes \beta_i : K \otimes M_{0,i} \cong K \otimes N_{0,i}$$
$$\text{for } \alpha_i \in O(K) \text{ and } \beta_i : M_{0,i} \cong N_{0,i}. \tag{20}$$

In particular, $M_0 \cong N_0$.

(iii) *$\sigma(K \otimes m) = K \otimes n$ $(m \in M, n \in N)$ implies $m = 0$; moreover there exists a basis $\{u_i\}$ of K, submodules M_i (resp. N_i) of M (resp. N) and an isometry $\mu : N \cong M$ such that for every i*

$$\sigma(K \otimes M_i) = u_i \otimes N, \; \sigma^{-1}(K \otimes N_i) = u_i \otimes M, \tag{21}$$

$$\mu(N_i) = M_i \text{ and } \sigma^{-1}(u_i \otimes n) = u_i \otimes \mu(n) \text{ for } n \in N_i. \tag{22}$$

Proof. Suppose $Y \neq N_0 \perp N$, and fix an arbitrary element $y \in Y$ such that

$$y \notin N_0 \perp N.$$

From (19), (21) we have $\sigma^{-1}(K \otimes N) = K \otimes M$ and $\sigma^{-1}(K \otimes N_i) = u_i \otimes M$ which imply

$$N = \oplus N_i \text{ and similarly } M = \oplus M_i. \tag{23}$$

Hence we can write, using $\mathbb{Q}Y = \mathbb{Q}N_0 \perp \mathbb{Q}N = \perp_i \mathbb{Q}N_{0,i} \perp (\oplus \mathbb{Q}N_i)$

$$y = \sum_i y_{0,i} + \sum_i y_i, \; (y_{0,i} \in \mathbb{Q}N_{0,i}, \; y_i \in \mathbb{Q}N_i). \tag{24}$$

By (21), (22) and $y_i \in \mathbb{Q}N_i$, we have, for $w \in K$

$$\sigma^{-1}(w \otimes y_i) = u_i \otimes \mu(w, y_i) \; (\mu(w, y_i) \in \mathbb{Q}M), \tag{25}$$

$$\mu(u_i, y_i) = \mu(y_i). \tag{26}$$

Then we have, for $w \in K$

$$\sigma^{-1}(w \otimes y) = \sigma^{-1}(\sum_i w \otimes y_{0,i} + \sum_i w \otimes y_i) \quad \text{by (24)}$$

$$= \sum_i \alpha_i^{-1}(w) \otimes \beta_i^{-1}(y_{0,i}) + \sum_i u_i \otimes \mu(w, y_i), \tag{27}$$

by (20) and (25). Recalling that $\{u_i\}$ is a basis of K and $y \in Y$, we can put, for $w \in K$,

$$\sigma^{-1}(w \otimes y) = \sum u_i \otimes x_i(w), \; x_i(w) \in X. \tag{28}$$

Now (27) and $\beta_i^{-1}(y_{0,i}) \in \mathbb{Q}M_{0,i} \subset \mathbb{Q}M_0$ (by (24), (20)) for every i imply $\sigma^{-1}(w \otimes y) - \sum_i u_i \otimes \mu(w, y_i) \in \mathbb{Q}(K \otimes M_0)$. Substituting (28) we then have

$$x_i(w) - \mu(w, y_i) \in \mathbb{Q}M_0 \quad \text{for } w \in K. \tag{29}$$

Suppose $\mu(y_i) \in M$ for all i; then $\mu(\sum y_i) \in M$ and hence

$$\sum y_i \in \mu^{-1}(M) = N \subset Y,$$

which implies

$$\sum_i y_{0,i} = y - \sum_i y_i \in Y \cap \mathbb{Q}N_0 = N_0$$

by (24), $y \in Y$ and (i). This gives $y = \sum_i y_{0,i} + \sum y_i \in N_0 \perp N$, contradicting the choice of y. Hence $\mu(y_i) \notin M$ for some i. We may assume $i = 1$ without loss of generality; then we have

(30) $$x_0 := \mu(y_1) \notin M.$$

Putting

(31) $$x := x_1(u_1) \in X$$

by the notation (28), $x - x_0 = x_1(u_1) - \mu(y_1) = x_1(u_1) - \mu(u_1, y_1) \in \mathbb{Q}M_0$ by (26), (29), and then we can write, using (ii)

(32) $$x - x_0 = \sum x_i \quad (x_i \in \mathbb{Q}M_{0,i}).$$

Suppose $x_i \in M_{0,i}$ for all i; then $x - x_0 = \sum x_i \in \perp M_{0,i} = M_0 \subset X$ implies $x_0 \in X$ by (31) and with it $x_0 = \mu(y_1) \in \mu(\mathbb{Q}N_1) = \mathbb{Q}M_1$ (by (24), (22)) implies $x_0 \in X \cap \mathbb{Q}M_1$. Since $M = \oplus M_i$ (by (23)) is primitive in X, M_1 is also primitive in X and then $x_0 \in X \cap \mathbb{Q}M_1 = M_1 \subset M$ holds. This contradicts (30) and hence for some h

(33) $$x_h \notin M_{0,h}.$$

Put, for the isometry $\alpha_i \in O(K)$,

(34) $$\alpha_i(u_k) = \sum_j a_{ikj} u_j \ (a_{ikj} \in \mathbb{Z}).$$

On the other hand, by (24), (22) $x_0 = \mu(y_1) \in \mathbb{Q}M_1$ which implies, by the left side of (21)

$$\sigma(u_k \otimes x_0) = u_1 \otimes \eta(u_k, x_0)$$

for some $\eta(u_k, x_0) \in \mathbb{Q}N$. Thus we have

$$\begin{aligned}
&\sigma(u_k \otimes x) \\
&\quad = \sigma(u_k \otimes x_0) + \sum_i \sigma(u_k \otimes x_i) \quad \text{by (32)} \\
&\quad = u_1 \otimes \eta(u_k, x_0) + \sum_i \alpha_i(u_k) \otimes \beta_i(x_i) \quad \text{by } x_i \in \mathbb{Q}M_{0,i} \text{ and (20)} \\
&\quad = u_1 \otimes \eta(u_k, x_0) + \sum_j u_j \otimes (\sum_i a_{ikj}\beta_i(x_i)) \quad \text{by (34).}
\end{aligned}$$

By (31), $u_k \otimes x \in K \otimes X$ and so $\sigma(u_k \otimes x) \in K \otimes Y$, and hence the above implies, for $j \geq 2$

$$\sum_i a_{ikj}\beta_i(x_i) \in Y.$$

Since $N_0 = \perp_i N_{0,i}$ and $\beta_i(x_i) \in \mathbb{Q}N_{0,i}$ by (32), (20), we have

$$\begin{aligned}\sum_i a_{ikj}\beta_i(x_i) &\in Y \cap \perp_i \mathbb{Q}N_{0,i} \\ &= Y \cap \mathbb{Q}N_0 = N_0 \quad \text{(by (i))} \\ &= \perp_i N_{0,i}\end{aligned}$$

and so we have $a_{ikj}\beta_i(x_i) \in N_{0,i}$, which yields by (20)

(35) $$a_{ikj}x_i \in M_{0,i} \text{ for } j \geq 2 \text{ and every } i, k.$$

Define a natural number s by $s\mathbb{Z} = \{a \in \mathbb{Z} \mid ax_h \in M_{0,h}\}$; then (33) implies $s > 1$, and from (35) follows

$$a_{hkj} \equiv 0 \bmod s \quad \text{for} \quad j \geq 2 \text{ and } k \geq 1.$$

Therefore (34) yields $\alpha_h(u_k) \equiv a_{hk1}u_1 \bmod sK$ for every $k \geq 1$. This contradicts the assumption $\operatorname{rank} K > 1$. Thus we have shown that the assumption $Y \neq N_0 \perp N$ induces a contradiction; hence $Y = N_0 \perp N$ and (19) yields $X = M_0 \perp M$. By (ii) (resp. (iii)), $M_0 \cong N_0$ (resp. $N \cong M$), and so $X \cong Y$. □

Proof of (B) ⇒ (A).

We use induction on $\operatorname{rank} X$. If $\operatorname{rank} X = 1$, then the assertion (A) is trivial. Assume that the assertion (A) is true if $\operatorname{rank} X \leq k$. Suppose $\operatorname{rank} X = k + 1$.

First we show (A2) by verifying the assumption in Lemma 7.2.8 is valid for $K := L$. Let Y be positive lattices and suppose $\sigma : L \otimes X \cong L \otimes Y$. Applying Lemma 7.2.4 to $\sigma : L \otimes X \cong L \otimes Y$, there exist submodules M_0, M of X and N_0, N of Y satisfying the conditions (i), (ii) in Lemma 7.2.8 for $K := L$ and

(36) $$\sigma(L \otimes m) = L \otimes n \ (m \in M, n \in N) \text{ implies } m = 0.$$

Since $\sigma(L \otimes M) = L \otimes N$ by (19), and (36) is just the assumption in (B), Lemma 7.2.1 implies $M \cong N$. On the other hand, $M_0 \cong N_0$ follows from (ii) of Lemma 7.2.8. If $M = 0$, then the condition (i) in Lemma 7.2.8 yields $M_0 = X$. So $L \otimes Y = \sigma(L \otimes X) = \sigma(L \otimes M_0) = L \otimes N_0$

yields $Y = N_0$. Hence $X = M_0 \cong N_0 = Y$. Similarly, $M_0 = 0$ implies $X = M \cong N = Y$. Thus we may assume $M \neq 0, M_0 \neq 0$ and hence $1 \leq \operatorname{rank} M$, $\operatorname{rank} M_0 \leq k = \operatorname{rank} X - 1$. To show $X \cong Y$, we have only to verify the condition (iii) in Lemma 7.2.8. Since $M \cong N$, we may assume $M = N$ and write

$$M = \perp K_i, \quad K_i = \otimes^{r_i} L \otimes H_i \tag{37}$$

where K_i is indecomposable and H_i is not divisible by L. Since $\operatorname{rank} K_i \leq \operatorname{rank} M \leq k$, using the assumption (A1) of the induction for K_i we see that $L \otimes K_i$ is indecomposable. By $\sigma(L \otimes M) = L \otimes M$ and Theorem 6.7.1, we have $\sigma(L \otimes K_i) = L \otimes K_{s(i)}$ for some permutation s. Then, by the assumption (A2) of the induction, $K_i \cong K_j$ $(j = s(i))$ holds. Using (A2) repeatedly, we have $r_i = r_j, H_i \cong H_j$. By the assumption (A3) of the induction, $O(L \otimes K_i)$ is generated by $O(L), O(H_i)$ and interchanges of L. Then noting $H_i \cong H_j$, we see $\sigma : L \otimes K_i \cong L \otimes K_j$ is of form

$$\sigma(x_1 \otimes \cdots \otimes x_{r_i+1} \otimes x) = \sigma_1(x_{\mu(1)}) \otimes \cdots \otimes \sigma_{r_i+1}(x_{\mu(r_i+1)}) \otimes \beta(x)$$

for $x_k \in L$, $x \in H_i$, $\sigma_k \in O(L)$, $\beta : H_i \cong H_j$ and a permutation μ. (36) implies $\mu(1) \neq 1$. Hence for any fixed basis $\{v_i\}$ of L, we can define submodules $M_{i,h}$ of $K_i, N_{j,h}$ of K_j by

$$\begin{aligned} \sigma(v_h \otimes K_i) &= \sigma(v_h \otimes L \otimes \cdots \otimes L \otimes H_i) = L \otimes N_{j,h}, \\ \sigma^{-1}(v_h \otimes K_j) &= \sigma^{-1}(v_h \otimes L \otimes \cdots \otimes L \otimes H_j) = L \otimes M_{i,h}, \end{aligned} \tag{38}$$

that is

$$\begin{aligned} M_{i,h} &= L \otimes \cdots \otimes L \otimes \sigma_1^{-1}(v_h) \otimes L \otimes \cdots \otimes L \otimes H_i \text{ and} \\ N_{j,h} &= L \otimes \cdots \otimes L \otimes \sigma_{\mu^{-1}(1)}(v_h) \otimes L \otimes \cdots \otimes L \otimes H_j, \end{aligned}$$

where $\sigma_1^{-1}(v_h)$, resp. $\sigma_{\mu^{-1}(1)}(v_h)$ are on the $(\mu(1) - 1)$th, resp. $(\mu^{-1}(1) - 1)$th, coordinate, respectively. Hence we have

$$\sigma(v_h \otimes M_{i,h}) = v_h \otimes N_{j,h}. \tag{39}$$

Putting $M_h = \oplus_i M_{i,h} \subset \oplus K_i = M, N_h = \oplus_j N_{j,h} \subset \oplus K_j = M = N$, we have, upon summing (38), (39) on $i, j = s(i)$

$$\sigma(v_h \otimes M) = L \otimes N_h, \ \sigma^{-1}(v_h \otimes M) = L \otimes M_h, \ \sigma(v_h \otimes M_h) = v_h \otimes N_h,$$

which satisfy the conditions (B2) and (B3). Corollary 7.2.1 shows the existence of the isometry μ in the condition (iii) of Lemma 7.2.8 and then it shows $X \cong Y$. Thus the condition (A2) has been shown.

Next we show the condition (A3). Let X be an indecomposable positive lattice with $\operatorname{rank} X = k+1$ as before, and $X = \otimes^t L \otimes X'$ where X' is not divisible by L. Take $\sigma \in O(L \otimes X)$ and apply the above to $\sigma : L \otimes X \cong L \otimes X$; as above, there exist submodules M_0, M, N_0, N of X satisfying the conditions in Lemma 7.2.8. Since X is indecomposable, then only one of the conditions (ii) or (iii) can happen.

We show that

$$(*) : \sigma(L \otimes x) = L \otimes y \text{ for some non-zero } x, y \in X,$$

in particular the condition (ii) is valid, then σ is expressed by $O(L), O(X')$ and interchanges of L.

Suppose that $\sigma(L \otimes x) = L \otimes y$ for some non-zero $x, y \in X$; then $X \neq M$ and hence $X = M_0$, so we can write $\sigma = \alpha \otimes \beta$ for $\alpha \in O(L)$, $\beta \in O(X)$. Since $X = L \otimes (\otimes^{t-1} L \otimes X')$ and $\operatorname{rank}(\otimes^{t-1} L \otimes X') \leq k$, we can use the assumption (A3) of the induction. Then β is generated by $O(L)$, $O(X')$ and interchanges of L and hence $\sigma = \alpha \otimes \beta$ is also.

Next assume that the condition (iii) is satisfied and that σ is not in the subgroup G generated by $O(L), O(X')$ and interchanges of L; then by Lemma 7.2.1, there exist $x_1 (\neq 0) \in L$ and $X_1 \subset X$ such that $\sigma(L \otimes X_1) = x_1 \otimes X$. Define an isometry $\mu_2 \in O(L \otimes X)$ by

$$\mu_2(x_1 \otimes x_2 \otimes \cdots) = x_2 \otimes x_1 \otimes \cdots .$$

We have

$$\mu_2 \sigma(L \otimes X_1) = L \otimes (x_1 \otimes (\otimes^{t-1} L) \otimes X'). \tag{40}$$

If $\mu_2 \sigma(L \otimes x_2') = L \otimes y$ for $x_2' \in X_1$ $(x_2' \neq 0)$, then (*) shows $\mu_2 \sigma \in G$ and hence $\sigma \in G$, which contradicts the assumption $\sigma \notin G$. Hence the condition (iii) in Lemma 7.2.8 is valid for $\mu_2 \sigma$ in (40) by $\operatorname{rank} X_1 \leq k$, and by Lemma 7.2.1, there exist $0 \neq x_2 \in L$ and $X_2 \subset X_1$ satisfying $\mu_2 \sigma(L \otimes X_2) = x_2 \otimes (x_1 \otimes (\otimes^{t-1} L) \otimes X')$. Denoting by μ_3 the isometry of interchanging the first and the third coordinates, we have

$$\mu_3 \mu_2 \sigma(L \otimes X_2) = L \otimes (x_1 \otimes x_2 \otimes (\otimes^{t-2} L) \otimes X'),$$

which is similar to (40). Repeating the above argument, we have

$$\mu \sigma(L \otimes X_{t+1}) = x_1 \otimes \cdots \otimes x_{t+1} \otimes X'$$

for some interchange μ of L and some submodule $X_{t+1} \subset X$ and non-zero $x_1, \cdots, x_{t+1} \in L$. This contradicts the fact that X' is not divisible by L. Thus we have proved the assertion (A3).

Lastly we show (A1). Let X be an indecomposable positive lattice with $\operatorname{rank} X = k+1$, and write $X = \otimes^t L \otimes X'$ where X' is not divisible by L. By the proved condition (A3), $O(L\otimes X)$ is generated by $O(L), O(X')$ and interchanges of L and hence $O(L\otimes X) = O(\otimes^{t+1}L)\otimes O(X')$. Hence $O(\otimes^{t+1}L)$ is also generated by $O(L)$ and interchanges of L. The indecomposability of X implies that of $\otimes^t L$. Applying Lemma 7.2.7 to $m = t+1$, $\otimes^{t+1}L$ is indecomposable, and then applying Lemma 7.2.5 to $M = \otimes^{t+1}L, N = X'$, we have the indecomposability of $L\otimes X$ by virtue of $O(L\otimes X) = O(\otimes^{t+1}L)\otimes O(X')$.
□

7.3 Weighted graphs

In this section, we show the uniqueness of decompositions of a certain kind of graph, which is applied to the uniqueness of decompositions of a certain kind of positive lattice with respect to tensor product. Let A be a finite set and $[\,,]$ be a mapping from $A \times A$ to $\{t \mid 0 \le t \le 1\}$ satisfying the following conditions for $a, a' \in A$:

(i) $[a, a'] = 1$ if and only if $a = a'$,

(ii) $[a, a'] = [a', a]$.

We call $(A,[\,,])$ or simply A a *weighted graph.* A weighted graph A is called *connected* unless $A = A_1 \cup A_2$ (disjoint), $A_i \neq \emptyset$ and $[a_1, a_2] = 0$ for $a_i \in A_i$. It is easy to see that A is connected if and only if for any $x, y \in A$ there exist $a_1 := x, a_2, \cdots, a_m := y \in A$ such that $[a_i, a_{i+1}] \neq 0$ for $i = 1, 2, \cdots m-1$. If a subset B of A is maximal connected with respect to inclusion, then B is a *connected component* of A. So, if B is a connected component of A, then for $a \in A$ and $b \in B$, $[a, b] \neq 0$ implies $a \in B$. If A, B are weighted graphs, then $A \times B$ becomes a weighted graph by

$$[(a_1, b_1), (a_2, b_2)] := [a_1, a_2][b_1, b_2].$$

If a bijection σ from A to B satisfies $[x, y] = [\sigma(x), \sigma(y)]$ for $x, y \in A$, then we call σ an *isometry* and write $\sigma : A \cong B$.

We give an example which is important to us. Let L be a positive lattice and for $x, y \in \mathcal{M}(L)$, we put $[x, y] := |B(x, y)|/\min(L)$. Since

$$Q(tx + y) = Q(x)t^2 + 2B(x, y)t + Q(y) \ge 0 \text{ for all } t \in \mathbb{Q},$$

we get $(\min L)^2 = Q(x)Q(y) \ge B(x, y)^2$ and so $0 \le [x, y] \le 1$. If $[x, y] = 1$, then $B(x, y) = \pm Q(x)$ and so $Q(tx + y) = (t \pm 1)^2 \min L$, which yields $x = \pm y$. Thus $\mathcal{M}(L)/\pm 1$ becomes a weighted graph. We call it the weighted graph induced by a positive lattice L.

Let $\tilde{L}$ be the submodule of L spanned by the elements of $\mathcal{M}(L)$. We show that $\tilde{L}$ is indecomposable if and only if the graph $\mathcal{M}(L)/\pm 1$ is connected.

If $\tilde{L} = L_1 \perp L_2$, $L_1 \neq 0$, $L_2 \neq 0$, then $x \in \mathcal{M}(L)$ is in L_1 or L_2. Hence putting $G_i := \{x \in \mathcal{M}(L) \mid x \in L_i\}$, we find $\mathcal{M}(L) = G_1 \cup G_2$. If, for instance $G_1 = \emptyset$, then $\mathcal{M}(L) = G_2 \subset L_2$ and then $\tilde{L} \subset L_2$. This is a contradiction. Thus $G_1 \neq \emptyset$ and similarly $G_2 \neq \emptyset$. By definition of L_1, L_2, $[G_1, G_2] = 0$ is clear. Thus $\mathcal{M}(L)/\pm 1$ is not connected.

Conversely suppose $\mathcal{M}(L)/\pm 1$ is not connected; then there exist subsets H_1, H_2 of $\mathcal{M}(L)$ such that

$$H_i = -H_i\,, \mathcal{M}(L) = H_1 \cup H_2,\ B(H_1, H_2) = 0.$$

Let M_i be the submodule of L spanned by the elements of H_i; then $M_1 \perp M_2 = \tilde{L}$ and so $\tilde{L}$ is not indecomposable. □

Lemma 7.3.1. *Let $A = \{e_1, \cdots, e_n\}, A', B, C$ be weighted graphs and suppose $\sigma : A \times B \cong A' \times C$. Take and fix any element $b \in B$. We define $f_i \in A'$, $c_i \in C$, $g_{ij} \in A$, $b_{ij} \in B$ by*

$$\sigma(e_i, b) = (f_i, c_i),\ \sigma(g_{ij}, b_{ij}) = (f_i, c_j)\ \ \text{for}\ 1 \le i, j \le n.$$

Then $b \neq b_{ij}$ yields $[e_i, e_j] = 0$.

Proof. Put $a_{ij} = [e_i, e_j]$; then the property (ii) means $a_{ij} = a_{ji}$, and

$$\begin{aligned} a_{ij} =& [e_i, e_j] = [(e_i, b), (e_j, b)] = [(f_i, c_i), (f_j, c_j)] \\ =& [f_i, f_j][c_i, c_j]. \end{aligned} \tag{1}$$

We must show that $b \neq b_{ij}$ implies $a_{ij} = 0$. Choose $1 \le u, m \le n$ with $b \neq b_{um}$ such that $a_{um} \ge a_{ij}$ if $b \neq b_{ij}$. We have only to conclude the contradiction, supposing $a_{um} \neq 0$. Put $g_{um} = e_p$, $b_{um} = b'$; then

$$\sigma(e_p, b') = (f_u, c_m),\ b' \neq b. \tag{2}$$

$$\begin{aligned} a_{ip}[b, b'] &= [e_i, e_p][b, b'] = [(e_i, b), (e_p, b')] \\ &= [(f_i, c_i), (f_u, c_m)] = [f_i, f_u][c_i, c_m]. \end{aligned} \tag{3}$$

$$a_{ip}[b, b'] = [c_i, c_m] \quad \text{if } f_i = f_u \ \ (\text{by } (3))\,. \tag{4}$$

$$a_{mp}[b, b'] = [f_m, f_u] \quad (\text{putting } i = m \text{ in } (3)). \tag{5}$$

Hence we have

$$\begin{aligned} a_{um} &= [f_u, f_m][c_u, c_m] \quad \text{by } (1) \\ &= a_{up} a_{mp} [b, b']^2 \quad \text{by } (4), (5). \end{aligned} \tag{6}$$

We first show

$$f_p \neq f_u. \tag{7}$$

Suppose $f_u = f_p$; then (2) implies $\sigma(e_p, b') = (f_p, c_m)$. If $b = b_{pm}$, then it implies $\sigma(g_{pm}, b) = \sigma(g_{pm}, b_{pm}) = (f_p, c_m) = \sigma(e_p, b')$ and so $b = b'$ which contradicts (2). Thus we have $b \neq b_{pm}$ and then the definition of u, m implies $a_{um} \geq a_{pm}$, which yields

$$\begin{aligned} 0 < a_{um} &= a_{up}a_{mp}[b, b']^2 \quad \text{by (6)} \\ &\leq a_{up}a_{um}[b, b']^2 \leq a_{um}. \end{aligned}$$

Hence we get $[b, b'] = 1$, contradicting (2). Thus we have shown (7). If $b_{up} \neq b$, then $a_{up} \leq a_{um}$ and so

$$\begin{aligned} 0 < a_{um} &= a_{up}a_{mp}[b, b']^2 \quad \text{by (6)} \\ &\leq a_{um}a_{mp}[b, b']^2 \leq a_{um}, \end{aligned}$$

which also induces $[b, b'] = 1$ contradicting (2). Thus we have

(8) $$b_{up} = b \quad \text{and} \quad \sigma(g_{up}, b) = (f_u, c_p).$$

Now, we define t by $g_{up} = e_t$; then $\sigma(e_t, b) = (f_t, c_t)$ and (8) yield

(9) $$f_u = f_t, \ c_p = c_t;$$

then (4) implies

(10) $$a_{tp}[b, b'] = [c_t, c_m].$$

Substituting $i = p$ to (3), we have

(11) $$[b, b'] = [f_p, f_u][c_p, c_m].$$

Thus we get

$$\begin{aligned} [c_p, c_m] &= [c_t, c_m] \quad \text{by (9)} \\ &= a_{tp}[b, b'] \quad \text{by (10)} \\ &= a_{tp}[f_p, f_u][c_p, c_m] \quad \text{by (11)} \end{aligned}$$

and $[c_p, c_m](1 - a_{tp}[f_p, f_u]) = 0$. If $[c_p, c_m] \neq 0$, then $a_{tp}[f_p, f_u] = 1$ and hence $[f_p, f_u] = 1$, which contradicts (7). If $[c_p, c_m] = 0$, then $[b, b'] = 0$ by (11) and (6) induces $a_{um} = 0$. This contradicts the assumption $a_{um} \neq 0$, which completes the proof. □

Lemma 7.3.2. *Let A, A', B and C be weighted graphs and suppose that $A = \{e_1, \cdots, e_n\}$ is connected and $\sigma : A \times B \cong A' \times C$. Fix any element $b \in B$. Putting $\sigma(e_i, b) = (f_i, c_i)$, we have*

$$A \cong \{\sigma(e_i, b) \mid 1 \leq i \leq n\} = \{(f_i, c_j) \mid 1 \leq i, j \leq n\}.$$

Proof. For $1 \leq i \leq n$, we put $C_i = \{c_k \mid f_k = f_i\}$. Clearly $c_i \in C_i$; denote by C_i' the connected component of C_i containing c_i. Here we note that $x \in C_i', y \in C_i$ and $[x, y] \neq 0$ yield $y \in C_i'$.

We first show that $[e_i, e_j] = [f_i, f_j][c_i, c_j] \neq 0$ induces $C_i' = C_j'$. Suppose

$$(12) \qquad [e_i, e_j] = [f_i, f_j][c_i, c_j] \neq 0.$$

Then by Lemma 7.3.1, we have $\sigma(e_t, b) = (f_i, c_j)$ for some t. Hence $\sigma(e_t, b) = (f_t, c_t)$ implies $c_j = c_t, f_i = f_t$ and so $c_t \in C_i$. Now $[e_i, e_j] \neq 0$ implies $[c_i, c_j] \neq 0$ and hence $C_i' \ni c_j$ since $c_i \in C_i'$ and $c_j = c_t \in C_i$. Thus we get $(c_j \in) \, C_i' \cap C_j' \neq \emptyset$.

Next we show that $x \in C_i' \cap C_j', \;\; y \in C_j'$ and $[x, y] \neq 0$ imply $y \in C_i'$. Let x and b be above; then by definition $x \in C_i$ and $y \in C_j$ imply

$$(13) \qquad x = c_u, f_u = f_i, y = c_k \text{ and } f_k = f_j \text{ for some } u, k.$$

They give

$$\begin{aligned} [e_u, e_k] &= [(e_u, b), (e_k, b)] = [(f_u, c_u), (f_k, c_k)] \\ &= [f_u, f_k][c_u, c_k] = [f_i, f_j][x, y] \text{ by (13)}, \end{aligned}$$

which yields $[e_u, e_k] \neq 0$ by virtue of the assumptions $[e_i, e_j] \neq 0$ and $[x, y] \neq 0$. Applying Lemma 7.3.1 to $[e_u, e_k] \neq 0$, we have $\sigma(e_s, b) = (f_u, c_k)$ for some s. Then $\sigma(e_s, b) = (f_s, c_s)$ implies, by virtue of (13), $f_s = f_u = f_i$ and hence

$$(14) \qquad y = c_k = c_s \in C_i.$$

Now $[y, x] \neq 0$ by the assumption, $x \in C_i'$ and (14) imply $y \in C_i'$, which is what we wanted.

As C_j' is connected, for any fixed $y_1 \in C_j'$, there exist $y_2, \cdots, y_m \in C_j'$ such that

$$[y_i, y_{i+1}] \neq 0 \quad (i = 1, 2, \cdots, m-1), \; y_m = c_j \; (\in C_j').$$

As proved as above, $c_j \in C_i' \cap C_j'$, and applying what we have proved to $x = c_j, y = y_{m-1}$, we have $y_{m-1} \in C_i' \cap C_j'$; repeating this argument we get $y_{m-2}, \cdots, y_1 \in C_i' \cap C_j'$. Thus $C_j' \subset C_i'$. Changing i and j, we have the converse inclusion and have proved $C_i' = C_j'$ if $[e_i, e_j] \neq 0$.

Since A is connected, we get $C_1' = \cdots = C_n'$. For $1 \leq s, t \leq n$, we have $c_t \in C_t' = C_s' \subset C_s$ and then by definition of C_s we find $c_t = c_k, f_k = f_s$ for some k. Finally, we get $(f_s, c_t) = (f_k, c_k) = \sigma(e_k, b)$, which completes the proof of the lemma. □

Lemma 7.3.3. *Let A, B and C be connected weighted graphs and suppose $\sigma : A \times B \cong A \times C$. If there exist $b_0 \in B, c_0 \in C$ such that $\sigma(x, b_0) = (f(x), c_0)$ for any $x \in A$, where f is a mapping from A to A, then f is an isometry from A to A and there exists an isometry $g : B \cong C$ such that $\sigma(x, y) = (f(x), g(y))$.*

Proof. Since A is a finite set and σ is injective, f is injective and hence bijective. For $a, a' \in A$, $[a, a'] = [(a, b_0), (a', b_0)] = [f(a), f(a')]$ implies that f is an isometry. Composing $f^{-1} \times \mathrm{id}$ with σ, we may assume $f = \mathrm{id}$. We consider a subset B_0 of B having the property

$$\sigma(a, b) = (a, g(b)) \quad \text{for} \quad a \in A, b \in B_0, \tag{14}$$

where g is a mapping from B_0 to C. Now $B_0 = \{b_0\}$ is such a subset by hypothesis. We let B_0 be a maximal subset among all subsets having this property. Once $B_0 = B$ is shown, then g is clearly an isometry and the proof is finished.

Suppose $B_0 \neq B$. We have only to show the existence of $b' \notin B_0$ such that $\sigma(x, b') = (x, c)$ for every $x \in A$ and some $c \in C$, which would contradict the maximality of B_0. We put

$$C_0 := g(B_0), \text{ and } m = \max_{\substack{b \in B_0 \\ b' \notin B_0}} [b, b'].$$

Considering σ^{-1} if necessary, we may assume

$$m \geq \max_{\substack{c \in C_0 \\ c' \notin C_0}} [c, c']. \tag{15}$$

Since B is connected, $m > 0$. Fix $b \in B_0, b' \notin B_0$ such that $m = [b, b']$. We define $h(x) \in A, c(x) \in C$ for $x \in A$ by

$$\sigma(x, b') = (h(x), c(x)).$$

Let us show that $h(x) = x$ and $c(x)$ is independent of $x \in A$, which contradicts the maximality of B_0 will complete the proof. If $c(x) \in C_0$, then $c(x) = g(b_1)$ for some $b_1 \in B_0$, and from

$$\sigma(x, b') = (h(x), g(b_1)) = \sigma(h(x), b_1)$$

we get $b' = b_1$, contradicting the property $b' \notin B_0$. Thus $c(x) \notin C_0$, and

$$\begin{aligned} m &= [b, b'] = [(x, b), (x, b')] \\ &= [(x, g(b)), (h(x), c(x)] = [x, h(x)][g(b), c(x)] \\ &\leq [x, h(x)]m \quad \text{by } g(b) \in C_0, c(x) \notin C_0 \text{ and (15)} \\ &\leq m. \end{aligned}$$

Now $m > 0$ implies $[x, h(x)] = 1$ so $x = h(x)$. Thus we have

$$\sigma(x, b') = (x, c(x)) \text{ for } x \in A.$$

If $x, y \in A$ and $[x, y] \neq 0$, then

$$[x, y] = [(x, b'), (y, b')] = [(x, c(x)), (y, c(y))] = [x, y][c(x), c(y)]$$

yields $[c(x), c(y)] = 1$, i.e. $c(x) = c(y)$. Since A is connected, it means that c is constant on A. This contradicts the maximality of B_0 and completes the proof. □

We say that a weighted graph A is *indecomposable* unless $A \cong B \times C$ for weighted graphs B, C with $\sharp B > 1, \sharp C > 1$.

Theorem 7.3.1. *Let A_i $(1 \leq i \leq m)$ and B_j $(1 \leq j \leq n)$ be connected and indecomposable weighted graphs with $\sharp A_i > 1, \sharp B_j > 1$ for all i, j, and assume $\sigma : \prod_{i=1}^{m} A_i \cong \prod_{j=1}^{n} B_j$. Then we have $m = n$ with σ a product of $\sigma_i : A_i \cong B_{s(i)}$ for some permutation s of $\{1, \cdots, m\}$.*

Proof. Changing suffices and taking σ^{-1} if necessary, we may suppose $\sharp A_1 \geq \sharp A_i, \sharp B_j$ for all i, j. For each $i \geq 2$, take and fix an element $e_i \in A_i$. Changing suffices if necessary, we may suppose that the projection of $\sigma(A_1, e_2, \cdots, e_m)$ to B_1 contains at least two distinct elements by the assumption that $\sharp A_1 > 1$. Applying Lemma 7.3.2 to $A = A_1, B = \prod_{i \geq 2} A_i$, $A' = B_1$ and $C = \prod_{i \geq 2} B_i$, we have

$$(16) \qquad A_1 \cong \sigma(A_1, e_2, \cdots, e_m) = A_0 \times C_0 \quad (A_0 \subset B_1, C_0 \subset \prod_{i \geq 2} B_i).$$

By hypothesis, $\sharp A_0 > 1$ and the assumption that A_1 is indecomposable implies $\sharp C_0 = 1$. Therefore $\sharp A_1 = \sharp A_0 \leq \sharp B_1$ yields $\sharp A_1 = \sharp B_1$ and hence $A_0 = B_1$ since $\sharp A_1 \geq \sharp B_j$. So (16) induces an isometry from A_1 to B_1. Applying Lemma 7.3.3 to $\sigma : A_1 \times B \cong B_1 \times C$, we have $\sigma = f \times g$ for

$$f : A_1 \cong B_1, \quad g : \prod_{i \geq 2} A_i \cong \prod_{j \geq 2} B_j.$$

The theorem is now inductively proved. □

7.4 The tensor product of positive lattices

For a positive lattice L, $\tilde{L}$ denotes a submodule of L spanned by $\mathcal{M}(L)$. In general, $\operatorname{rank} \tilde{L} \leq \operatorname{rank} L$.

Lemma 7.4.1. *Let L be a positive lattice of E-type such that $[L:\tilde{L}] < \infty$ and $\tilde{L}$ is indecomposable. Then L is indecomposable, and if $\sigma : L \otimes M \cong L \otimes N$ for positive lattices M, N, then for $m \in \mathcal{M}(M)$, we have $\sigma(L \otimes m) = F \otimes G$ for some primitive submodules $F \subset L, G \subset N$ satisfying* $\min F = \min L, \min G = \min N$.

Proof. Suppose $L = L_1 \perp L_2$; then any minimal vector of L is in L_1 or L_2. Put $\mathcal{M}_i = \{v \in \mathcal{M}(L) \mid v \in L_i\}$ for $i = 1, 2$; then $\mathcal{M}(L) = \mathcal{M}_1 \cup \mathcal{M}_2$ and for a submodule K_i of L_i spanned by $\mathcal{M}_i$ we have $\tilde{L} = K_1 \perp K_2$. Since $\tilde{L}$ is indecomposable, K_1 or $K_2 = 0$, which means $\mathcal{M}_1$ or $\mathcal{M}_2 = \emptyset$. Thus $\mathcal{M}(L) \subset L_2$ or L_1 and hence $\tilde{L} \subset L_2$ or L_1. Since $[L : \tilde{L}] < \infty$ means $\operatorname{rank} L = \operatorname{rank} \tilde{L}$, we have L_1 or $L_2 = 0$. Therefore L is indecomposable.

Since L is of E-type, $\sigma(\mathcal{M}(L) \otimes \mathcal{M}(M)) = \mathcal{M}(L) \otimes \mathcal{M}(N)$ and hence σ canonically induces an isometry from $A \times B$ to $A \times C$ as weighted graphs where $A = \mathcal{M}(L)/\pm 1, B = \mathcal{M}(M)/\pm 1, C = \mathcal{M}(N)/\pm 1$. Since $\tilde{L}$ is indecomposable, A is connected. Applying Lemma 7.3.2, we have $\sigma(\mathcal{M}(L) \otimes m) = \{x \otimes y \mid x \in F_1, y \in G_1\}$ for some subsets $F_1 \subset \mathcal{M}(L), G_1 \subset \mathcal{M}(N)$. Denoting by F', G' the submodules spanned by F_1, G_1, we have $\sigma(\tilde{L} \otimes m) = F' \otimes G'$. Put $F = \mathbb{Q}F' \cap L, G = \mathbb{Q}G' \cap N$; then $F \otimes G$ is a primitive submodule of $L \otimes N$ and $[F \otimes G : F' \otimes G'] < \infty$. Also $L \otimes m$ is a primitive submodule of $L \otimes M$ and $[L \otimes m : \tilde{L} \otimes m] < \infty$. Then $\sigma(\tilde{L} \otimes m) = F' \otimes G'$ induces $\sigma(L \otimes m) = F \otimes G$. Now $F' \subset F \subset L$ and F' contains the subset F_1 of $\mathcal{M}(L)$ and so $\min L = \min F$, similarly $\min G = \min N$. □

Corollary 7.4.1. *Let L be a positive lattice of E-type such that $[L : \tilde{L}] < \infty$ and $\tilde{L}$ is indecomposable. If $L = L_1 \otimes L_2$ for positive lattices L_1, L_2, then L_1, L_2 are of E-type and $[L_i : \tilde{L}_i] < \infty$ and $\tilde{L}_i$ is indecomposable for $i = 1, 2$.*

Proof. We define an isometry $\sigma \in O(L_1 \otimes L_2 \otimes L_2)$ by

$$\sigma(x \otimes y \otimes z) = x \otimes z \otimes y \ (x \in L_1, y, z \in L_2).$$

For $m \in \mathcal{M}(L_2)$, $\sigma(L \otimes m) = (L_1 \otimes m) \otimes L_2$, and applying Lemma 7.4.1, we have $\min(L_1 \otimes m) = \min(L)$. From (iii) of Lemma 7.1.1 it follows that $L_1 \otimes m$ is of E-type and so L_1 is of E-type. Similarly L_2 is of E-type. Hence $\mathcal{M}(L) = \mathcal{M}(L_1) \otimes \mathcal{M}(L_2)$ and then $\tilde{L} = \tilde{L}_1 \otimes \tilde{L}_2$. Now the indecomposability of $\tilde{L}$ and the property $[L : \tilde{L}] < \infty$ induce those of L_1, L_2. □

For a positive lattice L, we say that L is *multiplicatively indecomposable* if $L \cong K_1 \otimes K_2$ for positive lattices K_1, K_2 implies $\operatorname{rank} K_1$ or $\operatorname{rank} K_2 = 1$.

Theorem 7.4.1. *Let L be a positive lattice of E-type such that $[L : \tilde{L}] < \infty$, $\tilde{L}$ is indecomposable and L is multiplicatively indecomposable. Then for any indecomposable positive lattice X,*

(i) *$L \otimes X$ is indecomposable,*
(ii) *$L \otimes X \cong L \otimes Y$ yields $X \cong Y$ for a positive lattice Y,*
(iii) *if $X = \otimes^t L \otimes X'$ and X' is not divisible by L, then $O(L \otimes X)$ is generated by $O(L), O(X')$ and interchanges of L.*

Proof. If $\operatorname{rank} L = 1$ then the assertion is clear. Assume $\operatorname{rank} L > 1$. We have only to prove that L is indecomposable and L satisfies the condition (B) in the Fundamental Lemma in 7.2.

Lemma 7.4.1 shows the indecomposability of L.

To prove the condition (B), we use induction on $\operatorname{rank} M$ for the lattice M in the condition. Put $\mathcal{M}(L) = \{\pm v_i\}_{i=1}^t$, and for positive lattices M, N assume $\sigma : L \otimes M \cong L \otimes N$ and that $\sigma(L \otimes m) = L \otimes n$ ($m \in M, n \in N$) yields $m = 0, n = 0$. Now $\{v_i\}$ satisfies the condition (B1). If $\operatorname{rank} M = 1$, then this condition is not satisfied and hence we have $\operatorname{rank} M > 1$. Put

$$M^1 := \tilde{M}^\perp, \;\; M^2 := (M^1)^\perp,$$
$$N^1 := \tilde{N}^\perp, \;\; N^2 := (N^1)^\perp.$$

Then M^i, N^i ($i = 1, 2$) are primitive submodules; moreover $\mathbb{Q}M^2 = \mathbb{Q}\tilde{M}$, and $\mathbb{Q}N^2 = \mathbb{Q}\tilde{N}$ are clear. Since L is of E-type, we have $\sigma(\tilde{L} \otimes \tilde{M}) = \tilde{L} \otimes \tilde{N}$, which implies

(1)
$$\sigma(L \otimes M^i) = L \otimes N^i \; (i = 1, 2), [M : M^1 \perp M^2] < \infty, [N : N^1 \perp N^2] < \infty.$$

$\tilde{M} \neq 0$ implies $M^2 \neq 0$.

First we assume $M^1 \neq 0$; then $\operatorname{rank} M^i < \operatorname{rank} M$ for $i = 1, 2$. Put

$$M_j^i = \{m \in M^i \mid \sigma(L \otimes m) \subset v_j \otimes N^i\},$$
$$N_j^i = \{n \in N^i \mid \sigma^{-1}(L \otimes n) \subset v_j \otimes M^i\}$$

for $i = 1, 2$ and $j = 1, 2, \cdots, t$. Using the induction hypothesis, we have, by (B2)

(2) $$\operatorname{rank} M_j^i = \operatorname{rank} N_j^i = \operatorname{rank} M^i / \operatorname{rank} L \; (i = 1, 2, \; j = 1, 2, \cdots, t).$$

It then follows from (B3) that

(3) $$\sigma(\mathbb{Q}(v_j \otimes M_j^i)) = \mathbb{Q}(v_j \otimes N_j^i) \; (i = 1, 2, \; j = 1, 2, \cdots, t).$$

Putting

$$M_j = \{m \in M \mid \sigma(L \otimes m) \subset v_j \otimes N\},$$
$$N_j = \{n \in N \mid \sigma^{-1}(L \otimes n) \subset v_j \otimes M\},$$

we have

(4) $$M_j^1 \perp M_j^2 \subset M_j,\ N_j^1 \perp N_j^2 \subset N_j,$$

and hence

(5) $$\operatorname{rank} M_j \geq \operatorname{rank} M_j^1 + \operatorname{rank} M_j^2 = \operatorname{rank} M / \operatorname{rank} L$$

by (1), (2). We show that the inequality in (5) is indeed an equality. For an arbitrarily fixed j, we take a subset S of $\{v_i\}$ such that $v_j \in S$ and S is a basis of $\mathbb{Q}L$. By definition, $\sigma(L \otimes M_i) \subset v_i \otimes N$ is clear and then $\sum_{i \in S} v_i \otimes N = \oplus_{i \in S} v_i \otimes N$ yields $\sum_{i \in S} M_i = \oplus_{i \in S} M_i$, which implies

$$\operatorname{rank} M \geq \operatorname{rank} \oplus_{i \in S} M_i \geq (\sharp S) \operatorname{rank} M / \operatorname{rank} L$$

by (5). By the choice of S, $\sharp S = \operatorname{rank} L$ is valid and then the inequality in (5) becomes equality.

Similarly we get $\operatorname{rank} N_j = \operatorname{rank} N / \operatorname{rank} L$ which is nothing but the condition (B2). Moreover (4) implies

$$\mathbb{Q}M_j = \mathbb{Q}M_j^1 \perp \mathbb{Q}M_j^2 \text{ and } \mathbb{Q}N_j = \mathbb{Q}N_j^1 \perp \mathbb{Q}N_j^2.$$

Then (3) implies $\sigma(\mathbb{Q}(v_j \otimes M_j)) = \mathbb{Q}(v_j \otimes N_j)$ and (B3) is proved.

Next we assume $M^1 = 0$; then $N^1 = 0$ and $[M : \tilde{M}] < \infty, [N : \tilde{N}] < \infty$. Note that the situation is similar for $\sigma^{-1} : L \otimes N \cong L \otimes M$. We begin by showing that for $m \in \mathcal{M}(M)$ the following holds:

(6) $$\sigma(L \otimes m) \subset w \otimes N \quad \text{for some } w \in \mathcal{M}(L).$$

By virtue of Lemma 7.4.1, there exist, for $m \in \mathcal{M}(M)$, submodules $F \subset L, G \subset N$ satisfying $\sigma(L \otimes m) = F \otimes G$, $\min F = \min L$ and $\min G = \min N$. Since L is multiplicatively indecomposable, either $\operatorname{rank} F$ or $\operatorname{rank} G$ must be one. If $\operatorname{rank} G = 1$, then putting $G = \mathbb{Z}g$ we have $\sigma(L \otimes m) = F \otimes G \subset L \otimes g$, and $Q(g) = \min G = \min N = \min M = Q(m)$ implies $\mathrm{d}(L \otimes m) = \mathrm{d}(L \otimes g)$ and so $\sigma(L \otimes m) = L \otimes g$, which contradicts the assumption. Therefore we get $\operatorname{rank} F = 1$, and putting $F = \mathbb{Z}w$, $\min L = \min F$ yields $w \in \mathcal{M}(L)$ which is simply (6).

For $u \in \mathcal{M}(L)$ and $M_u := \{m \in M \mid \sigma(L \otimes m) \subset u \otimes N\}$, we show

(7) $$u \otimes \tilde{N} \subset \sigma(L \otimes M_u) \subset u \otimes N.$$

For $n \in \mathcal{M}(N)$, write $\sigma^{-1}(u \otimes n) = v \otimes m$ $(v \in \mathcal{M}(L), m \in \mathcal{M}(M))$; then by virtue of the above $\sigma(L \otimes m) \subset v' \otimes N$ holds for $v' \in \mathcal{M}(L)$.

Then $\sigma(v \otimes m) = u \otimes n$ yields $v' = \pm u$ and hence $m \in M_u$. Thus we have $u \otimes n = \sigma(v \otimes m) \in \sigma(v \otimes M_u) \subset \sigma(L \otimes M_u) \subset u \otimes N$, which immediately yields (7).

The assumption $[N : \tilde{N}] < \infty$ then implies $\operatorname{rank} L \operatorname{rank} M_u = \operatorname{rank} N$. Similarly we have $\operatorname{rank} L \operatorname{rank} N_u = \operatorname{rank} M$ which yields the condition (B2).

In the above, $m \in M_u$ implies

$$\min(M_u) = \min(M) \text{ and } \mathcal{M}(M_u) \subset \mathcal{M}(M).$$

Similarly, by considering $\sigma^{-1} : L \otimes N \cong L \otimes M$, we have $\mathcal{M}(N_u) \subset \mathcal{M}(N)$, where $N_u := \{n \in N \mid \sigma^{-1}(L \otimes n) \subset u \otimes M\}$. Starting with $u \in \mathcal{M}(L), n \in \mathcal{M}(N_u) \subset \mathcal{M}(N)$, we put $\sigma^{-1}(u \otimes n) = v \otimes m$ for $v \in \mathcal{M}(L), m \in \mathcal{M}(M)$. Then $n \in N_u$ implies $\sigma^{-1}(L \otimes n) \subset u \otimes M$ and hence $\sigma^{-1}(u \otimes n) = v \otimes m$ gives $v = \pm u$ and $\sigma(L \otimes m) \subset u \otimes N$ by (6), i.e. $m \in M_u$. Thus we have $\sigma^{-1}(u \otimes \mathcal{M}(N_u)) \subset u \otimes \mathcal{M}(M_u)$ and

$$\sigma^{-1}(u \otimes \widetilde{N_u}) \subset u \otimes \widetilde{M_u}. \tag{8}$$

Since $\min(L) \min(M) = \min(u \otimes \tilde{N}) = \min(u \otimes N)$, taking submodules spanned by minimal vectors in (7) we have $u \otimes \tilde{N} \subset \sigma(\tilde{L} \otimes \widetilde{M_u}) \subset u \otimes \tilde{N}$. Hence $\sigma(\tilde{L} \otimes \widetilde{M_u}) = u \otimes \tilde{N}$, and then $\sigma(L \otimes M_u) \subset u \otimes N$ implies

$$\begin{aligned}[N : \tilde{N}] &= [u \otimes N : u \otimes \tilde{N}] \\ &= [u \otimes N : \sigma(L \otimes M_u)][\sigma(L \otimes M_u) : u \otimes \tilde{N}] \\ &= [u \otimes N : \sigma(L \otimes M_u)][\sigma(L \otimes M_u) : \sigma(\tilde{L} \otimes \widetilde{M_u})].\end{aligned}$$

Now $[N : \tilde{N}] < \infty$ implies that $[L \otimes M_u : \tilde{L} \otimes \widetilde{M_u}] < \infty$ so $[M_u : \widetilde{M_u}] < \infty$. Similarly $[N_u : \widetilde{N_u}] < \infty$ is valid and then (8) yields $\mathbb{Q}\sigma^{-1}(u \otimes N_u) \subset \mathbb{Q}(u \otimes M_u)$. The condition (B2) proved above implies $\operatorname{rank} N_u = \operatorname{rank} M_u$ and so $\mathbb{Q}\sigma^{-1}(u \otimes N_u) = \mathbb{Q}(u \otimes M_u)$, which yields the condition (B3). □

Example 1.

Let L be one of the positive lattices A_n, D_n, E_6, E_7, E_8 in Example 1 in 7.1. Then L satisfies the assumptions in Theorem 7.4.1. Except for L being multiplicatively indecomposable, the other conditions are obvious. Suppose $L \cong K \otimes H$ for positive lattices K, H. Assuming $\operatorname{rank} K > 1$ and $\operatorname{rank} H > 1$, we will induce a contradiction. Let us assume it. By Corollary 7.4.1, K, H are of E-type and $[K : \tilde{K}], [H : \tilde{H}] < \infty$. Since K is of E-type, we get $\tilde{L} = \widetilde{K \otimes H} = \tilde{K} \otimes \tilde{H}$ and then $L = \tilde{L}$ yields

$K = \tilde{K}, H = \tilde{H}$. By scaling, we may assume $\min K = 1, \min H = 2$. For $e \in \mathcal{M}(K), u, v \in \mathcal{M}(H)$, we have

$$B(u,v) = Q(e)B(u,v) = B(e \otimes u, e \otimes v) \in B(L,L) = \mathbb{Z}.$$

If $B(u,v)$ is even for all $u, v \in \mathcal{M}(H)$, then $H = \tilde{H}$ yields $\mathrm{s}\,H \subset 2\mathbb{Z}$ which implies $H = \mathbb{Z}u \perp u^{\perp}$ for $u \in \mathcal{M}(H)$. This means that H and hence $L \cong K \otimes H$ is decomposable, which is a contradiction. Therefore $B(u,v)$ is odd for some $u, v \in \mathcal{M}(H)$. Since $B(u,v)^2 \leq Q(u)Q(v) = 4, B(u,v) = \pm 1$ and hence we may assume $B(u,v) = 1$ for some $u, v \in \mathcal{M}(H)$. For $e, f \in \mathcal{M}(K)$, we have $B(e,f) = B(e \otimes u, f \otimes v) \in B(L,L) = \mathbb{Z}$. Then $K = \tilde{K}$ yields $\mathrm{s}\,K \subset \mathbb{Z}$. Since $Q(e) = 1$, $K = \mathbb{Z}e \perp e^{\perp}$ holds and it implies the decomposability of L. Thus we have shown that L is multiplicatively indecomposable.

Example 2.

Let L be the lattice in Exercise 1 in 7.1 and use the notation there. Now L is of E-type and $L = \tilde{L}$. If the weighted graph induced by L is connected, then L is indecomposable. If $L \cong K \otimes H$ and L is indecomposable, then Corollary 7.4.1 implies that K, H are of E-type and $\mathcal{M}(K) \otimes \mathcal{M}(H) \cong \mathcal{M}(L)$. Hence the indecomposability of the weighted graph induced by L implies that L is multiplicatively indecomposable. Thus L satisfies the assumption of Theorem 7.4.1 if and only if the weighted graph induced by L is connected and indecomposable.

Among the conditions on L in Theorem 7.4.1, the multiplicative indecomposability is necessary to the assertion (iii). The author does not know whether other conditions are really necessary or not. When $\operatorname{rank} L = 2$, the following gives a satisfactory answer.

Theorem 7.4.2. *Let E be an indecomposable binary positive lattice. Then the assertion (A) is true for $L := E$ in the Fundamental Lemma.*

By the Fundamental Lemma, we have only to verify the assertion (B). We need several lemmas. First note that E is of E-type by Theorem 7.1.1; we may assume $\min E = 1$ without loss of generality by scaling of E by $(\min E)^{-1}$. We fix a basis $\{e_1, e_2\}$ of E such that

$$Q(e_1) = 1, a := Q(e_2) \geq 1, b := B(e_1, e_2) \text{ and } 0 < b \leq 1/2.$$

This is possible since $B(e_1, \pm e_2 + ne_1) = \pm B(e_1, e_2) + n$.

If $a = 1$, then $\mathcal{M}(E) = \{\pm e_1, \pm e_2\}$ and so this case is contained in the previous theorem.

We may assume hereafter

$$Q(e_1) = 1, a = Q(e_2) > 1, 0 < b = B(e_1, e_2) \leq 1/2.$$

We verify $\mathcal{M}(E) = \{\pm e_1\}$. For $m, n \in \mathbb{Z}$, it is easy to see that

$$Q(me_1 + ne_2) = (m + nb)^2 + (a - b^2)n^2 = 1$$

yields $|n| \leq 1$ and

$$Q(me_1 + e_2) = (m + b)^2 + (a - b^2) \geq a > 1 \text{ for } m \in \mathbb{Z}.$$

Hence if $me_1 + ne_2 \in \mathcal{M}(E)$, then $n = 0$ must be valid and then $m = \pm 1$, i.e. $\mathcal{M}(E) = \{\pm e_1\}$.

Lemma 7.4.2. *Let M be a positive lattice with* $\min M = 1$. *Then we have, for $f \in \mathcal{M}(M)$*

$$\begin{aligned} &\{z \in E \otimes M \mid B(z, e_1 \otimes f) = b, Q(z) \leq a\} \\ &\quad \subset e_1 \otimes \{x \in M \mid B(f, x) = b, Q(x) \leq a\} \cup \{e_2 \otimes f,\ (e_1 - e_2) \otimes f\}. \end{aligned}$$

Here $(e_1 - e_2) \otimes f$ can occur only when $b = 1/2$.

Proof. Let $z = e_1 \otimes x_1 + e_2 \otimes x_2 \in E \otimes M$ $(x_1, x_2 \in M)$ satisfy $B(z, e_1 \otimes f) = b, Q(z) \leq a$; then we have

$$\begin{aligned} Q(z) &= Q(x_1) + aQ(x_2) + 2bB(x_1, x_2) \\ &= Q(x_1 + bx_2 - bf) + b^2 + (a - b^2)Q(x_2), \end{aligned} \tag{9}$$

noting $B(x_1, f) + bB(x_2, f) = B(z, e_1 \otimes f) = b$ and $Q(f) = 1$.

Suppose $x_2 = 0$; then $B(f, x_1) = b$ and $Q(x_1) = Q(e_1 \otimes x_1) = Q(z) \leq a$. This gives the first case.

Suppose $x_2 \neq 0$; then $\min M = 1$ implies $Q(x_2) \geq 1$, and $Q(z) \leq a$ yields, by (9), $a \geq Q(z) \geq b^2 + (a - b^2)Q(x_2) \geq a$, and hence

$$Q(z) = a, Q(x_2) = 1, Q(x_1 + bx_2 - bf) = 0, \text{ i.e. } x_1 + bx_2 = bf. \tag{10}$$

Since M is positive definite, we have $B(x_2, f)^2 \leq Q(x_2)Q(f) = 1$, and then, by (10)

$$Q(x_1) = b^2 Q(f - x_2) = b^2(2 - 2B(f, x_2)) \leq 4b^2 \leq 1. \tag{11}$$

The assumptions $\min M = 1$ and $x_1 \in M$ yield $Q(x_1) = 1$ or 0 by (11). If $Q(x_1) = 0$, then $x_1 = 0$ and $x_2 = f$ by (10), i.e. $z = e_2 \otimes f$. This is the second case.

Suppose $Q(x_1) = 1$; then (11) implies $b = 1/2$ and $B(f, x_2) = -1$. Hence $Q(f + x_2) = 0$ and $x_2 = -f$ are valid. From (10) we have $x_1 = b(f - x_2) = f$ and so $z = (e_1 - e_2) \otimes f$. □

Lemma 7.4.3. *Let M, N be positive lattices with $\min M = \min N = 1$. Assume we have an isometry $\sigma : E \otimes M \cong E \otimes N$ such that $\sigma(E \otimes m) = E \otimes n$ $(m \in M, n \in N)$ implies $m = 0$. Then for a given $f_1 \in \mathcal{M}(M)$ there exist $f_2 \in M$, $g_1, g_2 \in N$ such that*

$$\sigma(e_1 \otimes f_1) = e_1 \otimes g_1, \sigma(e_2 \otimes f_1) = e_1 \otimes g_2, \sigma(e_1 \otimes f_2) = e_2 \otimes g_1.$$

Proof. Since E is of E-type, $\sigma(\mathcal{M}(E) \otimes \mathcal{M}(M)) = \mathcal{M}(E) \otimes \mathcal{M}(N)$ is valid and hence there exists $g_1 \in \mathcal{M}(N)$ such that $\sigma(e_1 \otimes f_1) = e_1 \otimes g_1$. Then $B(\sigma(e_2 \otimes f_1), e_1 \otimes g_1) = b, Q(\sigma(e_2 \otimes f_1)) = a$ are clear. From Lemma 7.4.2, we have $\sigma(e_2 \otimes f_1) = e_1 \otimes g_2$ for $g_2 \in N$, $e_2 \otimes g_1$ or $(e_1 - e_2) \otimes g_1$. The latter two yield $\sigma(E \otimes f_1) = E \otimes g_1$ contradicting the assumption. Thus we have $\sigma(e_2 \otimes f_1) = e_1 \otimes g_2$. Applying a similar argument to $\sigma^{-1} : E \otimes N \cong E \otimes M$, and $g_1 \in \mathcal{M}(N)$, we have $\sigma^{-1}(e_2 \otimes g_1) = e_1 \otimes f_2$ for $f_2 \in M$. □

Lemma 7.4.4. *$O(E \otimes E)$ is generated by $O(E)$ and the interchange of E.*

Proof. Let $\sigma \in O(E \otimes E)$. Suppose that there exist $x(\neq 0), y(\neq 0) \in E$ such that $\sigma(E \otimes x) = E \otimes y$. Then $\sigma(E \otimes x^\perp) = E \otimes y^\perp$ and, applying Lemma 7.2.3 to $K = M = N = E$, we have $\sigma \in O(E) \otimes O(E)$. Suppose $\sigma(E \otimes x) = E \otimes y$ yields $x = y = 0$; applying Lemma 7.4.3 to $M = N = E, f_1 = e_1$, there exist $g_1, g_2 \in E$ such that $\sigma(e_1 \otimes e_1) = e_1 \otimes g_1$, $\sigma(e_2 \otimes e_1) = e_1 \otimes g_2$. Hence $\sigma(E \otimes e_1) \subset e_1 \otimes E$ is valid and so $\sigma(E \otimes e_1) = e_1 \otimes E$. Letting $\mu \in O(E \otimes E)$ be the interchange of E we have $\mu\sigma(E \otimes e_1) = E \otimes e_1$. By the above argument, we have $\mu\sigma \in O(E) \otimes O(E)$. □

Lemma 7.4.5. *Let M, N be positive lattices isometric to E and suppose that an isometry $\sigma : E \otimes M \cong E \otimes N$ satisfies $\sigma(E \otimes m) = E \otimes n$ only when $m = 0, n = 0$. Then there exist decompositions $M = M_1 \oplus M_2, N = N_1 \oplus N_2$ such that, for $j = 1, 2$*

$$\sigma(E \otimes M_j) = e_j \otimes N, \sigma^{-1}(E \otimes N_j) = e_j \otimes M, \sigma(e_j \otimes M_j) = e_j \otimes N_j.$$

Proof. We may suppose $M = N = E$. The proof of the previous lemma shows $\sigma(x \otimes y) = \sigma_1(y) \otimes \sigma_2(x)$ $(x, y \in E)$ for some isometries $\sigma_1, \sigma_2 \in O(E)$. Noting $\mathcal{M}(E) = \{\pm e_1\}$, clearly $\sigma(E \otimes e_1) = e_1 \otimes E$. Thus we have only to put $M_1 = \mathbb{Z}e_1, M_2 = \mathbb{Z}\sigma_1^{-1}(e_2), N_1 = \mathbb{Z}e_1$ and $N_2 = \mathbb{Z}\sigma_2(e_2)$. □

Lemma 7.4.6. *The assertion (B) for E follows from the following assertion*

(C): *Let M, N be positive lattices and suppose that an isometry $\sigma : E \otimes M \cong E \otimes N$ satisfying $\sigma(E \otimes m) = E \otimes n$ $(m \in M, n \in N)$ implies $m = 0$. Then there exist submodules M_1, N_1 of M, N respectively such that $M_1 \cong N_1 \cong$ (a scaling of E) and $\sigma(E \otimes M_1) = E \otimes N_1$.*

Proof. Let M, N be positive lattices with $\sigma : E \otimes M \cong E \otimes N$ and suppose that $\sigma(E \otimes m) = M \otimes n$ $(m \in M, n \in N)$ yields $m = 0$. Applying the assertion (C) repeatedly, there exist submodules $M_1, \cdots, M_t, N_1, \cdots, N_t$ of M, N respectively such that $M_i \cong N_i \cong E^{c_i}$ (which denotes the scaling of E by c_i), $\sigma(E \otimes M_i) = E \otimes N_i$ and

(12) $$[M : M_1 \perp \cdots \perp M_t] < \infty, \; [N : N_1 \perp \cdots \perp N_t] < \infty.$$

(B1) is satisfied for $v_i := e_i$ $(i = 1, 2)$.

By virtue of Lemma 7.4.5, there exist decompositions, $M_i = M_{i1} \oplus M_{i2}$, $N_i = N_{i1} \oplus N_{i2}$ such that, for $1 \le i \le t, j = 1, 2$

(13) $$\sigma(E \otimes M_{ij}) = v_j \otimes N_i, \sigma^{-1}(E \otimes N_{ij}) = v_j \otimes M_i, \sigma(v_j \otimes M_{ij}) = v_j \otimes N_{ij}.$$

Putting

$$M_{v_j} = \{m \in M \mid \sigma(E \otimes m) \subset v_j \otimes N\},$$
$$N_{v_j} = \{n \in N \mid \sigma^{-1}(E \otimes n) \subset v_j \otimes M\},$$

Now $M_{v_j} \supset \oplus_i M_{ij}, N_{v_j} \supset \oplus_i N_{ij}$ follow from (13), and hence $\operatorname{rank} M_{v_j} \ge t, \operatorname{rank} N_{v_j} \ge t$. On the other hand, $M_{v_1} \cap M_{v_2} = \{0\}, N_{v_1} \cap N_{v_2} = \{0\}$ yield $\operatorname{rank} M_{v_1} \oplus M_{v_2} \ge 2t, \operatorname{rank} N_{v_1} \oplus N_{v_2} \ge 2t$. Since (12) implies $\operatorname{rank} M = \operatorname{rank} N = 2t$, we have $\operatorname{rank} M_{v_j} = \operatorname{rank} N_{v_j} = t$ for $j = 1, 2$, which yields (B2) and $\mathbb{Q}M_{v_j} = \mathbb{Q}(\oplus_i M_{ij}), \mathbb{Q}N_{v_j} = \mathbb{Q}(\oplus_i N_{ij})$. Hence

$$\begin{aligned}\sigma(\mathbb{Q}(v_j \otimes M_{v_j})) &= \sigma(\mathbb{Q}(v_j \otimes (\oplus_i M_{ij}))) = \mathbb{Q}(v_j \otimes (\oplus_i N_{ij})) \quad \text{(by (13))} \\ &= \mathbb{Q}(v_j \otimes N_{v_j}).\end{aligned}$$

Thus (B3) is valid. □

Lemma 7.4.7. *Let K be a positive lattice with* $\min K = 1$. *Then a submodule H of $E \otimes K$ isometric to E is either $E \otimes f$ $(f \in \mathcal{M}(K))$ or $e_1 \otimes K_0$ $(K_0 \subset K)$.*

Proof. Let $\{u, z\}$ be a basis of H such that $Q(u) = 1, B(u, z) = b, Q(z) = a$. Since E is of E-type, $\min(E \otimes K) = 1$ and hence $u \in \mathcal{M}(E \otimes K) = \mathcal{M}(E) \otimes \mathcal{M}(K)$. Thus we can put $u = e_1 \otimes f$ for $f \in \mathcal{M}(K)$. Then from Lemma 7.4.2 it follows that $z = e_1 \otimes x$ $(x \in K, B(f, x) = b, Q(x) \le a), e_2 \otimes f$ or $(e_1 - e_2) \otimes f$. Thus, H is either $e_1 \otimes \mathbb{Z}[f, x]$ or $E \otimes f$. □

Hereafter we make the assumption

(*): for positive lattices M and N, an isometry $\sigma : E \otimes M \cong E \otimes N$ satisfying $\sigma(E \otimes m) = E \otimes n (m \in M, n \in N)$ implies $m = 0$, and moreover $\min M = 1$ and hence $\min N = 1$.

We have only to show that the assertion (C) in Lemma 7.4.6 holds under (*) which we assume henceforth.

Lemma 7.4.8. *For $x \in \mathcal{M}(M), y \in \mathcal{M}(N)$, there are submodules $M' \subset M, N' \subset N$ such that $\sigma(E \otimes x) = e_1 \otimes N', \sigma(e_1 \otimes M') = E \otimes y$.*

Proof. Since $\min M = \min N = 1$ by assumption, $Q(x) = Q(y) = 1$ implies $\sigma(E \otimes x) \cong \sigma^{-1}(E \otimes y) \cong E$. Hence Lemma 7.4.7 we have the assertion, noting that the assumption that $\sigma(E \otimes m) = E \otimes n$ is valid only for $m = 0$. □

Lemma 7.4.9. *Let N' be a submodule of N containing $n \in \mathcal{M}(N)$ and being isometric to E. Then either $\sigma^{-1}(e_1 \otimes N') = E \otimes m$ $(m \in \mathcal{M}(M))$ or $= e_1 \otimes M'$ $(M' \subset M)$; the former is possible for at most one N' when n is given.*

Proof. Since $E \cong \sigma^{-1}(e_1 \otimes N') \subset E \otimes M$, from Lemma 7.4.7 we have $\sigma^{-1}(e_1 \otimes N') = E \otimes m$ $(m \in \mathcal{M}(M))$ or $e_1 \otimes M'$ $(M' \subset M)$. Suppose that there exist submodules N_i $(i = 1, 2)$ of N such that $n \in N_i \cong E$ and $\sigma^{-1}(e_1 \otimes N_i) = E \otimes m_i$ $(m_i \in \mathcal{M}(M))$; then $E \otimes m_1 \cap E \otimes m_2 \ni \sigma^{-1}(e_1 \otimes n)$ implies $m_1 = \pm m_2$ since m_1, m_2 are primitive, and hence we have $\sigma^{-1}(e_1 \otimes N_1) = \sigma^{-1}(e_1 \otimes N_2)$, i.e. $N_1 = N_2$. □

For $m_1 \in \mathcal{M}(M)$, we put $\sigma(e_1 \otimes m_1) = e_1 \otimes n_1$ $(n_1 \in \mathcal{M}(N))$. From Lemma 7.4.7 and the assumption (*) we have $\sigma(E \otimes m_1) = e_1 \otimes N_0$ for some submodule N_0 of N. Define a basis $\{n_1, n_2\}$ of N_0 by $\sigma(e_i \otimes m_1) = e_1 \otimes n_i$ $(i = 1, 2)$; then $Q(n_1) = 1, B(n_1, n_2) = b$ and $Q(n_2) = a$ are obvious. Similarly we define a submodule M_0 of M by $\sigma^{-1}(E \otimes n_1) = e_1 \otimes M_0$ and a basis $\{m_1, m_2\}$ of M_0 by $\sigma^{-1}(e_i \otimes n_1) = e_1 \otimes m_i$ $(i = 1, 2)$; m_1 is already given. Then the following are clear.

(14) $$Q(m_1) = Q(n_1) = 1, B(m_1, m_2) = B(n_1, n_2) = b, \quad Q(m_2) = Q(n_2) = a,$$

(15) $$\sigma(e_1 \otimes m_1) = e_1 \otimes n_1,$$

(16) $$\sigma(e_2 \otimes m_1) = e_1 \otimes n_2,$$

(17) $$\sigma(e_1 \otimes m_2) = e_2 \otimes n_1,$$

(18) $$\sigma(e_2 \otimes m_2) = e_1 \otimes x_1 + e_2 \otimes x_2 \text{ for some } x_i \in N \ (i=1, 2).$$

Once we have $x_1 = 0, x_2 = n_2$, we obtain $\sigma(E \otimes \mathbb{Z}[m_1, m_2]) = E \otimes \mathbb{Z}[n_1, n_2]$ and $\mathbb{Z}[m_1, m_2] \cong \mathbb{Z}[n_1, n_2] \cong E$, which is nothing but the assertion (C) in Lemma 7.4.6 and we complete the proof of Theorem 7.4.2. We shall verify $x_1 = 0, x_2 = n_2$ below.

Lemma 7.4.10. *We have $B(n_1, x_1) = 0, B(n_1, x_2) = b, x_2 \neq 0, B(n_2, x_1 + bx_2) = ab, Q(x_1 + bx_2) + (a - b^2)Q(x_2) = a^2$ and $Q(x_2) \leq a$.*

Proof. From the inner product of (15) and (18) follows

(19) $$b^2 = B(n_1, x_1) + bB(n_1, x_2).$$

From the inner product of (17) and (18) we have

$$ab = bB(n_1, x_1) + aB(n_1, x_2). \tag{20}$$

Similarly, (16) and (18) imply

$$ab = B(n_2, x_1) + bB(n_2, x_2) = B(n_2, x_1 + bx_2). \tag{21}$$

(18) implies

$$\begin{aligned} a^2 &= Q(x_1) + 2bB(x_1, x_2) + aQ(x_2) \\ &= Q(x_1 + bx_2) + (a - b^2)Q(x_2). \end{aligned} \tag{22}$$

Then (19), (20) yield $B(n_1, x_1) = 0, B(n_1, x_2) = b$ which implies $x_2 \neq 0$ by $b \neq 0$. It remains to show $Q(x_2) \leq a$. We know the inequality

$$B(n_2, x_1 + bx_2)^2 \leq Q(n_2)Q(x_1 + bx_2)$$

and substituting (21), (14), (22) into it, we get $a^2b^2 \leq a(a^2 - (a - b^2)Q(x_2))$, which yields immediately $Q(x_2) \leq a$. □

Lemma 7.4.11. $\sigma^{-1}(e_1 \otimes x_2) \notin e_1 \otimes M$.

Proof. Suppose

$$\sigma^{-1}(e_1 \otimes x_2) = e_1 \otimes y \text{ for some } y \in M. \tag{23}$$

Then we will show $Q(x_2) = b^2 (< 1)$, which contradicts $x_2 (\neq 0) \in N$ and $\min N = 1$, and hence we complete the proof. So, let us verify $Q(x_2) = b^2$. Put

$$\sigma^{-1}(e_2 \otimes x_2) = e_1 \otimes y_1 + e_2 \otimes y_2 \ (y_i \in M). \tag{24}$$

The inner product of (15) and (23), and Lemma 7.4.10 imply

$$B(m_1, y) = B(n_1, x_2) = b, \tag{25}$$

and (16), (23) and (25) give

$$B(n_2, x_2) = bB(m_1, y) = b^2. \tag{26}$$

From (15), (24) and Lemma 7.4.10 we have

$$B(m_1, y_1) + bB(m_1, y_2) = bB(n_1, x_2) = b^2, \tag{27}$$

and (16), (24) and (26) give

$$bB(m_1, y_1) + aB(m_1, y_2) = bB(n_2, x_2) = b^3,$$

which implies with (27)

(28) $$B(m_1, y_1) = b^2, B(m_1, y_2) = 0.$$

From (23) and (24) we find

$$Q(x_2) = Q(y),$$
$$aQ(x_2) = Q(y_1) + 2bB(y_1, y_2) + aQ(y_2) = Q(y_1 + by_2) + (a - b^2)Q(y_2),$$
$$bQ(x_2) = B(y, y_1) + bB(y, y_2) = B(y, y_1 + by_2).$$

Those yield

$$\begin{aligned} b^2Q(x_2)^2 &= B(y, y_1 + by_2)^2 \leq Q(y)Q(y_1 + by_2) \\ &= Q(x_2)(aQ(x_2) - (a - b^2)Q(y_2)), \end{aligned}$$

which immediately gives, since $Q(x_2) \neq 0$

(29) $$Q(y_2) \leq Q(x_2).$$

From (18) and (23) we find $bB(m_2, y) = B(x_1, x_2) + bQ(x_2)$. Since, on the other hand

$$\begin{aligned} B(m_2, y) &= B(e_1 \otimes m_2, e_1 \otimes y) \\ &= B(e_2 \otimes n_1, e_1 \otimes x_2) \quad \text{(by (17), (23))} \\ &= b^2 \end{aligned}$$

by Lemma 7.4.10, we have

(30) $$B(x_1, x_2) = bB(m_2, y) - bQ(x_2) = b^3 - bQ(x_2).$$

The equalities

$$B(y_1, m_2) + bB(y_2, m_2) = aB(n_1, x_2) = ab,$$
$$bB(y_1, m_2) + aB(y_2, m_2) = bB(x_1, x_2) + aQ(x_2) = b^4 + (a - b^2)Q(x_2),$$

follow from (17), (24) and Lemma 7.4.10 and by (18), (24) and (30) respectively. Solving them, we have

(31) $$B(y_1, m_2) = b(a + b^2) - bQ(x_2), B(y_2, m_2) = Q(x_2) - b^2.$$

If $m_2 = y_2$, then (31) and Lemma 7.4.10 imply $Q(m_2) = Q(x_2) - b^2 \leq a - b^2$, which contradicts $Q(m_2) = a$ in (14) by $b \neq 0$. So, $m_2 \neq y_2$, giving $Q(y_2 - m_2) \geq \min M = 1$. From (29), (31) and (14) we have

$$\begin{aligned} 1 &\leq Q(y_2 - m_2) = Q(y_2) - 2B(y_2, m_2) + Q(m_2) \\ &\leq Q(x_2) - 2(Q(x_2) - b^2) + a = a + 2b^2 - Q(x_2). \end{aligned}$$

Thus we get, with (29)

$$Q(y_2) \leq Q(x_2) \leq a + 2b^2 - 1. \tag{32}$$

Now we put $\sigma(e_1 \otimes y_2) = e_1 \otimes w_1 + e_2 \otimes w_2$ $(w_i \in N)$. Then $Q(y_2) = Q(w_1 + bw_2) + (a - b^2)Q(w_2) \geq a - b^2$ if $w_2 \neq 0$. Hence $w_2 \neq 0$ yields $a - b^2 \leq Q(y_2) \leq a + 2b^2 - 1$ by (32); so $3b^2 \geq 1$ which contradicts $0 < b \leq 1/2$. Thus we have $w_2 = 0$, so

$$\sigma(e_1 \otimes y_2) = e_1 \otimes w_1, \tag{33}$$

and then by (28)

(34)

$$B(n_1, w_1) = B(e_1 \otimes n_1, e_1 \otimes w_1) = B(e_1 \otimes m_1, e_1 \otimes y_2) = B(m_1, y_2) = 0.$$

Finally we have

$$\begin{aligned} Q(x_2) - b^2 &= B(m_2, y_2) \text{ by (31)} \\ &= B(e_1 \otimes m_2, e_1 \otimes y_2) \\ &= B(e_2 \otimes n_1, e_1 \otimes w_1) \text{ by (17) and (33)} \\ &= bB(n_1, w_1) = 0 \text{ by (34)} \end{aligned}$$

which is the required contradictory equality. □

Lemma 7.4.12. *$Q(x_2) = a$ and $\mathbb{Z}[n_1, n_2] = \mathbb{Z}[n_1, x_2]$.*

Proof. We know $Q(x_2) \leq a$ by Lemma 7.4.10. Suppose $Q(x_2) < a$. Then from $B(\sigma^{-1}(e_1 \otimes n_1), \sigma^{-1}(e_1 \otimes x_2)) = b$ (by Lemma 7.4.10) and $Q(\sigma^{-1}(e_1 \otimes x_2)) < a$ we have $\sigma^{-1}(e_1 \otimes x_2) \in e_1 \otimes M$ by Lemma 7.4.2, which contradicts Lemma 7.4.11. Thus we obtain $Q(x_2) = a$; then with $Q(n_1) = 1, B(n_1, x_2) = b$ by Lemma 7.4.10 we get $\mathbb{Z}[n_1, x_2] \cong E$. Since

$$\sigma^{-1}(e_1 \otimes \mathbb{Z}[n_1, n_2]) = E \otimes m_1 \quad \text{(by (15), (16))},$$

Lemma 7.4.9 yields

$$\sigma^{-1}(e_1 \otimes \mathbb{Z}[n_1, x_2]) = E \otimes m \text{ or } \subset e_1 \otimes M.$$

Lemma 7.4.11 denies the latter inclusion, so $\sigma^{-1}(e_1 \otimes \mathbb{Z}[n_1, x_2]) = E \otimes m$ and $\sigma^{-1}(e_1 \otimes \mathbb{Z}[n_1, n_2]) = E \otimes m_1$ and Lemma 7.4.9 implies $\mathbb{Z}[n_1, x_2] = \mathbb{Z}[n_1, n_2]$. □

Now we can prove the theorem. As stated above, we have only to verify $x_1 = 0, x_2 = n_2$. To do it, we need one more equation

$$bn_2 = x_1 + bx_2, \tag{35}$$

which follows from

$$\begin{aligned} Q(x_1 + bx_2 - bn_2) &= Q(x_1 + bx_2) - 2bB(x_1 + bx_2, n_2) + b^2 Q(n_2) \\ &= \{a^2 - (a - b^2)Q(x_2)\} \\ &\qquad - 2b \cdot ab + b^2 Q(n_2) \text{ (by Lemma 7.4.10)} \\ &= 0 \text{ (by Lemma 7.4.12. and (14))}. \end{aligned}$$

By virtue of Lemma 7.4.12, we can put

$$n_2 = a_1 n_1 + a_2 x_2 \quad (a_i \in \mathbb{Z}).$$

We have to show $a_1 = 0, a_2 = 1$, which implies $n_2 = x_2$ and then $x_1 = 0$ by (35) and therefore completes the proof of Theorem 7.4.2.

Suppose $a_1 a_2 \neq 0$; then

$$\begin{aligned} a = Q(n_2) &= Q(a_1 n_1 + a_2 x_2) = a_1^2 + 2a_1 a_2 b + a_2^2 a \\ &= (1 - b)a_1^2 + b(a_1 + a_2)^2 + (a - b)a_2^2 \geq 1 - b + a - b = a - 2b + 1 \end{aligned}$$

hence $b \geq 1/2$ and so $b = 1/2$. Then the right-hand side is equal to a in the above inequality and $a_1 = \pm 1, a_1 + a_2 = 0, a_2 = \mp 1$. Then we have $n_2 = a_1(n_1 - x_2)$ and $b = B(n_1, n_2) = a_1 B(n_1, n_1 - x_2) = a_1(1 - b)$ by Lemma 7.4.10. Since $b = 1/2$, we have $a_1 = 1$ and then $n_2 = n_1 - x_2$. It follows from (35) that $x_1 = \frac{1}{2}(n_2 - x_2) = n_1/2 - x_2$. Thus we have $n_1/2 = x_1 + x_2 \in N$ contradicting the primitiveness of n_1 in N.

Suppose $a_2 = 0$; then we get $n_2 = a_1 n_1 + a_2 x_2 = a_1 n_1$ contradicting the linear independence of n_1, n_2.

Thus we have $a_1 = 0$ and then $n_2 = a_2 x_2$. Then

$$\begin{aligned} b &= B(n_1, n_2) && \text{(by (14))} \\ &= a_2 B(n_1, x_2) = a_2 b && \text{(by Lemma 7.4.10)} \end{aligned}$$

yields $a_2 = 1$ which is what we wanted. □

Next we give a result on the uniqueness of the decomposition.

Theorem 7.4.3. *Let L_i $(1 \le i \le m)$ be a positive lattice of E-type such that* $\operatorname{rank} L_i > 1, [L_i : \tilde{L}_i] < \infty$, *$\tilde{L}_i$ is indecomposable and L_i is multiplicatively indecomposable. Let M_i $(1 \le i \le n)$ be a multiplicatively indecomposable positive lattice with* $\operatorname{rank} M_i > 1$ *and suppose that $\sigma : \otimes_{i=1}^m L_i \cong \otimes_{i=1}^n M_i$. Then we have $m = n$ and changing suffices if necessary, $\sigma = \otimes \sigma_i, \sigma_i : L_i \cong M_i^{a_i}$ (scaling of M_i by a_i).*

Proof. Put $L = \otimes_{i=1}^m L_i$; then L is of E-type by Lemma 7.1.1, and $\tilde{L} = \otimes \tilde{L}_i$. Since the L_i's are of E-type, $[L_i : \tilde{L}_i] < \infty$ implies $[L : \tilde{L}] < \infty$. Next we show $\tilde{L}$ is indecomposable. By virtue of (iii) in Lemma 7.1.1, $\tilde{L}_i$ is of E-type. If $\tilde{L}_i \cong \otimes K_{ij}$ for multiplicatively indecomposable positive lattices K_{ij}, then applying Corollary 7.4.1 repeatedly to $\tilde{L}_i$, $K_{ij} = \widetilde{K}_{ij}$ is of E-type and indecomposable. Taking K_{ij} as L in Theorem 7.4.1, we obtain the indecomposability of $\tilde{L} = \otimes \tilde{L}_i = \otimes_{i,j} K_{ij}$. Now we can apply Corollary 7.4.1 to the above L to show M_i is of E-type, $[M_i : \tilde{M}_i] < \infty$ and $\tilde{M}_i$ is indecomposable; so L_i, M_j satisfy the same properties. Hence, without loss of generality we may assume by scaling that $\min(L_i) = \min(M_i) = 1$. Changing suffices and considering σ^{-1} if necessary, we may further assume

$$\text{(36)} \qquad \operatorname{rank} L_1 \ge \operatorname{rank} L_i, \operatorname{rank} M_j \quad (1 \le i \le m, 1 \le j \le n)$$

$$\text{(37)} \qquad \text{if } \operatorname{rank} L_1 = \operatorname{rank} M_j, \text{ then } \mathrm{d}\, L_1 \le \mathrm{d}\, M_j.$$

Since $\tilde{L}_i, \tilde{M}_j$ are indecomposable, the weighted graphs

$$G(L_i) := \mathcal{M}(L_i)/\pm 1, G(M_j) := \mathcal{M}(M_j)/\pm 1$$

induced by L_i, M_j are connected and σ induces an isometry

$$\tilde{\sigma} : \prod_i G(L_i) \cong \prod_j G(M_j).$$

Then decomposing $G(L_i), G(M_j)$ to indecomposable weighted graphs and applying Theorem 7.3.1 to $\tilde{\sigma}$, we have, for $e_i \in \mathcal{M}(L_i)$, $i \ge 2$,

$$\tilde{\sigma}(G(L_1) \times e_2 \times \cdots \times e_n) = \prod G_i \; (G_i \subset G(M_i)),$$

where we consider $e_i \in G(L_i)$. Denoting by M_i^0 the submodule of M_i spanned by elements in G_i, we have

$$\text{(38)} \qquad \sigma(\tilde{L}_1 \otimes e_2 \otimes \cdots \otimes e_m) = M_1^0 \otimes \cdots \otimes M_n^0.$$

Putting $\hat{M}_i = \mathbb{Q} M_i^0 \cap M_i$, we get

$$\sigma(L_1 \otimes e_2 \otimes \cdots \otimes e_m) = \sigma(\mathbb{Q}\tilde{L}_1 \otimes e_2 \otimes \cdots \otimes e_m \cap L) = \hat{M}_1 \otimes \cdots \otimes \hat{M}_n.$$

By the assumption that L_1 is multiplicatively indecomposable, there exists j $(1 \leq j \leq n)$ such that $\operatorname{rank} \hat{M}_i = 1$ if $i \neq j$. If $\operatorname{rank} \hat{M}_i = 1$, then $\operatorname{rank} M_i^0 = 1$ and so $\sharp G_i = 1$. Changing the suffix if $j \neq 1$, we may assume $j = 1$ without loss of generality. Representing G_i by $f_i \in \mathcal{M}(M_i)$ for $i \geq 2$, we have $M_i^0 = \mathbb{Z} f_i$ for $i \geq 2$ and (38) yields

$$\sigma(L_1 \otimes e_2 \otimes \cdots \otimes e_m) \subset M_1 \otimes f_2 \otimes \cdots \otimes f_n. \tag{39}$$

Thus we have $\operatorname{rank} L_1 \leq \operatorname{rank} M_1$ and then the assumptions (36), (37) imply $\operatorname{rank} L_1 = \operatorname{rank} M_1$ and $\mathrm{d}\, L_1 \leq \mathrm{d}\, M_1$. Since we assumed $\min(L_i) = \min(M_j) = 1$, we have $Q(e_i) = Q(f_j) = 1$ and hence (39) implies $\mathrm{d}\, L_1 \geq \mathrm{d}\, M_1$ so $\mathrm{d}\, L_1 = \mathrm{d}\, M_1$ which makes the inclusion in (39) an equality. Thus putting $\sigma(x \otimes e_2 \otimes \cdots \otimes e_m) = \sigma_1(x) \otimes f_2 \otimes \cdots \otimes f_n$, we have

$$\sigma_1 : L_1 \cong M_1, \tilde{\sigma}(x \times e_2 \times \cdots \times e_m) = \tilde{\sigma}_1(x) \times f_2 \times \cdots \times f_n,$$

where $\tilde{\sigma}_1$ is an isometry: $G(L_1) \cong G(M_1)$ induced by σ_1. Then by Lemma 7.3.3 there exists an isometry $\tilde{\sigma}_2 : \prod_{i \geq 2} G(L_i) \cong \prod_{j \geq 2} G(M_j)$ with

$$\tilde{\sigma} = \tilde{\sigma}_1 \times \tilde{\sigma}_2. \tag{40}$$

Hence for any $e_i \in \mathcal{M}(L_i)$, there exists some element $\sigma_2(e_2 \otimes \cdots \otimes e_m) \in \otimes_{j \geq 2} \mathcal{M}(M_j)$ which projects to $\tilde{\sigma}_2(e_2 \times \cdots \times e_m)$ and

$$\sigma(e_1 \otimes \cdots \otimes e_m) = \sigma_1(e_1) \otimes \sigma_2(e_2 \otimes \cdots \otimes e_m). \tag{41}$$

For a fixed e_1, noting $[L_i : \tilde{L}_i] < \infty$, (41) gives

$$\sigma(e_1 \otimes L_2 \otimes \cdots \otimes L_m) \subset \sigma_1(e_1) \otimes M_2 \otimes \cdots \otimes M_m.$$

Therefore we may consider σ_2 as an isometry from $L_2 \otimes \cdots \otimes L_m$ to $M_2 \otimes \cdots \otimes M_m$. Since $L_1 \cong M_1$, we have

$$\mathrm{d}(L_2 \otimes \cdots \otimes L_m) = \mathrm{d}(M_2 \otimes \cdots \otimes M_m)$$

and so σ_2 is an isometry on $M_2 \otimes \cdots \otimes M_m$. Let us verify that σ_2 is independent of $e_1 \in \mathcal{M}(L_1)$. Let $e_1' \in \mathcal{M}(L_1)$; then similarly there exists

$$\sigma_2' : \otimes_{i \geq 2} L_i \cong \otimes_{j \geq 2} M_j$$

such that $\sigma(e_1' \otimes x) = \sigma_1(e_1') \otimes \sigma_2'(x)$ for $x \in \otimes_{i \geq 2} L_i$. Suppose $B(e_1, e_1') \neq 0$. Let $e_i \in \mathcal{M}(L_i)$ $(i \geq 2)$; then (40) gives

$$\sigma_2(e_2 \otimes \cdots \otimes e_m) = \delta \sigma_2'(e_2 \otimes \cdots \otimes e_m) \text{ for } \delta = \pm 1.$$

Then

$$\begin{aligned} B(e_1, e_1') &= B(e_1 \otimes e_2 \otimes \cdots \otimes e_m, e_1' \otimes e_2 \otimes \cdots \otimes e_m) \\ &= B(\sigma_1(e_1) \otimes \sigma_2(e_2 \otimes \cdots \otimes e_m), \sigma_1(e_1') \otimes \sigma_2'(e_2 \otimes \cdots \otimes e_m)) \\ &= B(e_1, e_1')\delta Q(e_2 \otimes \cdots \otimes e_m) = \delta B(e_1, e_1') \end{aligned}$$

so $\delta = 1$. Thus we have $\sigma_2 = \sigma_2'$ on $\otimes_{i\geq 2}\mathcal{M}(L_i)$ and hence on $\otimes_{i\geq 2}L_i$ if $B(e_1, e_1') \neq 0$. As $\tilde{L}_1$ is indecomposable, for any given $e_1, e_1' \in \mathcal{M}(L_1)$, there exist $z_1 = e_1, z_2, \cdots, z_r = e_1'(z_i \in \mathcal{M}(L_1))$ such that $B(z_i, z_{i+1}) \neq 0$ for $i = 1, \cdots, r-1$. Hence σ_2 is independent of the choice of $e_1 \in \mathcal{M}(L_1)$. So we have proved $\sigma(x \otimes y) = \sigma_1(x) \otimes \sigma_2(y)$ for $x \in \mathcal{M}(L_1), y \in \otimes_{i\geq 2}L_i$ and then

$$\sigma_1 : L_1 \cong M_1, \sigma_2 : \otimes_{i\geq 2}L_i \cong \otimes_{j\geq 2}M_j$$

imply $\sigma = \sigma_1 \otimes \sigma_2$ on $L_1 \otimes (\otimes_{i\geq 2}L_i)$. Our theorem is now proved inductively. □

Corollary 7.4.2. *Let L_i be those in Theorem 7.4.3 and assume that if L_i is isometric to a scaling of L_j, then $L_i \cong L_j$. Then $O(\otimes L_i)$ is generated by $O(L_i)$ and interchanges of isometric components.*

Proof. We have only to put $M_i = L_i$ in the theorem. □

Remark.

Lattices in Example 1 satisfy the assumptions of Theorem 7.4.3.

Exercise 1.

Let L_i be binary indecomposable lattices with $\min(L_i) = 1$; show $O(\otimes L_i)$ is generated by $O(L_i)$ and interchanges of isometric components.
(Hint: Show that the binary submodule in $\otimes L_i$ with minimal discriminant is of the form $e_1 \otimes \cdots \otimes e_{j-1} \otimes L_j \otimes e_{j+1} \otimes \cdots$ for $e_i \in \mathcal{M}(L_i)$. Theorem 7.4.2 then yields that $\otimes L_i$ determines the components L_i. Suppose without loss of generality $\mathrm{d}\, L_1 \leq \mathrm{d}\, L_i$ for $i \geq 1$ and $L_1 \cong \cdots \cong L_k \not\cong L_i$ for $i > k$. If $\otimes_{i>k}L_i \cong L_1 \otimes K$ for some K then $\min K = 1$ so $\otimes_{i>k}L_i$ contains a submodule L_1' isometric to L_1 and hence the minimal discriminant of the binary submodules in $\otimes_{i>k}L_i$ is equal to $\mathrm{d}\, L_i$ for some $i > k$ and is greater than or equal to $\mathrm{d}\, L_1' = \mathrm{d}\, L_1$. By the choice of L_1, we have $\mathrm{d}\, L_1 = \mathrm{d}\, L_i$, which contradicts the assumption on k. Therefore $\otimes_{i>k}L_i$ is not divisible by L_1. Now use the condition (A3).)

Exercise 2.

We say a positive lattice is IR-type if for any positive lattice M, $u \otimes v$ $(u \in \mathcal{M}(L), v \in \mathcal{M}(M))$ is irreducible, that is $u \otimes v = x + y$ and $B(x, y) = 0$ $(x, y \in L \otimes M)$ imply x or $y = 0$. Show the following.

(a) If $t\mu_t^{-2} > 1/2$ for Hermite's constant μ_t for $t \le T$, then a positive lattice L with $\operatorname{rank} L \le T$ is of IR-type. ($T \ge 114$ follows from Theorem 2.2.1.)

(b) If L is an indecomposable positive lattice of IR-type and $[L : \tilde{L}] < \infty$, then $L \otimes M$ is indecomposable for any indecomposable positive lattice M.

(Hint: Imitate [Ki8].)

7.5 Scalar extension of positive lattices

Assuming elementary theory of algebraic number fields we treat the following problem:

Let L and M be quadratic modules over $\mathbb{Z}$ and F be an algebraic number field with the maximal order R_F. Then does $R_F L \cong R_F M$ imply $L \cong M$?

As we noted, this does not make sense unless L and M are positive lattices and F is a totally real algebraic number field. But under these conditions the problem becomes interesting. In this section we give some partial answers.

The maximal order of an algebraic number field F is denoted by R_F.

Lemma 7.5.1. *Let F be a totally real finite Galois extension of $\mathbb{Q}$. Then the following are equivalent:*

(i) *For positive lattices L and M, $\sigma : R_F L \cong R_F M$ yields $\sigma(L) = M$.*

(ii) *Let G be a finite group in $GL_n(R_F)$ such that $g(A) := (g(a_{ij})) \in G$, with $A := (a_{ij}) \in G$, holds for every $g \in \mathrm{Gal}(F/\mathbb{Q})$. Then $G \subset GL_n(\mathbb{Z})$.*

Proof. (i) $\Rightarrow$ (ii) Let G be as in (ii) and put $P = \sum_{A \in G} {}^tAA$. The assumption $\{g(A) \mid A \in G\} = G$ for every $g \in \mathrm{Gal}(F/\mathbb{Q})$ implies we have $g(P) = P$ for every $g \in \mathrm{Gal}(F/\mathbb{Q})$. Since ${}^tAA$ is positive definite for a real regular matrix A, P is positive definite with rational entries. Now we can define a positive lattice $L = \mathbb{Z}[u_1, \cdots, u_n]$ by $(B(u_i, u_j)) = P$. Using ${}^tAPA = P$ for $A \in G$, the mapping σ defined by

$$(\sigma(u_1), \cdots, \sigma(u_n)) = (u_1, \cdots, u_n)A$$

becomes an isometry of $R_F L$. Applying (i) to $L = M, \sigma$, we have $\sigma(L) = L$, that is $A \in GL_n(\mathbb{Z})$.

(ii) $\Rightarrow$ (i). Let L, M, σ be as in (i). We define an isometry

$$\eta \in O(R_F(L \perp M))$$

by $\eta = \sigma$ on $R_F L$ and σ^{-1} on $R_F M$. Let $\{u_i\}_{i=1}^m$ be a basis of $L \perp M$ over $\mathbb{Z}$, and define a matrix $A = A_\eta$ for $\eta \in O(R_F(L \perp M))$ by

(1) $$(\eta(u_1), \cdots, \eta(u_m)) = (u_1, \cdots, u_m)A.$$

Then the following are clear:

$$(2) \qquad A \in GL_m(R_F), (B(u_i, u_j))[A] = (B(u_i, u_j)).$$

Put $G := \{A \mid A \text{ satisfies } (2)\}$; since $B(u_i, u_j) \in \mathbb{Q}$, $g(A) \in G$ for $g \in \mathrm{Gal}(F/\mathbb{Q})$, $A \in G$. Moreover $(B(u_i, u_j))$ is positive definite and F is totally real. Therefore G is a finite group, and then (ii) implies $A \in G \subset GL_m(\mathbb{Z})$. This means $\eta(L \perp M) = L \perp M$, and so $\sigma(L) \subset (L \perp M) \cap R_F M$, yielding $\sigma(L) \subset M$, and similarly $\sigma^{-1}(M) \subset L$. Thus we get $\sigma(L) = M$. □

We note that for a linear algebraic group G defined over $\mathbb{Q}$, the points $G(R_F)$ of G with coordinates in R_F satisfy $gG(R_F) = G(R_F)$ for $g \in \mathrm{Gal}(F/\mathbb{Q})$.

Lemma 7.5.2. *Let $K/\mathbb{Q}$ be a finite Galois extension, p, q different rational primes, and $\tilde{p}, \tilde{q}$ prime ideals of K which lie on p, q respectively. Let G be a finite subgroup of $GL_n(R_K)$ such that G is fixed by $\mathrm{Gal}(K/\mathbb{Q})$ as a set where $\sigma \in \mathrm{Gal}(K/\mathbb{Q})$ acts entrywise. If $g \in G, g \equiv 1_n \bmod \tilde{q}$, then $\sigma g = g$ for every σ in the inertia group of $\tilde{p}$.*

Proof. First we claim that if $g \equiv 1_n \bmod \tilde{q}$ for $g \in G$, then the order of g is a power of q. Suppose not. Then considering g^{q^r} $(\equiv 1_n \bmod \tilde{q})$ instead of g, we may assume that the order h of g is greater than 1 and is relatively prime to q. We consider the completion field $K_{\tilde{q}}$ of K by $\tilde{q}$, and denote a prime element of $K_{\tilde{q}}$ by π. Write $g = 1_n + \pi^r g_1$ where $r \geq 1$ by assumption and g_1 is an integral matrix over $K_{\tilde{q}}$ such that $g_1 \not\equiv 0 \bmod \pi$. Then $1_n = g^h = 1_n + \sum_{j=1}^{h} \binom{h}{j} \pi^{rj} g_1^j$ yields

$$h\pi^r g_1 \equiv -\sum_{j=2}^{h} \binom{h}{j} \pi^{rj} g_1^j \equiv 0 \bmod \pi^{2r}$$

and hence $g_1 \equiv 0 \bmod \pi^r$ which contradicts the defining property of g_1. Thus the claim has been verified.

Take an element σ of the inertia subgroup

$$\Gamma_0(\tilde{p}) := \{\sigma \in \mathrm{Gal}(K/\mathbb{Q}) \mid \sigma x \equiv x \bmod \tilde{p} \text{ for every } x \in R_K\}$$

and $g \in G$ with $g \equiv 1_n \bmod \tilde{q}$. We put $g_0 = \sigma(g) g^{-1} \sigma(g)^{-1} g \in G$. Then $g_0 \equiv 1_n \bmod \tilde{q}$ follows from $g \equiv 1_n \bmod \tilde{q}$, and $\sigma \in \Gamma_0(\tilde{p})$ implies $\sigma(g) \equiv g \bmod \tilde{p}$ and hence $g_0 \equiv 1_n \bmod \tilde{p}$. Applying the above claim to g_0 and $\tilde{p}, \tilde{q}$, we see the order of g_0 is a power of both p and q. Since p and q are relatively prime, $g_0 = 1_n$, and hence $\sigma(g)$ and g^{-1} commute. Therefore the order of $\sigma(g)g^{-1}$ is a power of q since the order of g is so. On the other hand, $\sigma(g)g^{-1} \equiv 1_n \bmod \tilde{p}$ implies the order of $\sigma(g)g^{-1}$ is a power of p. Thus $\sigma(g)g^{-1} = 1_n$, which is what we want. □

Lemma 7.5.3. *Let $K/\mathbb{Q}$ be a finite Galois extension and denote by G a finite subgroup of $GL_n(R_K)$ which is invariant under the action of* $\mathrm{Gal}(K/\mathbb{Q})$ *as a set. If the following are satisfied, then G is in $GL_n(\mathbb{Z})$.*

(i) *$G \cap GL_n(R_F) \subset GL_n(\mathbb{Z})$ is valid for every proper subfield F which is a Galois extension of $\mathbb{Q}$.*

(ii) *In $K/\mathbb{Q}$, at least two different rational primes p and q ramify.*

Proof. We may suppose $q \neq 2$. Take a prime ideal $\tilde{q}$ of K which lies above q and take $g \in G$ and $\sigma \in \Gamma_0(\tilde{q})$, the inertia subgroup of $\tilde{q}$; then $\sigma(g)g^{-1} \equiv 1_n \bmod \tilde{q}$, and by the previous lemma, $\sigma(g)g^{-1}$ is fixed by elements of the inertia groups of prime ideals lying above p. Since such elements generate a normal subgroup H, we have $\sigma(g)g^{-1} \in GL_n(R_F)$ where F is the subfield of K corresponding to H. Since p ramifies, $F \neq K$ and using the condition (i) we have $\sigma(g)g^{-1} \in GL_n(\mathbb{Z})$. Since $\sigma(g)g^{-1} \equiv 1_n \bmod \tilde{q}$, we have $\sigma(g)g^{-1} \equiv 1_n \bmod q$ and then by virtue of Lemma 7.2.2 it follows $\sigma(g)g^{-1} = 1_n$. Thus we have shown that $\sigma(g) = g$ is valid for $g \in G$, and $\sigma \in \Gamma_0(\tilde{q})$ for every prime ideal $\tilde{q}$ lying above q. Therefore $\sigma(g) = g$ is valid for $g \in G$ and every element σ of the normal subgroup H' generated by inertia subgroups of prime ideals lying above q. Since q ramifies, $H' \neq \{1\}$ and then using the condition (i), we get $g \in GL_n(\mathbb{Z})$. □

Lemma 7.5.4. *Let $K/\mathbb{Q}$ be a totally real finite Galois extension such that only one rational prime ramifies and the Galois group* $\mathrm{Gal}(K/\mathbb{Q})$ *is nilpotent. Then* $\mathrm{Gal}(K/\mathbb{Q})$ *is cyclic.*

Proof. We use induction on $[K:\mathbb{Q}]$. If $[K:\mathbb{Q}] = 1$, there is nothing to do. Let p be a ramified prime. Put $\Gamma = \mathrm{Gal}(K/\mathbb{Q})$ and denote the centre of Γ by Z. Since Γ is nilpotent, $Z \neq \{1\}$ and Γ/Z is nilpotent. By the induction hypothesis, Γ/Z is cyclic and then Γ is abelian. Since p is a unique prime which ramifies in the totally real abelian extension $K/\mathbb{Q}$, K is contained in $\mathbb{Q}(\zeta + \zeta^{-1})$ where ζ is a primitive p^n-th root of unity for some n. It is known that $\mathbb{Q}(\zeta + \zeta^{-1})$ is a cyclic extension of $\mathbb{Q}$, and hence $K/\mathbb{Q}$ is cyclic. □

Lemma 7.5.5. *Let L be a positive lattice and K a totally real algebraic number field. We introduce a symmetric bilinear form $B_K(x,y)$ on R_K by $B_K(x,y) = \mathrm{tr}_{K/\mathbb{Q}}(xy)$. If (R_K, B_K) or L is of E-type, then $z \in R_K L$ satisfying*

$$Q(z) = \min\{Q(y) \mid y \in R_K L, 0 \neq Q(y) \in \mathbb{Q}\}$$

is already contained in L.

Proof. We show first $\mathcal{M}((R_K, B_K)) = \{\pm 1\}$. For a non-zero $v \in R_K$, we have

$$B_K(v,v) = \mathrm{tr}_{K/\mathbb{Q}} v^2 \geq [K:\mathbb{Q}](\mathrm{N}_{K/\mathbb{Q}} v^2)^{1/[K:\mathbb{Q}]} \geq [K:\mathbb{Q}].$$

If $B_K(v,v) = [K : \mathbb{Q}]$, then all conjugates of v^2 are equal hence $v^2 = 1$ and so $v = \pm 1$.

Since K is totally real, (R_K, B_K) is clearly a positive lattice. We define a bilinear form $\tilde{B}$ on $R_K L$ by $\tilde{B}(x,y) = \mathrm{tr}_{K/\mathbb{Q}} B(x,y)$ for $x, y \in R_K L$. Then for $a, b \in R_K, x, y \in L$, we have

$$\tilde{B}(ax, by) = \mathrm{tr}_{K/\mathbb{Q}} B(ax, by) = \mathrm{tr}_{K/\mathbb{Q}}(ab) \cdot B(x,y)$$

where we use the same letter B for bilinear forms on L and $R_K L$. Thus we get $(R_K L, \tilde{B}) \cong (R_K, B_K) \otimes L$. Since (R_K, B_K) or L is of E-type,

$$\mathcal{M}((R_K, B_K) \otimes L) = \mathcal{M}((R_K, B_K)) \otimes \mathcal{M}(L) = \{\pm 1\} \otimes \mathcal{M}(L)$$

and hence $\mathcal{M}(R_K L, \tilde{B}) = \mathcal{M}(L)$ follows.

Let z be as in the lemma. Since $Q(z) \in \mathbb{Q}$, we have $Q(z) \le \min L$. Then

$$\begin{aligned} \tilde{B}(z,z) &= \mathrm{tr}_{K/\mathbb{Q}} Q(z) = [K : \mathbb{Q}] Q(z) \le [K : \mathbb{Q}] \min L \\ &= \min((R_K, B_K) \otimes L) \\ &= \min(R_K L, \tilde{B}) \end{aligned}$$

and hence $z \in \mathcal{M}(R_K L, \tilde{B}) = \mathcal{M}(L) \subset L$. □

Theorem 7.5.1. *Let $K/\mathbb{Q}$ be a totally real extension. Then for positive lattices L and M, $\sigma : R_K L \cong R_K M$ yields $\sigma(L) = M$ provided that* $\mathrm{rank}\, L = \mathrm{rank}\, M \le 43$ *or (R_K, B_K) defined in the previous lemma is of E-type.*

Proof. We note that L is of E-type by Theorem 7.1.1 if $\mathrm{rank}\, L \le 43$. Take $x_1 \in R_K L$ such that

$$Q(x_1) = \min\{Q(y) \mid y \in R_K L, 0 \ne Q(y) \in \mathbb{Q}\};$$

then by the previous lemma, we have $x_1 \in L$ and similarly $\sigma(x_1) \in M$. Hence σ induces $\sigma : R_K(x_1^\perp) \cong R_K(\sigma(x_1)^\perp)$. Repeating the similar argument, we obtain $x_1, \cdots, x_n \in L$ such that $L \supset \mathbb{Z}x_1 \perp \cdots \perp \mathbb{Z}x_n$ (finite index) and $\sigma(x_i) \in M$. Hence we get $\sigma(\mathbb{Q}L) = \mathbb{Q}M$ and so

$$\sigma(L) = \sigma(R_K L \cap \mathbb{Q}L) = R_K M \cap \mathbb{Q}M = M.$$

□

Theorem 7.5.2. *Let $K/\mathbb{Q}$ be a totally real nilpotent Galois extension. Then for positive lattices L, M $\sigma : R_K L \cong R_K M$ yields $\sigma(L) = M$.*

Proof. We use induction on $[K : \mathbb{Q}]$. If $[K : \mathbb{Q}] = 1$, then we have nothing to do. By virtue of Lemma 7.5.1, we have only to verify the condition (ii) there. Noting that every subfield of K which is a Galois extension of $\mathbb{Q}$ is a nilpotent extension of $\mathbb{Q}$, Lemma 7.5.3 completes the proof of the theorem if at least two different rational primes ramify in $K/\mathbb{Q}$. Hence we may suppose that the only one prime can ramify; then by Lemma 7.5.4, K is a cyclic extension and from Theorem 7.1.3, (R_K, B_K) is of E-type. Theorem 7.5.1 completes the proof of the theorem. □

Exercise 1.

Let m be a natural number > 1. Show there exists a finite set E_m of algebraic numbers with $E_m \cap \mathbb{Q} = \emptyset$ depending only on m such that if K is a totally real algebraic number field with $[K : \mathbb{Q}] = m$ and $K \cap E_m = \emptyset$, then $\sigma : R_K L \cong R_K M$ yields $\sigma(L) = M$ for positive lattices L, M.
(Hint. Let $x \in R_K L$ satisfy $Q(x) = \min\{Q(y) \mid R_K L \ni y \neq 0, Q(y) \in \mathbb{Q}\}$ and write $x = \sum_{i=1}^{m} c_i u_i (c_i \in R_K, u_i \in L)$ where $(B(u_i, u_j)) = a[n] \in S_{4/3,1/2}$ (a Siegel domain) for a diagonal matrix a and an upper triangular matrix n with diagonals being 1. Then $Q(x) = a[nc] \leq a_1$ holds for $c = {}^t(c_1, \cdots, c_m)$ and $a = \mathrm{diag}(a_1, \cdots, a_m)$. Since $Q(x) \in \mathbb{Q}$, the inequality yields $a_i \times$(conjugate of i-th entry of $nc)^2 \leq a_1$. Hence conjugates of entries of nc are bounded and similarly for c, and so they are roots of a finite number of polynomials over $\mathbb{Z}$. Denote by E_m the set of non-rational roots. Then c is a rational vector, that is $x \in L$; finally use the argument of Theorem 7.5.1.)

Exercise 2.

Let E be a totally real Galois extension over $\mathbb{Q}$ and suppose that all Sylow subgroups of the Galois group $\mathrm{Gal}(E/\mathbb{Q})$ are cyclic. Show $O(R_E L) = O(L)$ holds for any positive lattice L.
(Hint. See [BK].)

Exercise 3.

Let K_1, K_2 be totally real algebraic number fields with their discriminants being relatively prime. Suppose that $O(R_{K_i} L) = O(L)$ holds for every positive lattice and $i = 1, 2$. Show $O(R_{K_1 K_2} L) = O(L)$ holds for every positive lattice where $K_1 K_2$ denotes a composite field of K_1, K_2.
(Hint. $R_{K_1 K_2} = R_{K_1} \otimes R_{K_2}$ if the discriminants of K_1, K_2 are relatively prime.)

The next result is concerned with indecomposability under the scalar extension.

Theorem 7.5.3. *Let E/F be a Galois extension of totally real algebraic number fields such that F is the unique unramified (over F) intermediate subfield of E. Let V be a quadratic space over F such that $V_v = F_v \otimes V$ is positive definite for the completion F_v of F at every infinite place v of F. Let L be a lattice on V, that is $FL = V$ an L is finitely generated over R_F. If L is indecomposable, then $R_E L$ is also indecomposable.*

Proof. Before the proof, we recall three facts:
(F1) $L, R_E L$ are positive definite for every infinite place, and then the generalization of Theorem 6.7.1 holds, that is if

$$L = L_1 \perp \cdots \perp L_n = M_1 \perp \cdots \perp M_m$$

for indecomposable submodules L_i, M_j, then $m = n$ and $\{L_i \mid 1 \le i \le n\} = \{M_i \mid 1 \le i \le n\}$. It is also true for $R_E L$.

(F2) Let U be a vector space over an algebraic number field K, and N a finitely generated submodule of U over R_K. Then there exist $u_1, \cdots, u_r \in U$ such that $N = R_K u_1 \oplus \cdots \oplus R_K u_{r-1} \oplus A u_r$ where A is an integral ideal of K such that $\mathrm{N}_{K/\mathbb{Q}}(A) := [R_K : A]$ is relatively prime to any given natural number.

(F3) The integral ideal of F generated by $(\det a_i^{(j)})^2$ where $a_1, \cdots, a_n \in R_E$ $(n = [E : F])$ and $x^{(j)}$ means the j-th conjugate of x, is called the relative discriminant $\mathrm{d}(E/F)$ of E/F, and E/F is unramified if and only if $\mathrm{d}(E/F)$ is R_F.

Let us begin the proof of the theorem.

Let L be a lattice on V. Suppose that L is indecomposable but $R_E L$ is decomposable. We have only to prove the existence of the intermediate field K such that $K \neq F$ and K is unramified over F, which contradicts the assumption. Write

(3) $$R_E L = L_1 \perp \cdots \perp L_m \ (m > 1)$$

where L_i's are indecomposable R_E-modules for $i = 1, \cdots, m$. We make $G := \mathrm{Gal}(E/F)$ act on EV by

$$g(av) = g(a)v \quad \text{for} \quad a \in E,\ v \in V \text{ and } g \in G.$$

We note that for $w \in EV$, $g(w) = w$ for all $g \in G$ yields $w \in V$. For $x = \sum a_i v_i, y = \sum b_i w_i$ $(a_i, b_i \in E, v_i, w_i \in V)$, we have

$$\begin{aligned} g(B(x,y)) &= g(\sum_{i,j} a_i b_j B(v_i, w_j)) \\ &= \sum_{i,j} g(a_i) g(b_j) B(v_i, w_j) = B(g(x), g(y)). \end{aligned}$$

Hence we have $R_E L = g(R_E L) = g(L_1) \perp \cdots \perp g(L_m)$ for $g \in G$. Since L_i is an R_E-module, $g(L_i)$ is so for $g \in G$. By (3),

$$\{g(L_i) \mid 1 \leq i \leq m\} = \{L_i \mid 1 \leq i \leq m\}.$$

We show that G acts transitively on $\{L_i \mid 1 \leq i \leq m\}$. Otherwise, changing suffices if necessary we may suppose that G acts on $\{L_i \mid 1 \leq i \leq t\}$ and $\{L_i \mid t+1 \leq i \leq m\}$ for some t $(1 \leq t < m)$. Put

$$L_1' = \perp_{1 \leq i \leq t} L_i,\ L_2' = \perp_{t+1 \leq i \leq m} L_i;$$

then $g(L_i') = L_i'$ for $i = 1, 2$ and $g \in G$. We decompose $v \in L$ as $v = v_1 + v_2, v_i \in L_i'$. Then for $g \in G$ we have $v = g(v) = g(v_1) + g(v_2)$ and $g(v_i) \in L_i'$ for $i = 1, 2$, so $v_i = g(v_i)$ and $v_i \in V$ for $i = 1, 2$. Thus we have

$$L \subset (L_1' \cap V) \perp (L_2' \cap V) \subset R_E L \cap V = L$$

and $L = (L_1' \cap V) \perp (L_2' \cap V)$. Since L is indecomposable, we may assume $L_1' \cap V = 0$; then $L = L_2' \cap V \subset L_2'$, which yields $L_1' \perp L_2' = R_E L \subset L_2'$ hence $L_1' = 0$, which is a contradiction. Thus G acts transitively on $\{L_i \mid 1 \leq i \leq m\}$. In particular, $\operatorname{rank} L_i = \operatorname{rank} L_1$ holds for $i \geq 1$.

Put

$$H := \{g \in G \mid g(L_1) = L_1\}$$

and fix an element $g_i \in G$ such that

(4) $$g_i(L_1) = L_i,\ g_1 = \mathrm{id}.$$

Then $G = \cup_{i=1}^m g_i H$ is clear. Put

$$K := \{a \in E \mid g(a) = a \text{ for } g \in H\}.$$

Then $[K : F] = [G : H] = m > 1$ implies $K \neq F$.

We will show that K is unramified over F, which contradicts the hypothesis and will complete the proof. Put

$$M_1 := \{v \in L_1 \mid h(v) = v \text{ for } h \in H\} = KV \cap L_1 \subset KV \cap R_E L = R_K L,$$

which is an R_K-module. Writing $v \in R_K L$ as

$$v = \tilde{v}_1 + \tilde{v}_2\ (\tilde{v}_1 \in L_1, \tilde{v}_2 \in \perp_{i \geq 2} L_i),$$

we have, for $h \in H$, $v = h(v) = h(\tilde{v}_1) + h(\tilde{v}_2)$, implying $h(\tilde{v}_i) = \tilde{v}_i$ since

(5) $$h(L_1) = L_1 \quad \text{and } \{h(L_i) \mid i \geq 2\} = \{L_i \mid i \geq 2\}.$$

Hence we have $\tilde{v}_1 \in M_1$ and hence $\tilde{v}_2 \in R_K L$. Thus

$$R_K L = M_1 \perp (R_K L \cap \perp_{i\geq 2} L_i)$$

yielding

(6) $$L_1 = R_E M_1.$$

Let

(7) $$M_1 = R_K v_1 \oplus \cdots \oplus R_K v_{r-1} \oplus A v_r,$$

where $v_i \in KV$ and A is an integral ideal of K such that $\mathrm{N}_{K/\mathbb{Q}}(A) := [R_K : A]$ and the norm of $\mathrm{d}(E/F)$ are relatively prime by virtue of (F2). Since $\{v_i\}_{i=1}^r \subset KV$ are linearly independent over K, they are also linearly independent over E in EV. Similarly we write

(8) $$L = R_F u_1 \oplus \cdots \oplus R_F u_{n-1} \oplus B u_n \ (n = mr),$$

where $\{u_i\}$ is a basis of V over F and B is an integral ideal of F such that $\mathrm{N}_{K/\mathbb{Q}} B = [R_F : B]$ is relatively prime to the norm of $\mathrm{d}(E/F)$. Put

(9) $$v_i = \sum_{j=1}^{n} a_{ij} u_j \ (a_{ij} \in K) \text{ and } u_i = \sum_{j=1}^{m} g_j(w_{ji}) \ (w_{ji} \in EL_1).$$

By virtue of $u_i \in V$, compare components in L_1 along the decomposition (3) of $u_i = g(u_i) = \sum_{j=1}^m g g_j(w_{ji})$ for $g = g_h^{-1}$, to get $w_{1i} = w_{hi}$ for $1 \leq h \leq m$. Putting $w_i = w_{1i} \in EL_1$, we can write

$$u_i = \sum_{j=1}^{m} g_j(w_i).$$

Substituting the above into $h(u_i) = u_i$ for $h \in H$ and comparing components of EL_1, we have $h(w_i) = w_i$ for $h \in H$ by (4) and hence $w_i \in KV \cap EL_1 = KM_1$ by definition of M_1. Therefore, putting

$$w_j = \sum_{t=1}^{r} b_{jt} v_t \ (b_{jt} \in K)$$

by virtue of (7), we have

$$v_i = \sum_{j=1}^{n} a_{ij} u_j = \sum_{j=1}^{n} a_{ij} \sum_{k=1}^{m} g_k(w_j)$$

(10) $$= \sum_{j=1}^{n} \sum_{k=1}^{m} \sum_{t=1}^{r} a_{ij} g_k(b_{jt}) g_k(v_t).$$

Since $v_t \in KM_1 \subset EL_1$ by (7), (6) and $g_k(v_t) \in EL_k$ by (4), comparing the components in (10) of EL_k, we have

$$\sum_{j=1}^{n}\sum_{t=1}^{r} a_{ij}b_{jt}v_t = v_i$$

$$\sum_{j=1}^{n}\sum_{t=1}^{r} a_{ij}g_k(b_{jt})g_k(v_t) = 0 \quad \text{for } k \geq 2,$$

which implies $\sum_{j=1}^{n}\sum_{t=1}^{r} a_{rj}b_{jt}v_t = v_r$ and

$$\sum_{t=1}^{r}\{\sum_{j=1}^{n} g^{-1}(a_{ij})b_{jt}\}v_t = 0 \text{ for } g \notin H$$

since $b_{jt} \in K$ $v_t \in KV$ are invariant by H. With $a_j := a_{rj}, b_j := b_{jr} \in K$, the above implies

$$(11) \qquad \sum_{j=1}^{n} a_j b_j = 1 \text{ and } \sum_{j=1}^{n} a_j g(b_j) = 0 \text{ for } g \notin H,$$

noting that $\{v_i\}$ are linearly independent over E.

Next we show

$$(12) \qquad a_j \in A^{-1}R_K \quad \text{and} \quad b_j \in B^{-1}R_K$$

From $M_1 \subset R_K L$ follows

$$A(\sum_j a_{rj}u_j) = Av_r \subset M_1 \subset R_K L \subset \oplus_{i=1}^{n} R_K u_i.$$

Thus $\sum_j a_{rj}u_j \subset \oplus_i A^{-1}R_K u_i$ and $a_j = a_{rj} \in A^{-1}R_K$. From $Bu_j \subset L$ we get $B(\sum_k g_k(w_j)) = Bu_j \subset L \subset\perp L_i$ and hence $Bw_j \subset L_1$. Thus we have $R_E M_1 = L_1 \supset Bw_j = B(\sum_{t=1}^{r} b_{jt}v_t)$, which implies $AR_E \supset Bb_{jr}$ by (7). Therefore $b_j = b_{jr} \in B^{-1}AR_E \cap K \subset B^{-1}R_E \cap K = B^{-1}R_K$.

Finally we show that the existence of a_j, b_j with the properties (11) and (12) implies that K is unramified over F as we wanted. Write

$$R_K = R_F z_1 \oplus \cdots \oplus R_F z_{m-1} \oplus C z_m$$

where C is an integral ideal of F such that $\mathrm{N}_{F/\mathbb{Q}}(C)$ is relatively prime to the norm of $\mathrm{d}(E/F)$. Put

$$a_i = \sum_{j=1}^{m} c_{ij}z_j, b_i = \sum_{t=1}^{m} d_{it}z_t \quad (c_{ij}, d_{it} \in F).$$

Then (11) implies

$$(13)\qquad \begin{cases} \sum_{j,t=1}^{m}(\sum_{i=1}^{n} c_{ij}d_{it})z_j z_t = 1, \\ \sum_{j,t=1}^{m}(\sum_{i=1}^{n} c_{ij}d_{it})z_j g(z_t) = 0 \text{ for } g \notin H. \end{cases}$$

Putting $f_{jt} := \sum_{i=1}^{n} c_{ij}d_{it} \in F$ and

$$z := (z_1, \cdots, z_m) \in K^m, z' = {}^t(\sum_{t=1}^{m} f_{1t}z_t, \cdots, \sum_{t=1}^{m} f_{mt}z_t) \in K^m,$$

(13) means

$$(14)\qquad zz' = 1, \quad g(z)z' = 0 \text{ for } g \notin H.$$

We note that $\{g_1, \cdots, g_m\}$ gives a complete set of conjugate mappings of K over F. The above means

$$Mz' = \begin{pmatrix} 1 \\ 0 \\ \vdots \\ 0 \end{pmatrix} \text{ for } M := \begin{pmatrix} g_1(z) \\ \vdots \\ g_m(z) \end{pmatrix} = (g_i(z_j)).$$

Let g_i act on both sides and noting that g_iM is the interchange of rows of M, we have

$$(15)\qquad Mg_i(z') = {}^t(0, \cdots, 0, 1, 0, \cdots, 0).$$

If $Mg_i(z') = Mg_j(z')$, then $g_i(z') = g_j(z')$ since M is a regular matrix, and $z' = g_i^{-1}g_j(z')$. If $i \neq j$, then $g_i^{-1}g_j \notin H$ and so by (14)

$$(g_i^{-1}g_j)^{-1}(z)z' = 0,$$

which implies $zz' = zg_i^{-1}g_j(z') = 0$ contradicting $zz' = 1$. Thus we have proved that $Mg_i(z') = Mg_j(z')$ implies $i = j$, and then the positions of 1 in the right-hand side of (15) are different if $i \neq j$. Thus $M(g_1(z'), \cdots, g_m(z'))$ is an orthogonal matrix by (15). So

$$(16)\qquad \det M \det(g_1(z'), \cdots, g_m(z')) = \pm 1.$$

Now $z_i \in R_K$ for $i \leq m-1$ and $Cz_m \subset R_K$ implies

$$(17)\qquad C^2 \det M^2 = C^2 \det(g_i(z_j))^2 \subset \mathrm{d}(K/F).$$

$\mathrm{N}_{K/\mathbb{Q}}(A) = [R_K : A] \in A$ and (12) imply $R_K \supset Aa_i \ni \mathrm{N}_{K/\mathbb{Q}}(A)a_i = \sum_t \mathrm{N}_{K/\mathbb{Q}}(A)c_{ij}z_j$ and hence

(18) $$\mathrm{N}_{K/\mathbb{Q}}(A)c_{ij} \in R_F.$$

Similarly, $R_K \supset Bb_i$ implies

(19) $$\mathrm{N}_{K/\mathbb{Q}}(B)d_{it} \in R_F.$$

Therefore, from (18) and (19) follows $\mathrm{N}_{K/\mathbb{Q}}(AB)f_{jt} = \mathrm{N}_{K/\mathbb{Q}}(AB)\sum_{i=1}^n c_{ij}d_{it} \in R_F$ and then $\mathrm{N}_{K/\mathbb{Q}}(AB)Cz' \in R_K^m$. Thus we have $(\mathrm{N}_{K/\mathbb{Q}}(AB)C)^{2m} \det(g_1(z'), \cdots, g_m(z'))^2 \subset \mathrm{d}(K/F)$. Multiplying by $C^2 \det M^2$, (16) and (17) yield $C^2(\mathrm{N}_{K/\mathbb{Q}}(AB)C)^{2m} \subset \mathrm{d}(K/F)^2$. Since $(\mathrm{N}_{K/\mathbb{Q}}(AB)C, \mathrm{d}(E/F)) = 1$ and $\mathrm{d}(K/F)$ divides $\mathrm{d}(E/F)$, $(\mathrm{N}_{K/\mathbb{Q}}(AB)C, \mathrm{d}(K/F)) = 1$ follows and then the above inclusion yields $\mathrm{d}(K/F) = R_F$, which means that K is unramified over F. □

Notes

Chapter 1

The theory of quadratic forms has a long and rich history, but it was done in terms of matrices before [Wi].

Treatment by matrices is nothing but fixing a coordinate. At the beginning it may be easy to understand, but an adherence to it turns out to be inconvenient and we might lose perspective. We have only to give convenient coordinates if necessary.

There are many generalizations of [Wi]. Theorem 1.2.2 is one of them, and is due to Kneser [Kn2]. It seems to be the most convenient one for us.

The author does not know any references which give the complete formula for the number of isometries from a quadratic space V to W over a finite field.

For more on Clifford algebras, see [Bou], [Kn3], [La].

In this book, we do not mention Hermitian forms. But the study of quadratic forms often involves Hermitian forms so it is necessary to know about them. However almost all results of quadratic forms are easily generalized.

For algebraic theory of quadratic forms over general fields, see [La]. For Hermitian forms, see [Bou] and [Scha].

Chapter 2

The main results on reduction theory are due to Lagrange for $n = 2$, Seeber for $n = 3$ and Hermite and Minkowski for general n. The version of 2.1 is due to Siegel.

$p \in P(m)$ is reduced in the sense of Minkowski if and only if

(i) $p[x] \geq p_{kk}$ for all integral $x = {}^t(x_1, \cdots, x_m)$ with $\gcd(x_k, \cdots, x_m) = 1$ $(1 \leq k \leq m)$,

(ii) $p_{k,k+1} \geq 0$ for $1 \leq k \leq m-1$

are satisfied.

Denoting the set of all $p \in P(m)$ reduced in the sense of Minkowski by M_m, the following are known:

(i) M_m is a fundamental domain, that is $P(m) = \cup_{g \in GL(m,\mathbb{Z})} M_m[g]$ and for $g \neq \pm 1$, $M_m \cap M_m[g]$ is contained at most in the boundary of M_m.

(ii) M_m is contained in $S_{t,u}$ for some $t, u > 0$.

For $p \in P(m)$, there is $p' \in M_m$ such that $p' = p[g]$, $g \in GL(m, \mathbb{Z})$ by (i), but p' is not unique in general unless it is an interior point of M_m. For $m \leq 3$, the conditions for the uniqueness are known [D]. For $m \geq 4$, they are not known. Unless being the fundamental domain comes into question, the Siegel domain is more convenient than the reduced domain M_m. Results in 2.1 are generalized over algebraic number fields and for the general theory of Siegel domains, see [Wel], [Bor].

The evaluation and the estimate from above and below of class numbers of quadratic forms in m-variables is an important problem. Here we give only a few references. When $m = 2$, the class number formula is due to Dirichlet and it is called Dirichlet's class number formula. It is the origin of various class number formulas. (See [BS].) If $m \geq 3$ and quadratic forms are indefinite, then the evaluation of class numbers is reduced to almost local problems (see Chapter 6). On the other hand, the definite case is not so easy. When m and the discriminant are relatively small, see[CS2I].

Many things are known about Hermite's constant. For example, $\mu_2 = \sqrt{4/3}, \mu_3 = \sqrt[3]{2}, \mu_4 = \sqrt[4]{4}, \mu_5 = \sqrt[5]{8}, \mu_6 = \sqrt[6]{64/3}, \mu_7 = \sqrt[7]{64}$ and $\mu_8 = 2$, but μ_9 is not known yet. μ_2 is taken by $\begin{pmatrix} 2 & 1 \\ 1 & 2 \end{pmatrix}$. The other cases (up to dimension 8) are also realized by explicit lattices on pp. 30-31 in [MH].

It is known that there is a positive definite matrix $p \in P(m)$ such that $\mu_m = \min p/(\det p)^{1/m}$. We may assume $\min p = 1$ and $p = a[n] \in S_{4/3,1/2}$. Put $\mathcal{M} := \{z \in \mathbb{Z}^m \mid p[z] = 1\}$; then p is a unique solution of $A[z] = 1$ for $z \in \mathcal{M}$. Theorem 2.1.2 gives $a[z] \leq c_1^{-1}$ for $z \in \mathcal{M}$ for an explicitly given constant c_1. Hence $a_i \geq (3/4)^{i-1}a_1 = (3/4)^{i-1}$ implies $z_i^2 \leq c_1^{-1}(4/3)^{i-1}$ and so $\mathcal{M} \subset \mathcal{M}' := \{z \in \mathbb{Z} \mid z_i^2 \leq c_1^{-1}(4/3)^{i-1}\}$. So, if we can solve

$A[z] = 1$ for $z \in \mathcal{N}$ in subsets $\mathcal{N}$ of $\mathcal{M}'$, we get μ_m. Is the computer not useful for $m = 9$?

The estimate $\mathrm{vol}(K_m)^{-2/m} < \mu_m \le 4\,\mathrm{vol}(K_m)^{-2/m}$ was known by Minkowski. Theorem 2.2.1 states

$$\mu_m \le 2(1 + m/2)^{2/m}\,\mathrm{vol}(K_m)^{-2/m}.$$

Rogers gave the estimate

$$\mu_m \le 4(\sigma_m/\,\mathrm{vol}(K_m))^{2/m} \sim 2\{(me^{-1}\,\mathrm{vol}(K_m)^{-1}\}^{2/m},$$

for $\sigma_m = \mathrm{vol}(B)/\,\mathrm{vol}(S)$, where S is a regular m–dimensional simplex of edge length 2 and B is the subset consisting of all points in S with distance from some vertex of S at most 1. Kabatiansky and Levenshtein gave a better estimate which involves the smallest positive zero of the Bessel function $J_{m/2}$.

Now the construction of a positive definite quadratic form A, which has a bigger value of $\min A/\det A^{1/m}$, and estimating μ_m involve other branches of mathematics, for example the geometry of numbers, sphere packing, coding theory, the least positive eigenvalue of the Laplacian on certain kind of manifolds and even algebraic geometry (see [CS1]).

For a brief introduction to Hermite's constant see §7 in [MH]. More details are in [CS1]. For the geometry of numbers see [C1], [Le]. Section 2.2 was taken from [Le].

Chapter 5

One of the basic problems is the following:
Given regular quadratic lattices L and M, give good sufficient and/or necessary conditions so that L is represented by M, and then give a complete set of representatives for the $O(M)$-equivalence classes of submodules isometric to L.
There are many papers related to this, including ones on the generalization of Witt's theorem over R. Taking account of global problems, i.e. ones over $\mathbb{Z}$, additional conditions, which are hard to imagine from the local view point, are sometimes put on it.

The calculation of the spinor norm of $O^+(M)$ is important to the global problem through the strong approximation theorem Theorem 6.2.2.

The evaluation of local densities is important and equally so is the study of their behaviour. There are only a few cases where the explicit values are known. The explicit values seem to be very complicated and their behaviour is still another story. The infinite product of local densities turns out to be a global quantity via Siegel's formula. The quantitative behaviour of the infinite product of local densities gives global qualitative results or perspectives. The infinite product of local densities appears as arithmetic parts in the Fourier coefficients of Eisenstein series of the symplectic group. Hence relations among local densities give corresponding relations among Fourier coefficients of Eisenstein series, and vice versa. For example, the functional equation of the real analytic Eisenstein series of the symplectic group implies the following:

Let H_k be the $2k$-dimensional hyperbolic lattice $\perp_k \langle \begin{pmatrix} 0 & 1 \\ 1 & 0 \end{pmatrix} \rangle$ over R and L be a regular quadratic lattice over R with $\operatorname{rank} L = n$ and $\mathrm{n}(L) \subset 2R$. Then by Theorem 5.6.1 and Exercise 4, $\beta_p(L, H_k)$ is a polynomial in p^{-k} dependent on L. So there is a polynomial $f(x; L)$ in x such that $\beta_p(L, H_k) = f(p^{-k}; L)$. The functional equation of Eisenstein series shows that for $M \subset L$ with $\operatorname{rank} M = \operatorname{rank} L = n$

$$\frac{f(x; M)}{f(x; L)} x^{-\operatorname{ord}_p [L:M]}$$

is invariant under $x \to p^{-n-1}x^{-1}$. (Conversely this gives the functional equation of Eisenstein series with one of the generalized hypergeometric functions as analytic parts of the Fourier coefficients.)

[Y] gives relations among Fourier coefficients of Eisenstein series and it should imply something about local densities.

Can one verify these identities directly without global theory?

We know $\beta_p(M, H_k)$ for $p \neq 2$ and $\operatorname{rank} M = 3$ ([Ki13]). The case of $p = 2$ remains. The explicit formula of this case gives the explicit formula of the Eisenstein series of degree 3 and level 1.

There are many relations among local densities $\beta_p(M, N)$ when M varies. One concrete description is: $\sum \beta_p(p^t M, N)x^t$ is a rational function of x. This is generalized to a power series in many variables, taking

$$p^{t_1} M_1 \perp \cdots \perp p^{t_n} M_n$$

and $x_1^{t_1} \cdots x_n^{t_n}$ instead of $p^t M, x^t$, respectively. (See [BöSa], [Hi], [K], [Ki14], [Ki16].)

Chapter 6

One of the basic problems is

Problem 1.

Let M, N be quadratic lattices over $\mathbb{Z}$. Give sufficient and/or necessary conditions so that M is (primitively) represented by N.

From this view point, the following is natural.

Problem 2.

Is the condition $\dim W \geq 2\dim V + 3$ in Theorem 6.6.2 the best possible? Can one give $c(N)$ explicitly?

(see [W2] for the latter question.) For convenience we consider the following condition

($\natural$): Let V, W be positive definite quadratic spaces over $\mathbb{Q}$; let M, N be lattices on V, W with $\mathrm{s}(M), \mathrm{s}(N) \subset 2\mathbb{Z}$, $m = \operatorname{rank} M$, $n = \operatorname{rank} N$ respectively and M_p is represented by N_p for every prime p.

It is known that the answer to the first question in Problem 2 is affirmative when $\dim V = 1$. Taking account of Corollary 5.6.2, readers may conjecture that this is best. But under the condition ($\natural$) we know the following [Ki19], [Ki17]:

(C1): In the case of $n = 2m+2 = 6$, there is a positive constant $c(N)$ such that if $\min M > c(N)$ and either $\mathrm{d}\, M > (\min M)^{32.2}$ or $M \cong \langle rB \rangle$ for any given integral matrix B in advance and $(r, \mathrm{d}\, N) = 1$, then M is represented by N.

Moreover we know the following [Ki18]:

(C2): Assume the condition ($\natural$) with $n = 2m + 2$. If $\min M$ is sufficiently large, then there exists a lattice $M' \supset M$ such that $\min M'$ is still large and M'_p is primitively represented by N_p for every prime p.

This is false if $m = 1$, and it leads to the best possibility of Theorem 6.6.2 in the case of $m = 1$. So, if the following problem is true, then Theorem 6.6.2 is true for $\dim W = 2\dim V + 2 \geq 6$.

Problem 3.

Assume the condition ($\natural$) with $n = 2m + 2$, and suppose that M_p is primitively represented by N_p for every prime p and $\min M$ is sufficiently large. Then is M (primitively) represented by N ?

The local primitive representation implies the global primitive representation if $n \geq 3m + 3$ [HKK] or $m = 1, n = 4 = 2m + 2$. (C1), (C2) may suggest that Theorem 6.6.2 is true for $\dim W = 2\dim V + 2 > 4$ and moreover some experiments by a computer suggest that the following is likely.

Problem 4.

Assume the condition ($\natural$). If $n = 2m + 1 \geq 5$ and $M \cong \langle tT \rangle$ for any given fixed T and a sufficiently large integer t, is M represented by N ?

If $n = 2m \geq 6$ and $M \cong \langle t^2 T \rangle$ for any given fixed T and a sufficiently large integer t, is M represented by N ?

When $n < 2m$, the author has no perspectives.

The genus of unimodular positive definite quadratic lattices N of rank $N = 16$ and $\mathrm{n}(N) = 2\mathbb{Z}$ contains two classes N_1, N_2.

Problem 5.

Study the representation of M by N_i for a lattice M with rank $M \geq 8$.

As a variation of Corollary 6.4.1, we propose

Problem 6.

Assume the condition ($\natural$) with $n > m$. When does the total information of representation numbers $r(M, N)$ determine the isometry class of N, where M runs over all possibilities ?

Let us translate this in terms of modular forms. For an integral positive definite matrix S of size n, we define the degree m theta function of S by

$$\begin{aligned}\theta_m(z, S) := &\sum_{g \in M_{n,m}(\mathbb{Z})} exp(2\pi i \,\mathrm{tr}\, S[g]z)\\ = &\sum_T r(T, S) exp(2\pi i \,\mathrm{tr}\, Tz)\end{aligned}$$

where z belongs to Siegel's upper half space

$$H_m := \{z = {}^t z \in M_m(\mathbb{C}) \text{ with } \mathrm{Im}\, z > 0\},$$

and T runs over all symmetric integral non-negative matrices of size m. Then the theta function $f(z) := \theta_m(z, S)$ is a modular form of degree m, and weight $n/2$, that is it satisfies the following three properties:

(i) $f(z)$ is holomorphic on H_m.

(ii)

$$(f| \begin{pmatrix} a & b \\ c & d \end{pmatrix})(z) := \det(cz + d)^{-n/2} f((az + b)(cz + d)^{-1}) = \kappa f(z),$$

where $\begin{pmatrix} a & b \\ c & d \end{pmatrix}$ is in

$$Sp(m, \mathbb{Z}) := \{M \in SL(2m, \mathbb{Z}) \mid {}^t MJM = J\}$$

for $J = \begin{pmatrix} 0 & 1_m \\ -1_m & 0 \end{pmatrix}$, with $c \equiv 0 \bmod q$ where $q := 4|S|$. (Here a, b, c, d are $m \times m$ matrices and $cz + d$ is always regular in this case; κ is a complex number with $|\kappa| = 1$ dependent on a, b, c, d).

(iii) For any $M \in Sp(m,\mathbb{Z})$, $(f|M)(z)$ has a Fourier expansion

$$\sum_T h(T,M) exp(2\pi i \operatorname{tr} Tz/q)$$

for some complex number $h(T,M)$ where T runs over all $m\times m$ non-negative symmetric matrices such that the scale $\mathrm{s}(\langle T\rangle)\subset\mathbb{Z}$.

A function which satisfies these three conditions is called a modular form of weight $n/2$, degree m and level q.

Then Problem 6 is equivalent to

Problem 6′.

When does the theta function $\theta_m(z,S)$ determine the isometry class of $\langle S\rangle$?

Corollary 6.4.1 shows that this is true if $\det S$ is fixed and $n=m+1$. Is it always true with $n=m+1$? A related problem is

Problem 7.

Let S_i be integral positive definite matrices of degree n and assume that quadratic lattices $\langle S_i\rangle$ are not mutually isometric. When are $\theta_m(z,S_i)$ linearly independent ?

This is true if $n=m+1$ and $\det S_i$ are equal by virtue of Corollary 6.4.1. Schulze-Pillot taught me that this is false when $n=m+2$ and $\langle S_i\rangle$ runs over a complete set of representatives of the classes in the genus of $L=\langle 1\rangle\perp\langle 17^2\rangle\perp\cdots\perp\langle 17^{2k}\rangle$ $(2k=n)$. That is based on the Remark in 6.8 and

$$\theta(O^+(L_p))=\begin{cases}\mathbb{Q}_2^\times & \text{for } p=2,\\ (\mathbb{Q}_{17}^\times)^2 & \text{for } p=17,\\ \mathbb{Z}_p^\times(\mathbb{Q}_p^\times)^2 & \text{for } p\neq 2,17,\end{cases}$$

because these imply that $gen(L)$ contains two different spinor genera, and Siegel's weighted sum of theta functions in a spinor genus is independent of the choice of a spinor genus. What about the case when $\langle S_i\rangle$ runs over a fixed spinor genus ?

Incidentally it is a problem to generalize the Siegel formula for spinor genus (c.f. Remark in 6.8 and [SP2]).

The number of isometry classes of positive definite integral matrices with $\det = D$, degree $=n$ is bounded by $D^{(n-1)/2}$ from below and by $D^{(n-1)/2+\epsilon}$ from above up to a constant, which is a corollary of Theorem 6.8.1. [Ki2], [Bi]. The main contribution in the dimension formula of the above modular forms is supposed to be the volume of the fundamental domain of H_m by

$$\Gamma_0(q) := \{\begin{pmatrix} a & b\\ c & d\end{pmatrix}\in Sp(m,\mathbb{Z}) \mid c\equiv 0 \bmod q\}.$$

So, (up to a constant) the main part of this should be $[Sp(m,\mathbb{Z}) : \Gamma_0(q)]$, which is almost equal to $q^{m(m+1)/2}$. If the number of non-isometric quadratic forms is greater than the dimension of the modular forms, then we get a linear relation of theta functions. Hence, if $n-1 > m(m+1)$, then we will get ample linear relations among these theta functions. This is the case if $m = 1$. What happens in the case of $n-1 \leq m(m+1)$?

For example, as a refinement of Problem 6′.

Problem 8.

If $n-1 \leq m(m+1)$, then does the theta function $\theta_m(z,S)$ determine the isometry class of S ? If $n-1 > m(m+1)$, then are there non-isometric positive definite quadratic forms which have the same theta function ?

A finite number of examples such that non-isometric positive definite quadratic lattices have the same theta function is known when $n = 4, m = 1$ ([Schi], [Shio]). Are there infinitely many such examples and if so, what characterizes them ?

When $n = 3, m = 1$, the above problem seems to be answered affirmatively.

Problem 9.

Is there any example of a positive definite lattice N with $\operatorname{rank} N = n$ such that for an m given in advance, M is represented by N (with finitely many exceptions) if M is locally represented by N and $\operatorname{rank} M = m$?

Of course, if $gen(N) = cls(N)$, then the answer is yes. But it is known that the number of such N is finite. If $n = 3$ and $m = 1$, then it is known that there are only a finite number of lattices N which do not have exceptions above.

Suppose that N is a quinary positive definite quadratic form with $Q(N) \subset 2\mathbb{Z}$ and $\mathrm{d}N = 20$. Then $gen(N)$ contains two classes N_1, N_2 whose quadratic forms are:

$$\begin{aligned} Q_1(\textstyle\sum x_i v_i) &= 2(x_1^2 + x_2^2 + x_3^2 + x_4^2 + 2x_5^2 - x_1x_2 - x_2x_3 - x_3x_4), \\ Q_2(\textstyle\sum x_i v_i) &= 2(x_1^2 + x_2^2 + x_3^2 + x_4^2 + 3x_5^2 - x_1x_2 - x_1x_3 - x_1x_4 \\ &\quad - x_1x_5 + x_2x_5 + x_3x_5) \end{aligned}$$

For these, the exceptional binary quadratic forms seems to be finite.

To show the existence of an isometry from M to N, it is important to study the totality of all the numbers $r(K,N)$ of isometries from K to N. Because the generating function $\theta_m(z,S)$ is a modular form, we can make use of the theory of modular forms. If $n \geq 4m+4$, then we can give the asymptotic formula for $r(M,N)$ and in particular $r(M,N) > 0$ under the assumption of local representability and min M being sufficiently large. The second case in (C1) is based on the theory of modular forms.

Siegel's formula for indefinite quadratic forms is valid with some modifications. The orthogonal group of an indefinite quadratic form N is an infinite group unless $\mathbb{Q}N$ is a hyperbolic plane. So, the left-hand sides of the equalities of Theorem 6.8.1 are meaningless; however, they hold under some modifications. This involves some geometry, which we do not pursue. For modern approaches to Siegel's theory, see [We2], [S]. A generalization of Siegel's formula given in [Bl], [Ki15] is false when the ranks are equal. Siegel's formula Theorem 6.8.1 implies that the number of different classes in $gen(N)$ is almost the same as $2w(N)$ when rank N is fixed, [Ki2], [Ki3], [Bi]. But we do not know much when rank N varies. For example,

Problem 10.

Let N_n be a unimodular positive definite lattice with rank $N_n = n$. Is the class number of N_n almost the same as $2w(N_n)$?

There are many lattices in $gen(N_n)$ with only trivial isometries if n is large [Ba]. But it is not known whether this is true or not if $n = 32$ and $\mathrm{n}(N_n) = 2\mathbb{Z}$.

A general principle to evaluate the class number is given in [A]. It really works, at least for rank ≤ 5.

These problems are also considered over algebraic number fields. One problem on relative algebraic number fields is:

Let E be a Galois extension of an algebraic number field F and V be a quadratic space over E. When the Galois group G acts on V, how can one evaluate the class number of quadratic lattices over the ring R_E of algebraic integers in E whose isometry class is invariant under the action of G ? (c.f. [Ki5].)

Lastly, we briefly refer to an analytic approach to Theorem 6.6.2. Let S be a positive definite integral matrix of degree n and $\{S_i\}$ a complete set of representatives of the classes in the genus of S. Put

$$f(z) := \theta_m(z, S) - w(S)^{-1} \sum_i E(S_i)^{-1} \theta_m(z, S).$$

Then $f(z)$ is a modular form for which $h(0, M) = 0$ for every $M \in Sp(m, \mathbb{Z})$ with h as in the third condition of the definition of a modular form. Then $f(z)$ is a so-called cusp form if $m = 1$, but not so if $m > 1$. Put

$$a(T) := w(S)^{-1} \sum_i E(S_i)^{-1} r(S_i, T),$$

and

$$f(z) = \sum b(T) \exp(2\pi i \operatorname{tr} Tz).$$

A basic strategy is to show $b(T)/a(T) \to 0$, which implies an asymptotic formula for $r(S,T) = a(T)(1 + b(T)/a(T))$.

When $m = 1$, we get another proof of Theorem 6.6.2. This works even for $n = 3, 4$. The case of $n = 4 = 2m + 2$ is due to Kloosterman through introducing so-called Kloosterman sums, and to Eichler by solving Ramanujan's conjecture. The case of $n = 3 = 2m + 1$ involves Linnik-Selberg's conjecture [DS]. So, Theorem 6.6.2 is still valid with some modifications in the case $m = 1, n = 3, 4$.

The case of $m = 2$ is also successful and we have another proof of Theorem 6.6.2, and a partial result for $n = 6 = 2m + 2$. But we do not have a complete solution for $n = 6$ and any results for $n = 5 = 2m + 1$.

In the case of $m \geq 3$ we do not have a complete analytic proof of Theorem 6.6.2. This strategy is hard (in particular for the case $n \leq 2m + 2$) but very attractive ([Ki21]).

Chapter 7

The results in 7.1 show that there exist ample numbers of positive lattices of E-type. However an example which is not E-type is known at rank = 292 thanks to Steinberg (see p.47 in [MH]).

Problem 1.

Improve the number 43 in Theorem 7.1.1, and construct and study more examples not of E-type.

Related to 7.5,

Problem 2.

Let E be a finite Galois extension over $\mathbb{Q}$ and suppose that the complex conjugate induces an element of the centre of the Galois group $\mathrm{Gal}(E/\mathbb{Q})$. Then (R_E, B_E) $(B_E(x,y) := \operatorname{tr}_{E/\mathbb{Q}}(x\bar{y}))$ is a positive lattice. Is it of E-type?

So far, there is no counter-example.

Let us give several problems related to the tensor product. The basic one is

Problem 3.

Is there any positive lattice L which does not satisfy the condition (A) in Fundamental Lemma? In particular, what about the case of rank $L = 3$?

Problem 4.

Is there any example for which the uniqueness of decomposition does not hold? Even for $L = M \otimes N$ with $\operatorname{rank} M = 2, \operatorname{rank} N = 3$, it is not known whether M and N are determined uniquely by L up to scaling.

Problem 5.

Characterize a lattice which does not satisfy the condition (A), if it exists.

As we saw, Lemma 7.5.5 is critical. We propose a natural problem.

Problem 6.

Is Lemma 7.5.5 true without the assumption that (R_K, B_K) or L is of E-type?

The most basic question on scalar extension is

Problem 7.

Does $O(R_E L) = O(L)$ hold for any positive lattice L and any totally real algebraic number field E?

We considered the case that the base ring is the rational integer ring $\mathbb{Z}$. If the base ring is not $\mathbb{Z}$, then the problems are negatively answered in general. Let us explain this. Let E/F be an unramified Galois extension of totally real algebraic number fields with Galois group $G = \{g_i\}_{i=1}^n$. Put $V := F[G]$ (a group ring) and define an inner product $B(\,,)$ by $B(g, g') = \delta_{g,g'}$ (the Kronecker delta) for $g, g' \in G$. Then V is positive definite at all infinite places of F. We extend the operation of G to $EV = E[G]$ by

$$g(\sum a_i g_i) = \sum g(a_i) g g_i \ (g \in G, a_i \in E),$$

and put

$$L' := \perp_{g \in G} R_E g, L := \{\sum_{g \in G} g(a) g \mid a \in R_E\}$$

where R_E is the maximal order of E. Then for $a_1, \cdots, a_n \in R_E$ we have $\sum_{g \in G} g(a_j) g \in L$ and

$${}^t(\sum_{g \in G} g(a_1) g, \cdots, \sum_{g \in G} g(a_n) g) = (g_j(a_i))^t (g_1, \cdots, g_n).$$

Since E/F is unramified, $\det(g_i(a_j))^2$ generates R_F as an ideal. Hence

$$L' = R_E L.$$

Let us show that L is indecomposable. Suppose $L = L_1 \perp L_2$ ($L_1 \neq 0$), then $L' = R_E L_1 \perp R_E L_2$ and since $R_E L_1$ is an orthogonal sum of $R_E g$ ($g \in G$) by a generalization of Theorem 6.7.1, $GR_E L_1$ contains G so $GR_E L_1 = L'$. However, G operates trivially on L and so $L' = GR_E L_1 = R_E L_1$. This means $L_2 = 0$, that is L is indecomposable.

Put $M = \perp_{g \in G} R_F g$; then $R_E M = L' = R_E L$ but $M \not\cong L$ since M is decomposable. Let (R_E, B_E) be the quadratic lattice over R_F with $B_E(x, y) = \mathrm{tr}_{E/F}\, xy$; then $R'_E := (R_E, B_E)$ is positive definite at every infinite place and taking the trace of $R_E M = R_E L$, we have $R'_E \otimes M \cong R'_E \otimes L$. On the other hand, $R'_E \ni a \mapsto \sum_{g \in G} g(a) g \in L$ gives an isometry of R'_E and L. Thus we have an indecomposable lattice L and a decomposable one M such that $L \otimes L \cong L \otimes M$ but $L \not\cong M$.

Problem 8.

In the above example, we constructed lattices L, M and N ($= L$) such that $L \otimes M \cong L \otimes N$ does not imply $M \cong N$, making use of the existence of unramified extensions of the base field. Is there any such example independent of the existence of unramified extensions of the base field?

Let E/F be an unramified Galois extension of totally real algebraic number fields; then as above, we can construct an indecomposable lattice L over R_F such that $R_E L$ is decomposable and $V := FL$ is a quadratic space over F which is positive definite at every infinite place of F. Write $R_E L = L_1 \perp L_2$ ($L_i \neq 0$) and define an isometry $\sigma \in O(R_E L)$ by $\sigma = \mathrm{id}$ on L_1, $= -\mathrm{id}$ on L_2. Since L is indecomposable, $\sigma \notin O(L)$ holds. Hence $O(R_E L) = O(L)$ is false.

If conversely there is a special isometry $\sigma \in O(R_E L)$ such that $\sigma^2 = 1, \sigma(x) \equiv x \bmod 2R_E L$ for $x \in R_E L$ and $\sigma \neq \pm 1$, then

$$R_E L = \{x \in R_E L \mid \sigma x = x\} \perp \{x \in R_E L \mid \sigma x = -x\}$$

is clear and by virtue of Theorem 7.5.3, there exists an intermediate field $K (\neq F)$ which is unramified over F.
So a natural problem is

Problem 9.

What is the role of the existence of an unramified subfield ($\neq F$) to the problem $O(R_E L) = O(L)$?

Problem 10.

Does the Fundamental Lemma hold over R_F for a totally real algebraic number field F and lattices which are positive definite at every infinite place of F?

Problem 11.

When a totally real algebraic number field F does not have any totally real unramified extension, is there any example of L which is positive definite at every infinite place such that L does not satisfy the condition (A)? In particular, to what extent can one generalize the results in this chapter to $\mathbb{Q}(\sqrt{5})$? (It is known that F has no unramified extension.)

Problem 12.

$F = \mathbb{Q}(\sqrt{10})$ has a non-trivial totally real unramified extension (the maximal (not necessarily abelian) unramified totally real extension is $F(\sqrt{2})$, which the author learned from L. Washington). Let E/F be a Galois extension of totally real algebraic number fields. As shown above, some of the results in this chapter are false in general. Look for any examples independent of the unramified extensions for this special field.

Problem 13.

If L is a multiplicatively indecomposable positive lattice, is $R_F L$ also for a totally real algebraic number field F? In particular, what about the lattices in Example 1 in 7.1?

Problem 14.

If a positive lattice L satisfies the condition (A) in the Fundamental Lemma, does $R_F L$ satisfy the similar condition over R_F?

And so on.

Here we considered the scalar extension of positive definite quadratic forms to totally real algebraic number fields. But it is also possible to do this for positive definite Hermitian forms. (See [Ki10].)

If we confine ourselves to scalar extensions which are totally real, then the following holds:

Let $S = \{K\}$ be a set of totally real Galois extensions over $\mathbb{Q}$ such that every subfield of K in S which is normal over $\mathbb{Q}$ is also in S. Suppose that

(*): For any field K in S such that only one rational prime, say p ramifies, the $[\frac{e}{p-1}]$-th ramification subgroup is not trivial, where e is the ramification index of p and $[z]$ is the greatest integer not exceeding z.

Then $O(R_K L) = O(L)$ holds for any positive lattice L and $K \in S$. (See [Bar].)

If S is a set of totally real abelian fields, then (*) is valid. The author knows no example of a set S which does not satisfy the condition (*). But if we do not impose the assumption that fields are totally real, then there are examples of sets where the condition (*) is false. So this approach is not applicable to the generalization to positive definite Hermitian forms.

References

[AK] G.E.Andrews and Y.Kitaoka, *A result on q-series and its application to quadratic forms*, J. of Number Theory **34** (1990), 54–62.

[A] T.Asai, *The class numbers of positive definite quadratic forms*, Japan. J. Math. **3** (1977), 239–296.

[Ba] E.Bannai, *Positive definite unimodular lattices with trivial automorphism groups*, Memoirs of A.M.S. **85** (1990).

[Bar] H.-J.Bartels, *Zur Galoiskohomologie definiter arithmetischer Gruppen*, J. Reine Angew. Math. **298** (1978), 89–97.

[BK] H.-J.Bartels and Y.Kitaoka, *Endliche arithmetische Untergruppe der GL_n*, J. Reine Angew. Math. **313** (1980), 151–156.

[Bi] J.Biermann, *Gitter mit kleiner Automorphismengruppe in Geschlechtern von* **Z**-*Gittern mit positiv-definiter quadratischen Formen*, Dissertation Göttingen (1981).

[Bl] F. van der Blij, *On the theory of quadratic forms*, Ann. of Math. **50** (1949), 875–883.

[Bor] A.Borel, "Introduction aux groups arithmétiques," Hermann, 1969.

[Bou] N.Bourbaki, "Algebra, Ch.9," Hermann, 1959.

[BöSa] S.Böcherer and F.Sato, *Rationality of certain formal power series related to local densities*, Comment. Math. Univ. St. Paul. **36** (1987), 53–86.

[BöSP] S.Böcherer and R.Schulze-Pillot, *On a theorem of Waldspurger and on Eisenstein series of Klingen type*, Math.Ann. **288** (1990), 361–388.

[BS] Z.I.Borevich and I.R.Shafarevich, "Number theory," Academic Press, 1966.

[C1] J.W.S.Cassels, "An introduction to the geometry of numbers," Springer-Verlag, 1959.

[C2] ————————, "Rational quadratic forms," Academic Press, 1978.

[CS1] J.H.Conway and N.J.A.Sloane, "Sphere packings, lattices and groups," Springer-Verlag, 1988.

[CS2I] ——————, *Low-dimensional lattices. I. Quadratic forms of small determinant*, Proc. R. Soc. London A **418** (1988), 17–41.

[CS2II] ——————, *Low-dimensional lattices. II. Subgroups of $GL(n, \mathbf{Z})$*, Proc. R. Soc. London A **419** (1988), 29–68.

[CS2III] ——————, *Low-dimensional lattices. III. Perfect forms*, Proc. R. Soc. London A **418** (1988), 43–80.

[CS2IV] ——————, *Low-dimensional lattices. IV. The mass formula*, Proc. R. Soc. London A **419** (1988), 259–286.

[CS2V] ——————, *Low-dimensional lattices. V. Integral coordinates for integral lattices*, Proc. R. Soc. London A **426** (1989), 211–232.

[D] L.E.Dickson, "Studies in the theory of numbers," Chelsea, 1930.

[DS] W.Duke and R.Schulze-Pillot, *Representation of integers by positive ternary quadratic forms and equidistribution of lattice points on ellipsoids*, Invent. Math **99** (1990), 49–57.

[E] M.Eichler, "Quadratische Formen und orthogonal Gruppen," Springer-Verlag, Berlin, 1974.

[G] L.J.Gerstein, *Splitting quadratic forms over integers of global fields*, Amer.J.Math. **91** (1969), 106–133.

[Hi] Y.Hironaka, *On a denominator of Kitaoka's formal power series attached to local densities*, Comment. Math. Univ. St. Paul. **37** (1988), 159–171.

[H] J.S.Hsia, *Representations by sinor genera*, Pac. J. Math. **63** (1976), 147–152.

[HKK] J.S.Hsia, M.Kneser and Y.Kitaoka, *Representations of positive definite quadratic forms*, J. Reine und Ang. Math. **301** (1978), 132–141.

[K] H.Katsurada, *A certain formal power series of several variables attached to local densities of quadratic forms I.*

[Ki1] Y.Kitaoka, *On the relation between the positive definite quadratic forms with the same representation numbers*, Proc. of Japan Acad. **47** (1971), 439–441.

[Ki2] ————, *Two theorems on the class number of positive definite quadratic forms*, Nagoya Math.J. **51** (1973), 79–89.

[Ki3] ————, *On the asymptotic behaviour of Brauer-Siegel type of class numbers of positive definite quadratic forms*, Proc. of Japan Akad. **49** (1973), 789–790.

[Ki4] ————, *Representations of quadratic forms and their application to Selberg's zeta functions*, Nagoya Math. J. **63** (1976), 153–162.

[Ki5] ————, *Class numbers of quadratic forms over real quadratic fields*, Nagoya Math. J. **66** (1977), 89–98.

[Ki6] ————, *Scalar extension of quadratic lattices*, Nagoya Math. J. **66** (1977), 139–149.

[Ki7] ————, *Scalar extension of quadratic lattices II*, Nagoya Math. J. **67** (1977), 159–164.

[Ki8] ————, *Tensor products of positive definite quadratic forms*, Göttingen Nachr. (1977), 1–14.

[Ki9] ————, *Modular forms of degree n and representation by quadratic forms*, Nagoya Math. J. **74** (1979), 95–122.

[Ki10] ————, *Finite arithmetic subgroups of GL_n II*, Nagoya Math. J. **77** (1980), 137–143.

[Ki11] ————, *Modular forms of degree n and representation by quadratic forms II*, Nagoya Math. J. **87** (1982), 127–146.

[Ki12] ————, *A note on local densities of quadratic forms*, Nagoya Math. J. **92** (1983), 145–152.

[Ki13] ———, *Fourier coefficients of Eisenstein series of degree 3*, Proc. of Japan Akad. **60** (1984), 259–261.

[Ki14] ———, *Local densities of quadratic forms and Fourier coefficients of Eisenstein series*, Nagoya Math. J. **103** (1986), 149–160.

[Ki15] ———, *Modular forms of degree n and representation by quadratic forms IV*, Nagoya Math. J. **107** (1987), 25–47.

[Ki16] ———, *Local densities of quadratic forms*, Investigations in Number Theory (Advanced Studies in Pure Mathematics) **13** (1988), 433-460.

[Ki17] ———, *Modular forms of degree n and representation by quadratic forms V*, Nagoya Math. J. **111** (1988), 173–179.

[Ki18] ———, *Some remarks on representations of positive definite quadratic forms*, Nagoya Math. J. **115** (1989), 23–41.

[Ki19] ———, *A note on representation of positive definite binary quadratic forms by positive definite quadratic forms in* 6 *variables*, Acta Arith. **LIV** (1990), 317–322.

[Ki20] ———, "Lectures on Siegel modular forms and representation by quadratic forms," Springer-Verlag, Heidelberg, 1986.

[Ki21] ———, *Representation of positive definite quadratic forms and analytic number theory*, to appear in English, in "Sugaku," Iwanami-Shoten, Tokyo, 1991, pp. 115–127.

[Kn1] M.Kneser, *Darstellungsmasse infefiniter quadratischer Formen*, Math.Zeit. **77** (1961), 188–194.

[Kn2] ———, *Witts Satz fur qüadratische Formen über lokalen Ringen*, Nachr. die Akad.der Wiss.Göttingen,Math.-Phys.II **9** (1972), 195—203.

[Kn3] ———, "Quadratische Formen," Vorlesungs-Ausarbeitung, Göttingen, 1973/74.

[Kn4] ———, *Representations of integral quadratic forms*, Canadian Math.Soc. (1983), 159–172. In "Quadratic and Hermitian forms".

[Ko] O.Körner, *Die Masse der Geschlechter quadratischer Formen vom Range* ≤ 3 *in quadratischen Zahlkörpern*, Math.Ann. **193** (1971), 279–314.

[La] T.Y.Lam, "The algebraic theory of quadratic forms," Benjamin, 1973.

[Lan] E.Landau, "Primzahlen I," Chelsea, 1953.

[Le] C.G.Lekkerkerker, "Geometry of Numbers," North-Holland, Amsterdam-London, 1969.

[MH] J.Milnor and D.Husemoller, "Symmetric bilinear forms," Springer-Verlag, Berlin-Heidelberg-New York, 1973.

[MS] K.A.Morin-Strom, *A Witt theorem for non-defective lattices*, Can. J. Math **30** (1978), 499–511.

[O] O.T.O'Meara, "Introduction to quadratic forms," Springer-Verlag, Berlin, 1971.

[P] M.Peters, *Darstellungen durch definite ternäre quadratische Formen*, Acta Arith. **XXXIV** (1977), 57–80.

[S] F.Sato, *Siegel's main theorem of homogeneous spaces.*

[Scha] W.Scharlau, "Quadratic and Hermitian forms," Springer-Verlag, Berlin, 1985.

[SP1] R.Schulze-Pillot, *Darstellung durch Spinorgeschlechter ternärer quadratischer Formen*, J. of Number Theory **12** (1980), 529–540.

[SP2] ———, *Darstellungsmasse von Spinorgeschlechtern ternärer quadratischer Formen*, J. Reine und Ang. Math. **352** (1984), 114–132.

[SP3] ______________, *A linear relation between theta series of degree and weight 2*, in "Number theory, Lect. Notes in Math.," Springer-Verlag, Berlin-Heidelberg-New York, 1989, pp. 197–201.

[Schi] A.Schiemann, *Ein Beispiel positiv definiter quadratischer Formen der Dimension 4 mit gleichen Darstellungszahlen*, Arch. Math. **54** (1990), 372–375.

[Shim] G.Shimura, *Confluent hypergeometric functions on tube domains*, Math.Ann **260** (1982), 269–302.

[Shio] K.Shiota, *On theta series and the splitting of* $S_2(\Gamma_0(q))$.

[Si] C.L.Siegel, "Gesammelte Abhandlungen," Springer-Verlag, Berlin, 1966.

[W1] G.L.Watson, "Integral quadratic forms," Cambridge Univ. Press, London, 1960.

[W2] ______________, *Quadratic diohantine equations*, Philos. Trans. Roy. Soc. London Ser.A **253** (1960/61), 227–254.

[W3] ______________, *The 2-adic density of a quadratic form*, Mathematika **23** (1976), 94–106.

[We1] A.Weil, "Discontinuous subgroups of classical groups," University of Chicago, 1958.

[We2] A.Weil, *Sur la theorie des forms quadratiques*, in "collected works," Springer-Verlag, Berlin-Heidelberg-New York, 1979, pp. 471–484.

[Wi] E.Witt, *Theorie der quadratischen Formen in beliebigen Körpern*, J. Reine und Ang. Math. **176** (1937), 31–44.

[Y] T.Yamazaki, *Jacobi forms and a Maass relation for Eisenstein series*, J. Fac. Sci. Univ. Tokyo **33** (1986), 295–310.

Index

adele group, 132
anisotropic, 10
Arf invariant, 14
associated symmetric bilinear form, 3

canonical involution, 22
characteristic submodule, 153
class, 131
Clifford algebra, 20
connected, 217
connected component, 217

decomposable, 169
discriminant, 1, 70
divisible, 199
dual lattice, 71

finite place, 64

genus, 132

Hasse invariant, 56
Hermite's constant, 37
Hilbert symbol, 54
hyperbolic basis, 11
hyperbolic plane, 11
hyperbolic space, 11

idele group, 132
indecomposable, 169, 222
indefinite, 129
indefinite (quadratic) lattice, 129
infinite place, 64
irreducible, 170
isometric, 5
isometry, 5, 217
isotropic, 10
Iwasawa decomposition, 34

Jordan component, 79
Jordan decompositon, 79

Legendre symbol, 52
lattice, 70, 129
linked, 57
local field, 47
local density, 94

majorant, 41
maximal, 72
minimal vector, 190
modular, 71
multiplicatively indecomposable, 223

negative definite, 129
negative (definite quadratic) lattice, 129
norm, 25, 71

not divided, 199

orthogonal basis, 2
orthogonal sum, 2

p-adic integer, 48
p-adic number, 47
place, 64
positive definite, 129
positive (definite quadratic) lattice, 129
positive lattice, 190
primitive local density, 100
proper class, 131
proper spinor genus, 132
pure, 24

quadratic form, 3
quadratic module, 3
quadratic residue symbol, 52
quaternion algebra, 24

reducible, 169
regular, 2
regular quadratic lattice, 70, 129
represented, 5

scalar extension, 32
scale, 71
scaling, 28
second Clifford algebra, 22
Siegel domain, 34
Siegel-Eichler transformation, 12
spinor genus, 132
spinor norm, 29
split, 190
symmetric bilinear form, 1
symmetric bilinear module, 1
symmetry, 6

tensor algebra, 20
tensor product, 27
totally isotropic, 10
trace, 25

weight, 173
weighted graph, 217
Witt index, 11